鹤发医养卷

——北京曜阳国际老年公寓环境改造设计

MEDICAL AND HEALTH CARE FOR THE ELDERLY VOLME—BEIJING YAOYANG INTERNATIONAL SENIOR CITIZEN APARTMENT RECONSTRUCTION DESIGN

CIID室内设计教学参考书 CIID“室内设计6+1”2016（第四届）校企联合毕业设计

中国建筑学会室内设计分会
北京建筑大学
编

中国水利水电出版社
www.waterpub.com.cn

·北京·

内 容 提 要

国务院总理李克强在2015年政府工作报告中首次提出“健康中国”概念，指出：“健康是群众的基本需求，我们要不断提高医疗卫生水平，打造健康中国。”李克强总理在2016年政府工作报告中提出，要织密织牢社会保障安全网，将开展养老服务业综合改革试点，推进多种形式的医养结合作为2016年的工作重点。2016年6月16日，国家卫生计生委办公厅、民政部办公厅提出确定北京市东城区等50个市（区）作为第一批国家级医养结合试点单位。

为响应国家号召，探索新形势下对老年人的关怀和医养建筑的多种可能，中国建筑学会室内设计分会主办了CIID“室内设计6+1”2016（第四届）校企联合毕业设计活动，同济大学、华南理工大学、哈尔滨工业大学、西安建筑科技大学、北京建筑大学、南京艺术学院、浙江工业大学等作为参加高校，北京城建设计发展集团股份有限公司建筑院作为“鹤发医养——北京曜阳国际老年公寓环境改造设计”联合毕业设计命题的总指导企业。来自建筑学、工业设计、环境设计、艺术与科技等4个专业的2016届毕业班师生们，围绕北京曜阳国际老年公寓环境改造的医养园区环境与景观设计、医养建筑改造与室内设计、医养建筑设施与展示设计3个设计专题，联合开展了综合性实践教学活动。

中国建筑学会室内设计分会和北京建筑大学共同编写的《鹤发医养卷——北京曜阳国际老年公寓环境改造设计》，记录了这次实践教学活动的命题研讨、开题仪式、现场踏勘、中期检查、答辩评审与表彰奖励、编辑出版、专题展览、对外交流等过程，并以6+1个“活动脚步”板块作为编辑体例，全面展现了校企联合毕业设计活动情况和丰富设计成果。本书中英双语，图文并茂、内容翔实、可供建筑学、历史建筑保护工程、城乡规划、风景园林、工业设计、环境设计、产品设计、视觉传达设计、公共艺术、艺术与科技等专业人员及设置相关专业的院校师生参考借鉴。

出版资助：北京建筑大学北京市财政专项
学科建设——硕士学位授权学科点——设计学（含工业设计工程）

图书在版编目（CIP）数据

鹤发医养卷 ： 北京曜阳国际老年公寓环境改造设计 ： CIID室内设计教学参考书 CIID“室内设计6+1”2016（第四届）校企联合毕业设计 / 中国建筑学会室内设计分会，北京建筑大学编. -- 北京 ： 中国水利水电出版社，2016.10
ISBN 978-7-5170-4836-7

Ⅰ. ①鹤… Ⅱ. ①中… ②北… Ⅲ. ①老年人住宅－室内装饰设计－作品集－中国－现代 Ⅳ. ①TU241.93

中国版本图书馆CIP数据核字(2016)第247226号

书　名	CIID室内设计教学参考用书 CIID“室内设计6+1”2016（第四届）校企联合毕业设计 鹤发医养卷——北京曜阳国际老年公寓环境改造设计 HEFA YIYANG JUAN—BEIJING YAOYANG GUOJI LAONIAN GONGYU HUANJING GAIZAO SHEJI
作　者	中国建筑学会室内设计分会 北京建筑大学 编
出版发行	中国水利水电出版社 （北京市海淀区玉渊潭南路1号D座 100038） 网址：www.waterpub.com.cn E-mail：sales@waterpub.com.cn 电话：（010）68367658（营销中心）
经　售	北京科水图书销售中心（零售） 电话：（010）88383994、63202643、68545874 全国各地新华书店和相关出版物销售网点
排　版	中国建筑学会室内设计分会
印　刷	北京博图彩色印刷有限公司
规　格	210mm×285mm 16开本 17.25印张 409千字
版　次	2016年10月第1版 2016年10月第1次印刷
印　数	0001—1000册
定　价	120.00元

CIID
China Institute of
Interior Design
中国建筑学会室内设计分会

CIID 6+1
校企联合毕业设计

Preface

Since 2013, the CIID "Interior Design 6+1"University and Enterprise Joint in Graduation Design activity, which was hosted by CIID, cofounded and participated by universities and famous design companies, has been successfully held for four sessions. Each year, the topic is selected from the actual subjects needed to be solved by famous design companies. Related activities are carried out on the theme of Post Olympic Commercial Opportunities: Remoulding Interior Design in National Stadium after Games, The Volume of the New Environment of City Track: The Environmental Design of Shanghai Subway Rebuilding, The Ordnance Factory Relic: The Environmental Design of Nanjing Chenguang 1865 Creativity Industry Park and Healthy Aging: Beijing Yaoyang International Retirement Community Environmental Redesign respectively. From the opening site exploration, to the mid-term review, and then to the graduation reply, students from different areas and teaching systems have expanded their perspectives and ideas through communication with the judge teachers as well as specialists, thus have benefited a lot and achieved good results. These activities are geared to the needs of society and market, which help the students master knowledge through the way of combining theory with practice, and change to their work role in the future, in order to adapt to the rapid development of current society.

One of the characteristics of CIID "Interior Design 6+1"University and Enterprise Joint in Graduation Design is the formation of an education platform co-founded by universities and famous design companies, aiming to promote the development of interior design education. Each year, seven universities and famous design companies participate in the activities, to communicate and learn from each other, and to improve together.

During the past four years, CIID "Interior Design 6+1"University and Enterprise Joint in Graduation Design activities have achieved good results, which help to develop innovative and entrepreneurial talent, contribute to school education reform. As the only academic group in Chinese interior design, CIID has been committed to promoting Chinese interior design cause, and paid great attention to the growth of young students. It's our responsibility to constantly work hard, but not for the false reputation and vanity. In the future, the 6+1 activities will continue to rely on enterprises in guiding joint graduation design with the principle of education being offered in needs of talent cultivation, learn lessons and constantly explore, to better host the activity and facilitate the reform and uplift of interior design education.

Zou Huying
July , 2016

邹瑚莹
中国建筑学会室内设计分会理事长
清华大学　教授
Zou Huying
President of CIID
Professor of Tsinghua University

序

自 2013 年以来，由中国建筑学会室内设计分会（简称 CIID）主办、高校与知名设计企业共同创建并参与的 CIID“室内设计 6+1”校企联合毕业设计活动，至今年已经成功举办四届了。每一届设计题目的选定都来自知名设计企业当前需要解决的实际课题，分别以“赛后商机——国家体育场赛后改造室内设计”“城轨新境——上海地铁改造环境设计”“兵工遗产——南京晨光 1865 创意产业园环境设计”“鹤发医养卷——北京曜阳国际老年公寓环境改造设计”为题开展相关活动。从开题的现场勘踏，到中期检查，再到毕业答辩，来自不同地域、不同教学体系的同学们通过与不同院校评委老师、专家们的交流，视角和思路都得到了拓展，受益良多，取得了良好成绩。活动既面向社会，面向市场，又让学生在毕业之前能理论联系实际地掌握知识，为今后尽快转变角色进行工作打下基础，以适应当前社会的高速发展。

“室内设计 6+1”校企联合毕业设计的特色之一是打造了一个高校与知名设计企业共同创建的教育平台，这是一个推动室内设计教育发展的平台，每年由 7 所高校师生和知名设计企业组成并参与活动，七所高校师生在这个平台上互相交流、互相学习、共同提高。

四年来，室内设计 6+1”校企联合毕业设计活动取得了良好的成果，这有助于创新、创业型人才的培养，有助于学校教育改革的推进。CIID 作为中国室内设计界唯一的学术团体，致力于推动我国室内设计事业的发展。CIID 一直以来重视年青学生的成长，不为虚名浮华，踏踏实实坚持不懈地努力工作是我们应尽的责任。今后 6+1 的活动将继续坚持依托行（企）业开展联合毕业设计教学，体现教育服务于人才培养需求的原则，还要总结经验，不断摸索，把这项活动更好地开办下去，以促进室内设计教育的改革与提高。

邹瑚莹
2016 年 7 月

Forword

Under the guidance of the strategy of building an innovation-oriented country and strengthening the country with talented people, based on the new environment of China's university majors as well as facing the new target of training interior design talents, CIID Interior Design 6+1, a school-enterprise cooperated Graduation Project platform came into being, which aims to explore new ways for training interior architects under the "outstanding program". In 2013, sponsored by the Interior Design Branch of China Architectural Society, the 6 universities, Tongji University, University of Technology, Harbin Institute of Technology, Xi'an University of Architecture and Technology, Beijing Architecture University, Nanjing Art Institute, which are located in different areas, setting up interior design courses under Architecture and Design majors, co-created the CIID Interior Design 6+1 platform with some famous designing companies. The school-enterprise cooperated CIID "interior design 6+1" image has been constantly strengthened by 6+1 units, 6+1 months, 6+1 steps, 6+1 sections in each event.

During the four consecutive activities of school-enterprise cooperated CIID "Interior Design 6+1", the graduation projects are always following the hot issues of industry and society. In 2013, the first year of this activity, the project is "Business Opportunities After the Games---Interior Design for National Stadium", which studies on the operation of large stadiums after big sports events. In 2014, "New Environment for Urban Rail--Environment Design for Shanghai Subway Reconstruction", discussed the environmental design of subway station with city updating. In 2015, "Military Industry Heritage---Environment Design of Chenguang 1865 Creative Industry Park, aimed protection and utilization of modern industrial heritage in historic and cultural cities. In 2016, "Medical and Health Care for the Elderly--Environment Improvement Design of Beijing YaoYang International Senior Citizen Apartment", dug out the the new needs of aging society, to adapt the existing medical and health care building to livable needs of the elderly.

China is the most populous country in the world and also the one has the most number of senior citizens. China's aging population is not only a problem of China itself, but also a problem of the world. As facing the great test and pressure of "getting old before getting rich ", the rapid aging of the population structure will cause the "suffocate" of China's economic and social development if we do not speed up the improvement of the service system of "active aging" and "healthy aging" . Therefore, to solve the aging problem in China has become an unavoidable issue.

Li Keqiang, premier of the State Council, put forward the concept of "Healthy China" for the first time in the government work report in 2015. He pointed out that health is the basic needs of the masses and we must continuously improve the level of health care and build a healthy China. In the 2016 government work report, premier Li stressed that to weave a tight and strong social security safety net, we will focus on carrying out pilot pension service comprehensive reform and promoting various forms of medical and health care integration. June 16, 2016, the National Health and Family Planning Commission Office and the General Office of the Ministry of Civil Affairs announced the first batch of national medical and health care pilot units, 50 cities (districts), including Beijing Dongcheng District.

Beijing YaoYang International Senior Citizen Apartment construction site is located 10km north of Miyun County, north to the Miyun Reservoir, east of Jingmi diversion canal head, west of Xiejia mountain, south of the Gongzizhuang Village, Tiangezhuang County. The total land area is 137461m2, with a total construction area of 67230m2 (including underground 5390m2) and the construction height is 18m. This project is one of the few large-scale development projects in the suburbs of Beijing, with the theme of living and health care for the elderly. The project area is surrounded by mountain on 3 sides, which is 3km away from Miyun Reservoir on the north, and on its south is plain for agriculture with unique natural conditions such as topography, water source and groundwater. Since the apartment is at the foot of the mountain, the climate here is cool in summer and warm in winter. Also, with rich vegetation and fresh air, it is a wonderful health resort in summer in suburb of Beijing.

This event continued the cooperated graduation project teaching way, reflecting the principle of education serves personnel training needs. After selection, the institute commissioned the Architectural Institute, Beijing Urban Construction

前言

陈静勇
执行主编
中国建筑学会室内设计分会副理事长
北京建筑大学　教授

Chen JingYong
Executive Editor
Vice President of China Architecture Society Interior Design Branch
Professor of Beijing Architecture University

在建设创新型国家和人才强国战略的指引下，立足我国学科专业的新环境，面对建筑室内设计人才培养的新目标，2013 年，由中国建筑学会室内设计分会主办，由同济大学、华南理工大学、哈尔滨工业大学、西安建筑科技大学、北京建筑大学、南京艺术学院等 6 所地处不同区域、设置建筑学或设计学等学科室内设计方向的高校与知名设计企业，共同创建了 CIID“室内设计 6+1”校企联合毕业设计活动新平台，探索“卓越计划”目标下的室内建筑师教育培养之路。6+1 个单位、6+1 个月、6+1 个环节、6+1 个板块等也使得 CIID“室内设计 6+1”校企联合毕业设计的每届活动在内涵凝练上不断丰富的传播形象。

CIID“室内设计 6+1”校企联合毕业设计活动连续开展 4 届以来，毕业设计选题始终是以行业、社会密切关注的热点问题为导向。2013（首届）的“赛后商机——国家体育场赛后改造室内设计”课题，涉足了大型体育赛事之后大型体育场馆经营利用问题；2014（第二届）的“城轨新境——上海地铁改造环境设计”课题，探讨了城市更新与地铁站点的环境设计问题；2015（第三届）的“兵工遗产——南京晨光 1865 创意产业园环境设计”研习了历史文化名城近代工业遗产保护与利用问题；2016（第四届）的“鹤发医养——北京曜阳国际老年公寓环境改造设计”，挖掘了既有医养建筑空间环境适应老龄社会宜居新需求问题。

中国是世界上人口最多的国家，也是老龄人口最多的国家。中国的人口老龄化问题，不仅仅是中国自身的问题，也是世界的问题，备受关注。面对“未富先老”的巨大考验和压力，如果不加快健全“积极老龄化”“健康老龄化”的制度和服务体系，人口结构迅速老化将使未来中国经济社会的发展活力“窒息”。因此，解决中国老龄化问题已经成为不可回避的问题。

国务院总理李克强在 2015 年政府工作报告中首次提出“健康中国”概念，指出：健康是群众的基本需求，我们要不断提高医疗卫生水平，打造健康中国。李克强总理在 2016 年政府工作报告中提出，要织密织牢社会保障安全网，将开展养老服务业综合改革试点，推进多种形式的医养结合作为 2016 年的工作重点。2016 年 6 月 16 日，国家卫生计生委办公厅、民政部办公厅提出确定北京市东城区等 50 个市（区）作为第一批国家级医养结合试点单位。

北京曜阳国际老年公寓建设用地位于密云县城北 10km 处，北临密云水库，东临京密引水渠渠首，西侧为卸甲山，南侧为西田各庄镇龚庄子村。总用地面积为 137461 m²，总建筑面积为 67230 m²（含地下 5390 m²），建设高度 18m。该项目是北京市城郊区县范围内为数不多的以老年人居住和养生为主题的大型开发项目。该项目园区三面环山，山北侧 3km 外为密云水库，南侧为平原农地，地形地貌、水源大气和地下水等自然条件独特。由于园区处于京郊山脚下，园区内冬暖夏凉、植被丰富、空气清新，成为京郊的避暑养生胜地。

本届活动继续坚持依托行（企）业开展联合毕业设计教学，体现教育服务于人才培养需求的原则。经学会遴选，委托北京城建设计发展集团股份有限公司建筑院负责本届联合毕业设计活动命题和总指导工作。

Design and Development Group Co., Ltd. to be responsible for this graduation project design propositions and general guidance. Medical and Health Care for the Elderly Volume--Beijing YaoYang International Senior Citizen Apartment Reconstruction Design---CIID Interior Design 6+1, 2016, the 4th school-enterprise cooperated Graduation Project, will be published by China Water&Power Press. The book is written in English and Chinese to record remarkable processes of the activities, including the proposition study, opening ceremony and on-site survey, mid-term check, dissertation defense and award, editing and publishing, special exhibitions, foreign exchange,etc.

The activity includes four teaching sessions: on Oct 20th, 2015, CIID Interior Design 6+1 proposition seminar was held in Lanzhou, in 25th annual meeting of the Society; on Mar 5th-6th, 2016, opening ceremony in Beijing Architecture University and Beijing YaoYang International Senior Citizen Apartments on-site reconnaissance; Apr 16th-17th, 2016, mid-term check in Tongji University, Shanghai; Jun 18th-19th, dissertation defense and award ceremony in Harbin Institute of Technology, Harbin. The institutes, colleges, enterprises and experts jointly carried out a comprehensive practical teaching activity about Beijing YaoYang International Senior Citizen Apartment. There are 3 Design Themes: reconstruction of medical and health care building and interior design, apartment environment and landscape design, medical and health care facilities and display design. Teachers and students discussed how to adapt the existing medical buildings to the livable needs of the elderly and have in-depth communication which promotes the upgrading of the teaching level of graduation project.

The next two sessions of the event are also scheduled to proceed. The Medical and Health Care Volume is being co-edited by the Institute and the Beijing Architecture University. The manuscripts about the cooperation results of Institutes, universities, enterprises and experts have been collected. The thematic exhibition will be present from October 27th to 30th in 2016, the 26th (Hangzhou) Annual Conference and International Academic Exchange Symposium.

The Medical and Health Care Volume take 6+1 activity steps as editorial system, put together the activities "Constitution" (2015 revision), the current activity "outline", "graduation project framework", "working rules of dissertation defense, Awards and Recognition"and other basic material of cooperated graduation project teaching in the" General Principles of Activities "to facilitate future search. In 2015, we added 3 parts, on site research survey, process solution, teaching and research paper. This year, we created “lecture summary” to record the key points and theme of the lectures. Classified according to the 3 “Design theme”, We compiled the college works, listed the works by the design concept, highlighted the problem-solving focus, attached expert reviews, student reflections, award certificates, teachers and students photos and expert message. Finally, the photos of each session during the activities will be shown in the "activity impression". This book continues to be listed as the Chinese Architectural Society of Interior Design Branch recommended teaching reference.

The Medical and Health Care for the Elderly Volume will be published at the same time with (Hangzhou) annual meeting. As a case study of interior design education exchange among AIDIA, IFI and other associations, it is also an extensive session of the activity.

In short, I hope that more activities can be included, but considering the limited words, the list is far from being complete. So your suggestions and comments are welcomed here for us to improve.

Thanks to the Architecture Discipline Steering Committee, National Higher Education Institution, Design Discipline Teaching Committee, the Ministry of Education, for your long time guidance to the related subjects construction!

Thanks to the relevant local professional committees, participated colleges and universities, Beijing Urban Construction Design and Development Group Co., Ltd. lectures and assessment experts, for your help and guidance on this event!

Thanks to the volunteers and contractor colleges and universities, for their considerate preparation.

Thanks to thank China Water&Power Press, for your great support and hard work of editors.

The 2016, fourth CIID Interior Design 6+1 activity has been successfully concluded. Graduates who have come from Reconstruction Design of Medical and Health Care Building, Interior Design, Display Design, and Landscape Design graduation projects, now are the Bachelors of Engineering, Bachelors of Engineering and Bachelors of Arts, who will set foot on the new journey of meeting industry needs and going forward to seek excellence.

This year is the 27th anniversary of the establishment of Interior Design Branch, China Architectural Design Institute. Here, we present Medical and Health Care for the Elderly Volume to you as one of the innovative achievements of the Institute. In response to the national innovation-driven development strategy and serving economic and social development needs, cultivating a large number of qualified interior designers who are innovative, is always the objective industry authorities, associations, colleges, enterprises and society face.

Chen Jingyong

July , 22nd , 2016

《鹤发医养卷——北京曜阳国际老年公寓环境改造设计（CIID“室内设计6+1”2016（第四届）校企联合毕业设计）》由中国水利水电出版社出版；全书采用中英文对照方式记载了活动的命题研讨、开题踏勘、中期检查、答辩评审与表彰奖励、编辑出版、专题展览、对外交流等的过程印记。

本届活动的前4个教学环节，从2015年10月20日［兰州市，学会二十五届（甘青）年会］召开CIID“6+1室内设计”校企联合毕业设计命题研讨会，经历2016年3月5日～6日（北京市，北京建筑大学）开题仪式和北京曜阳国际老年公寓现场踏勘、4月16日～17日（上海市，同济大学）中期检查，到6月18日～19日（哈尔滨市，哈尔滨工业大学）辩评审会和表彰奖励仪式结束，学会、学校、企业、专家等多方协同，就北京曜阳国际老年公寓的医养建筑改造与室内设计、医养园区环境与景观设计、医养建筑设施与展示设计3个“设计专题”，联合开展了综合性实践教学活动。师生们对既有医养建筑空间环境适应老龄社会宜居新需求问题的协同探讨感同身受，交流深入，促进了毕业设计教学水平的提升。

本届活动的后2个展示环节也在按计划进行。《鹤发医养卷》由学会和北京建筑大学联合主编，满载着学会、高校、企业、专家等协作成果的书稿从东西南北中汇集起来；专题展览将于10月27日～30日在学会2016年二十六届（杭州）年会暨国际学术交流会上呈现。

《鹤发医养卷》以6+1个“活动脚步”板块作为编辑体例，将活动《章程》（2015修订版）、本届活动《纲要》、《毕业设计框架任务书》、《答辩、评奖、表彰工作细则》等联合毕业设计教学基础资料归集在“活动总则”，以利今后检索；在2015届扩充了“调研踏勘”“过程方案”“教研论文”等3个板块的基础上，本届再新增“讲座提要”板块，以记录专家讲座的主旨和要点；按3个“专题设计”分别汇编各校作品，列比作品设计概念，突出解决问题侧重点，附上专家点评、学生感言、获奖证书、师生照片等；进而安排“专家寄语”；最后将反映活动各环节的影像定格在“活动印记”；该书继续列作中国建筑学会室内设计分会推荐教学参考用书。

《鹤发医养卷》将于（杭州）年会同期出版发行，也作为与亚洲室内设计联合会（AIDIA）和国际室内建筑师/设计师联盟（IFI）等开展室内设计教育交流的特色案例，成为活动的1个拓展环节。

总之，希望编入的活动印记较多，限于篇幅，编辑之中难免挂一漏万，敬请指正，以利改进。

感谢全国高等学校建筑学学科专业指导委员会、教育部高等学校设计学类专业教学委员会等长期以来对高校相关学科专业建设工作的指导！

感谢学会相关地方专业委员会、参加高校、北京城建设计发展集团股份有限公司建筑院、专题讲座与评审专家等对本届活动的指导和帮助！

感谢本届活动承办高校的悉心筹备，以及甘当志愿者的朋友们！

感谢中国水利水电出版社的大力支持和编辑们的辛勤工作！

CIID“室内设计6+1”2016（第四届）校企联合毕业设计活动已圆满结束，从医养建筑改造设计、室内设计、展示设计、景观设计等协同设计教学活动中走过来的建筑学、工业设计、环境设计、艺术与科技4个专业的毕业生们，已成为了新一届的建筑学学士（专业学位）、工学学士、艺术学学士，他们即将踏上面向行业需求、走向卓越的新征途。

今年是中国建筑学会室内设计分会成立27周年。在此，以《鹤发医养卷》作为学会室内设计教育工作创新成果之一，呈献朋友们。响应国家创新驱动发展战略和服务经济社会发展需求，培养造就一大批创新能力强、适应需要的高质量室内设计师，始终是行业主管部门、学（协）会、高校、企业、社会等共同面对的命题。

陈静勇

2016年7月22日

中国建筑学会室内设计分会（CIID）2016年（

届）“室内设计6+1”校企联合毕业设计答辩

2016年6月18日

CIID“室内设计 6+1”2016（第四届）校企联合毕业设计 全体师生合影

目　录

Preface
序 005
Forword
前言 007

Introduction of the Activity
01　活动总则 015
Activity Outline
2016届活动纲要 016
The Artides of Association
2016届活动章程 019
The Framework of the Task
2016届框架任务书 023
Report of Building Plan
附件：建筑与场地图 024
Defense Rules
答辩规则 027

Site Survey
02　调研踏勘 029
Design process
03　过程方案 045

Aged care district environmental and landscape design
04　医养园区环境与景观设计 065
Communal Living
团聚 066
To Live, To Learn
十方黉舍 072
Awakening the Senses
唤醒感知 076
Seniors & Flowers – Beijing Yaoyang International Retirement Community Landscape Renovation
老人与花海 -- 北京曜阳国际老年公寓环境改造设计 080
Innovative Retirement Community
飞越老人院 088
Six Harmonies
六合・宓馆 094
Opera and Clouds
梨舍・云居 098
Memories in Courtyard Houses
四方合院 102
Regional Self-Sufficient Life in Groups
居住组团区域自足 106

Aged care building renovation and interior redesign
05　医养建筑改造与室内设计 113
Innovative Retirement Community
飞越老人院 114
Memories in Courtyard Houses
四方合院 122
Regional Self-Sufficient Life in Groups
居住组团区域自足 128
Living in a Fairyland
仚居行 136
Station of Interaction
交互・驿站 146
To Live, To Learn
十方黉舍 154
Six Harmonies
六合・宓馆 160
Lotus Garden
藕耕苑 164
Opera and Clouds
梨舍・云居 172
Awakening the Senses
唤醒感知 178
Communal Living
团聚 186

Aged care facility and exhibition design
06 医养建筑设施与展示设计 193
Photosynthesis – Regrowth
光合作用——再生长 194
Opera and Clouds
梨舍·云居 202
Six Harmonies
六合·宓馆 208
The speaker spoke
07 演讲人发言 213
Ying Zhi Operation Mode with the Combination of Rehabilitation Treatment and Aged Care
医康养结合的英智运营模式 215
Speech from Professor Hao Luoxi at Tongji University
同济大学郝洛西教授 215
Speech from Wang Chuanshun
王传顺老师 217

Teaching Graduation Design
08 联合毕业设计的教学探讨 219
Tutorial Process of the Retirement Community Environmental Redesign Project: Reflections and Explorations
适老环境改造设计教学中的反思与探索
左琰 221
Interdisciplinary Teaching of Basic Architectural Design Knowledge under the Environmental Design Major: An Exploration
环境设计专业建筑设计基础知识课程群集教学探索
李莉 225
How to Become a Professional Designer
——To students participating in the 4th CIID "Interior Design 6+1" University-Business Graduation Projects Competition
如何成为职业设计师
——写给CIID"室内设计 6+1"2016（第四届）校企联合毕业设计同学们
马辉 231
Some Thoughts on the Education Reform of the Environmental Design Major
谈环境艺术专业教学改革
西安建筑科技大学 235
Bathroom Well-design Exploration
适老化卫生间设计探索
杨琳、朱宁克 239
Professional Competencies Required to be Exhibition Designers in China: A Study
我国展示设计人才职业要求研究
朱飞 245
Research-Oriented Graduation Projects Under the "Meridian Theory – Survey throughout the Entire Project" Innovative Teaching Model
"经线论——调研贯穿设计始终"创新教学模式下的研究型毕业设计课题实践
任彝 249

Activity Mark
09 活动印记 253
Message
寄语 254
Tidbits
花絮 260

Unit Introduction
单位介绍 263

零壹

01

活動總則

Introduction of the Activity

CIID“室内设计 6+1”2016（第四届）
校企联合毕业设计

CIID"Interior Design 6+1"2016(Fourth Session)
University and Enterprise Joint in Graduation Design

鹤发医养卷
——北京曜阳国际老年公寓环境改造设计

White hair volume medical support
—Beijing Yao Yang International Apartments
for the elderly environmental reconstruction design

2016 届活动纲要

一、课　　题：鹤发医养——北京曜阳国际老年公寓环境改造设计

二、项目地点：

北京曜阳国际老年公寓建设用地位于密云县城北 10 公里处，北临密云水库，东临京密引水渠渠首，西侧为卸甲山，南侧为西田各庄镇龚庄子村。以改造园区内 C-1 栋、C-2 栋建筑及其周边空间环境为主。

三、主办单位：中国建筑学会室内设计分会（CIID）

四、承办高校：

CIID2015 年第二十五届（甘青）（命题研讨）
北京建筑大学（开题仪式、现场踏勘）
同济大学（中期检查）
哈尔滨工业大学（答辩评审、表彰奖励）
北京建筑大学（《鹤发医养卷》总编）

五、参加高校（学院 \ 专业）：

同济大学（建筑与城市规划学院 \ 建筑学）
华南理工大学（建筑学院 \ 环境设计）
哈尔滨工业大学（建筑学院 \ 环境设计）
西安建筑科技大学（艺术学院 \ 环境设计）
北京建筑大学（建筑与城市规划学院 \ 工业设计）
南京艺术学院（工业设计学院 \ 艺术与科技）
浙江工业大学（艺术学院 \ 环境设计）

六、命题企业：北京城建设计发展集团股份有限公司建筑院

七、出版企业：中国水利水电出版社

八、媒体支持：中国室内设计网 http://www.ciid.com.cn

九、时　　间：2015/10-2016/10

十、活动安排：

序号	阶段	时间	地点	活动内容	相关工作
1	命题研讨	2015/10/16-2015/10/20	学会二十五届（甘青）年会	• 10/16 报到，场外活动 • 10/19 学会二十五届（甘青）年会开幕式及学术交流活动 • 10/20 2015（第三届）总结与 2016（第四届）活动命题研讨会	• 命题企业负责起草《2016（第四届）校企联合毕业设计框架任务书》 • 研讨《框架任务书》 • 商讨确定承办高校工作
2	教学准备	2015/10/21-2016/03/03	各高校	• 校企联合毕业设计教学工作准备	• 各高校报送参加毕业设计师生名单 • 各高校反馈对《活动纲要》的修改意见和建议 • 各高校结合参加专业实际，依据《框架任务书》，分别编制本校该专业《毕业设计详细任务书》 • 各高校安排文献检索与毕业实习 • 开题仪式、现场踏勘等活动准备 • 学会和各高校邀请支持企业 • 开题仪式和现场踏勘准备
3	开题踏勘	2016/03/04-2016/03/06	北京建筑大学	• 03/04　报到 • 03/05　开题仪式专题讲座 • 03/06　现场踏勘调研交流	• 举行开题仪式 • 学会安排专题讲座；颁发讲座专家聘书 • 安排现场踏勘，核对图纸、补充测绘、拍照、访谈等 • 编制调研 PPT，进行调研工作交流 • 商议中期检查、编辑出版等相关工作

序号	阶段	时间	地点	活动内容	相关工作
4	方案设计	2016/03/07-2016/04/14	各高校	• 方案设计	• 各高校安排相关讲授、辅导、设计、研讨等 • 选择设计专题，明确方案设计目标，完成相关文案、图表等 • 完成调研报告、方案设计基本图示、效果图、模型、分析图表及等相应成果 • 中期检查准备；编制中期检查方案设计汇报 PPT
5	中期检查	2016/04/15-2016/04/17	同济大学	• 04/15　报到 • 04/16　中期检查 • 04/17　专题讲座调研交流	• 中期检查汇报、专家点评 • 研讨方案设计深化重点 • 商议答辩评审、表彰奖励、编辑出版、展览等相关工作
6	深化设计	2016/04/18-2016/06/16	各高校	• 方案深化设计	• 完成方案深化设计图示、效果图、模型、分析图表、设计说明等相应成果，制做展板（A0 竖版，3 张 / 方案组），提交电子版 • 编制毕业设计答辩汇报 PPT，提交电子版 • 《鹤发医养卷》分配页面排版（每一方案组每一设计专题 4P 或 6P）；提交电子版 • 答辩评审、表彰奖励准备
7	答辩评审	2016/06/17-2016/06/19	哈尔滨工业大学	• 06/17　报到 • 06/18　答辩、评审 • 06/19　表彰奖励	• 答辩布展与观摩 • 毕业答辩，等级奖评审 • 表彰毕业设计等级奖、优秀毕业设计指导教师、毕业设计最佳组织单位、毕业设计突出贡献单位等 • 2016（第四届）活动总结与 2017（第五届）活动命题研讨会
8	编辑出版	2016/06/20-2016/10/15	学会、参加高校、命题企业、支持企业、出版企业	• 07/15 前 高校完成《鹤发医养卷》作品书稿部分编辑工作，提交学会秘书处 • 08/31 前 完成书稿总编工作，提交出版企业校审 • 09/15 前 完成书稿校审工作，形成清样，送印厂 • 10/15 前 《鹤发医养卷》出版	• 各高校和相关专家提交《鹤发医养卷》书稿 • 学会秘书处和北京建筑大学负责书稿总编 • 出版企业负责书稿校审 • 学会二十六届（杭州）年会“室内设计 6+1”总结研讨会《鹤发医养卷》发行式准备
9	展览交流	2016/10/27-2016/10/30	学会二十六届（杭州）年会室内设计教育论坛	• 10/27　报到 • 10/28 2016（第四届）校企联合毕业设计活动总结与 2017（第五届）校企联合毕业设计命题研讨 • 10/28　《鹤发医养卷》发行式 • 10/29 CIID“室内设计 6+1”校企联合毕业设计教学组在 AIDIA（杭州）教育专场上做报告交流 • 10/27 ～ 30 学会年会专题展览	• 展板布展与观摩 • 学会二十六届（杭州）年会“室内设计 6+1”总结研讨会《鹤发医养卷》发行式 • 学会室内设计教育成果交流活动

The Artides Of Association

In order to meet the need of cultivating special talents of interior design in the fi eld of urban and rural construction,strengthen the pertinence of indoor architect cultivation, promote the relevant institutions of higher education's major construction on interior design discipline and communications on teaching and guide the relevant majors conduct teaching work on graduation design towards architecture, the institute of interior design of Architectural Society of China(CIID, Association for short in the following) acted as the director of advocacy and the colleges and universities who have interior design and the relating disciplines and well-known architectural and interior design enterprises carry out joint graduation design together.

In order to make the graduation design to be normative and to form its brand and characteristics, the CIID made regulations based on the opinions and advices of the relevant schools. The regulations passed on the CIID "interior design 6+1" 2013 (fi rst) school and enterprise joint graduation design proposition meeting and are announced now.

Ⅰ The background, aim and significance of the school and enterprise joint graduation design

In 2010, the Ministry of Education launched the "Excellent Engineer education program", and announced the subject and major list added to the plan of the "education program for excellent Engineer" from 2011 to 2013. The Committee of degree of the State Council and the Ministry of education publish the "discipline catalogue of degree awarding and talent cultivating (2011)" and added the discipline of "Arts (13)", making "Design Science (1305)" as the fi rst class discipline under Arts. The environmental design is advised as the second class discipline under design science and the interior design is advised as the second class discipline under design science and the interior design is advised as the second class discipline under Architecture. In 2012, the Ministry of Education announced the "catalog of undergraduate majors of colleges and universities" (2012), setting the major of "Design Science (1305)" under" Arts science" and making "environment design" (130503) as its core majors. The independent setting of the "Arts", the setting of the fi rst class principle of design science and the setting and adjusting of interior design and other majors have formed the new professional pattern of the environment design education and the cultivating of special talents of interior design.

The holding of the interior design school and enterprise joint graduation design has a great effect on strengthening the characteristic of relevant disciplines, deepening the teaching exchanges of the graduation design, promoting the innovation of interior design education and cultivating the interior design special talents needed by the service industry during the conduction of the "Excellent Engineer education program" held by the ministry of education.

Ⅱ The organizations of the school and enterprise joint graduation design

1. The host unit

School and enterprise joint graduation design activity was hosted by the CIID (or the authorized local professional committee), and obtained the guidance and support of the professional high school science teaching guidance committee of architecture, the Ministry of education teaching and guidance committee of design and the Ministry of education teaching and guidance committee of industry design.

2. The participant and host schools

Usually, the 6 universities and colleges with the near discipline and specialty conditions and the relevant majors of interior design became the participant schools through consulting and organizing. There was some area span between the locations of the participant schools, and formed a certain interdisciplinary and collaborative design conditions in subject between.

Every year, there will be one to four schools be chosen as the host schools for the opening ceremony and site survey, the mid-term defense, the judge and award of the graduation defense and the editing and publishing of the school and enterprise joint graduation design of that session. The proper number of students participating in the joint graduation design for each school is six and the students must be companied with one to two directors. Among the teachers, there must be at least one has senior professional title and be familiar with environment design, interior design and other industry practice business and has a wide communication with the relevant enterprises.

3. The proposition enterprise

The schools hosting the school and enterprise joint graduation design opening ceremony and site survey are responsible for recommending the CIID (or the authorized local professional committee) one famous architecture and interior design enterprise in their provinces (cities) to be as the proposition enterprise. The proposition enterprise are responsible for creating the proposition, making the sketch task book of the graduation design, participating in the opening ceremony and site survey, mid-term examination, the judge and award of the graduation defense, editing and publishing, exhibitions, external exchanges.

4. The supporting enterprise

The participant schools are responsible for recommending the CIID (or the authorized local professional committee) one famous architecture and interior design enterprise in their provinces (cities) to be as the supporting enterprise during every session of joint graduation design. The CIID (or the authorized local professional committee) and the supporting enterprises should sign the agreement of activity supporting and feedback and arrange the supporting enterprise to participate in the graduation design. The CIID (or the authorized local professional committee) is responsible for communicating the selected experts of the supporting enterprise, giving special lecture around the topic of the graduation design, participating in the in the opening ceremony and site survey, mid-term examination, the judge and award of the graduation defense, editing and publishing, exhibitions, external exchanges.

5. The pressing enterprises

Based on every session of school and enterprise joint graduation design, the CIID (or the authorized local professional committee) should select one famous pressing enterprise to take the responsibility of the publishing work of the school and enterprise joint graduation design.

Ⅲ The process of the school and enterprise joint graduation design

(1) According the graduation teaching work of the participant schools, the school and enterprise joint graduation design is held once every year (according to the session of the graduates).

(2) The sections of the school and enterprise joint graduation design mainly include the researching of the topic of the graduation design, the opening ceremony and site survey, mid-term examination, the judge and award of the graduation defense, editing and publishing of the school and enterprise joint graduation design of that session and exhibitions. It usually takes 6 months for the teaching section of the school and enterprise joint graduation design. External exchange is regarded as one extending section of the school and enterprise joint graduation design.

(3) The CIID (or the authorized local professional committee), as the host unit of this activity, is responsible for the overall planning, publicity, coordinating the participant schools, the relevant enterprises and the exhibition institutions, employing the relevant experts to hold special academic lectures, organizing annual award and the selection and awarding of the excellent guiding teachers, organizing units and special contribution of graduation design as well as the education and international exchange of interior design.

(4) The participant schools, together with the relevant enterprises, should draw out the outline of the activity of the school and enterprise joint graduation design under the guidance of the CIID (or the authorized local professional committee) and then the outline is sent to the CIID

2016 届活动章程

为服务城乡建设领域室内设计专门人才培养需求，加强室内建筑师培养的针对性，促进相关高等学校在室内设计学科专业建设和教育教学方面的交流，引导相关专业面向建筑行（企）业需求组织开展毕业设计教学工作，由中国建筑学会室内设计分会（CIID，以下简称学会）倡导、主管，国内设置室内设计相关学科专业的高校与知名建筑与室内设计企业开展联合毕业设计。

为使联合毕业设计活动规范、有序，形成活动品牌和特色，学会在征求相关高等学校意见和建议的基础上形成本章程，并于学会（CIID）"室内设计6+1"2013（首届）校企联合毕业设计命题会上审议通过，公布试行。

一、校企联合毕业设计活动设立的背景、目的和意义

2010 年教育部启动了"卓越工程师目录（2011 年）"，增设了"艺术学（13）"学科门类，将"设计学（1305）"设置为"艺术学"学科门类中的一级学科。"环境设计"建议作为"设计学"一级学科下的二级学科，"室内设计"建议作为新调整的"建筑学（0813）"一级学科下的二级学科。2012 年教育部公布了《普通高等学校本科专业目录（2012 年）》，在"艺术学"学科门类下设"设计学类（1305）"专业，"环境设计（130503）"等成为其下核心专业。"艺术学"门类的独立设置，设计学一级学科以及环境设计、室内设计等学科专业的设置与调整，形成了我国环境设计教育和室内设计专门人才培养新的学科专业格局。

举办室内设计校企联合毕业设计活动，对在教育部"卓越工程师教育培养计划"实施中加强相关学科专业特色建设，深化毕业设计各教学环节交流，促进室内设计教育教学协同创新，培养服务行（企）业需求的室内设计专门人才，具有十分重要的意义。

二、校企联合毕业设计活动组织机构

1．主办单位

校企联合毕业设计活动由学会（或经授权的地方专业委员会）主办，得到了全国高等学校建筑学学科专业指导委员会、教育部高等学校设计学类专业教学指导委员会、教育部高等学校工业设计专业教学指导分委员会等的指导和支持。

2．参加高校、承办高校

校企联合毕业设计活动一般由学科专业条件相近，设置室内设计方向的相关专业的6所高校间通过协商、组织成为参加高校。参加高校间所处地理区域具有一定的距离，在学科专业间形成一定的交叉性和协同设计条件。

每年在参加高校中推选 1 ～ 4 所高校分别作为毕业设计开题仪式与现场踏勘、中期检查、答辩评审与表彰奖励及当届《校企联合毕业设计》（主题卷）编辑出版等活动的承办高校。

每所高校参加联合毕业设计学生一般以6人为宜，要求配备 1 ～ 2 名指导教师，其中至少有 1 名指导教师具有高级职称，熟悉环境设计、室内设计等工程实践业务，与相关领域企业联系较广泛。

3．命题企业

承办校企联合毕业设计开题仪式与现场踏勘的高校，负责向学会（或经授权的地方专业委员会）推荐所在省（市）的（1 家）行业知名建筑与室内设计企业作为毕业设计命题企业。命题企业负责毕业设计课题命题，编制毕业设计框架任务书，参与开题仪式与现场踏勘、中期检查、答辩评审与表彰奖励、编辑出版、专题展览、对外交流等工作。

4．支持企业

参加高校在每届校企联合毕业设计活动中，分别向学会（或经授权的专业地方委员会）推荐（1 家）行业知名建筑与室内设计企业作为毕业设计支持企业，由学会（或经授权的地方专业委员会）负责联系支持企业选派专家，围绕毕业设计课题进行专题讲座，参与毕业设计开题仪式与现场踏勘、中期检查、答辩评审与表彰奖励、编辑出版、专题展览、对外交流等活动。

5．出版企业

学会（或经授权的地方专业委员会）就每届校企联合毕业设计活动，遴选（1 家）行业知名出版企业，负责承担当届《校企联合毕业设计》（主题卷）的出版工作。

三、校企联合毕业设计活动组织流程

（1）校企联合毕业设计活动按照参加高校毕业设计教学工作安排在每个年度(按毕业生届次)举行1次。

（2）校企联合毕业设计活动主要教学环节包括：毕业设计命题研讨、毕业设计开题仪式与现场踏勘、毕业设计中期检查、毕业设计答辩评审与表彰奖励、当届《校企联合毕业设计》（主题卷）编辑出版、专题展览等（6 个）主要环节。校企联合毕业设计活动的教学环节时间跨度一般为 6 个月，对外交流作为联合毕业设计活动的（1 个）扩展环节。

（3）学会（或经授权的地方专业委员会）作为活动主办单位，负责活动总体策划、宣传，协调参加高校、相关企业、展览机构等，聘请有关专家举办专题学术讲座，组织毕业设计学年奖、毕业设计优秀指导教师、毕业设计优秀组织单位、毕业设计特殊贡献奖等的评选、表彰，以及室内设计教育国际交流工作。

（4）参加高校在学会（或经授权的地方专业委员会）的指导、协调下，联合相关企业等，共同拟定校企联合毕业设计活动纲要，报学会（或经授权的地方专业委员会)审定。联合毕业设计开题仪式与现场踏勘、中期检查、答辩评审与表彰奖励、《校企联合毕业设计》（主题卷）编辑出版等工作由承办高校分别落实。

（5）命题研讨。参加高校的毕业设计指导教师参加联合毕业设计命题研讨会。毕业设计课题由命题企业与参加高校，着眼城市设计、建筑设计、环境设计、室内设计等领域发展前沿和行业热点问题，结合参加高校毕业设计教学实际商讨形成，报学会（或经授权的地方专业委员会）审定。毕业设计命题要求具备相关设计资料收集和现场踏勘等条件。命题研讨会一般安排在高校秋季学期中（每年 11 月左右），结合当年学会学术年会的教育论坛（或安排专题研讨会）进行。

（6）开题仪式与现场踏勘。参加高校的毕业设计师生参加联合毕业设计开题仪式和现场踏勘活动，命题企业提供毕业设计必要的设计基础资料和设计任务需求，安排现场踏勘等活动等。开题活动一般安排在高校春季学期开学初（3 月上旬）进行。

（7）中期检查。参加高校的毕业设计师生参加联合毕业设计中期检查活动。参加高校毕业设计指导教师和相关企业专家等，对毕业设计中期成果进行检查、评审，开展教学交流。中期检查一般安排在春季学期期中（4 月下旬）进行。中期检查活动中，每所参加高校优选不超过 3 个方案组进行陈述与答辩；其中陈述不超过 10 分钟，问答不超过 10 分钟。

（8）答辩评审与表彰奖励。参加高校的毕业设计师生参加联合毕业设计答辩评审与表彰奖励活动。学会（或经授权的地方专业委员会）专家、命题企业专家、

(or the authorized local professional committee) for approving. The host schools should fi nish the work of the opening ceremony of the joint graduation and site survey, mid-term examination, the judge and award of the graduation defense, editing and publishing, exhibitions.

(5)The research of the proposition.The guiding teacher of the participant schools should attend the research meeting of the proposition of the joint graduation design. The graduation design topic was formed by the proposition enterprise and participant schools by focusing on city design, architectural design, environmental design, the frontier of the interior design development and hot issues in industry and combining with the teaching reality of graduation design of the participant schools. Then the topic is sent to the CIID (or the authorized local professional committee) for approving. The graduation design proposition is required to have the approaches to collecting some relevant design data and site survey. The proposition research is generally arranged in the fall semester of college (around November each year), and be held with the education forum of the academic annual meeting of the CIID (or arrangement of seminars).

(6)The opening ceremony and site survey.The graduate teachers and students should attend the opening ceremony and site survey of the joint graduation design. The proposition enterprise provides the essential basic materials and task of the graduation design and arranges site survey and other activities. The opening activity is generally arranged in the spring semester of college (around the fi rst ten days of March).

(7)The mid-term examination.The graduate teachers and students should attend the mid-term examination of the joint graduation design. The guiding teacher of the participant schools of the joint graduation design and the experts of the relevant enterprises should check and review the mid-term results of the graduation design and conduct teaching exchanges. The mid-term examination is generally arranged in the spring semester of college (in the last ten days of April).

During the mid-term examination, each university can choose not more than 3 defense project to have the statement and achievements exhibition; each graduation of the program group should choose their designing theme for the defense. Each statement time is 10 minutes; question-and-answer is not more than 10 minutes.

(8)The judge and award of the graduation defense.The graduate teachers and students should attend the judge and award of the graduation defense. The experts of the CIID (or the authorized local professional committee), guiding teacher of the participant schools of the joint graduation design and the experts of the relevant enterprises should attend the judge and award of the graduation defense. The judge and award of the graduation defense is generally arranged in the end of the spring semester of college (in the fi rst ten days of June).

During the graduation defense, each university can choose not more than 2 defense project to have the statement and achievements exhibition; each graduation of the program group should choose their designing theme for the defense. Each statement time is 15 minutes; question-and-answer is not more than 15 minutes. In addition, each university can choose not more than 2 recommendation schemes to have the achievements exhibition, which can not have the statement and defense.

On the base of defense and judge, the CIID (or the authorized local professional committee) organizes the selection of the annual award, the excellent guiding teachers, the excellent organizing units and special contribution of graduation design and so on and gives awards to them(The proportion of awards can be set at 1:2:4). The graduation design annual award is set according to the design project and the level award is set.

(9)The exhibitions.The CIID (or the authorized local professional committee) arranges one annual award works thematic exhibition of the graduation design (November-December every year) in the CIID academic annual meeting in the end year of each session of the graduation design (or the authorized local professional committee activities); At the end of the thematic exhibition the school year awarding works of the graduation can be exhibited in the relevant schools at home abroad.

(10)The editing and publishing.Based on the graduation design activities, one volume of "school and enterprise joint graduation design" should be edited per session. The "school and enterprise joint graduation design" is edited by the CIID (or the authorized local professional committee) and the relevant host schools. They are responsible for soliciting and typesetting; the participant schools are as the participating organization. The guiding teachers of the graduation design of the participant schools are responsible for the examination of the manuscripts; as the editor, the publishing enterprise is responsible for proofreading and publishing.

(11)External exchange.The "school and enterprise joint graduation design" usually host the issue ceremony in the CIID academic annual meeting in the end year of each session of the graduation design. The CIID carries out the exchange of international interior design by communicating the Asian Interior Design Association (AIDIA) and international interior architect / designer alliance (1F1). In this way, it can open a new door for publicizing the education of Chinese interior design and carrying out international exchange.

IV. The fund of the school and enterprise joint graduation design

(1)The CIID (or the authorized local professional committee) is responsible for raising money for the selection of the annual award, the excellent guiding teachers, the excellent organizing units and special contribution of graduation design as well as the thematic exhibition of the annual meeting of the CIID.

(2)The participant schools should afford the travel fees of the relevant personnel participating in the school and enterprise joint graduation design and the site fees and relevant fees for the exhibition of the annual awarding works of the graduation design.

(3)The host schools afford the site fees, conversation fees, organization fees of the opening ceremony of the joint graduation and site survey, mid-term examination, the judge and award of the graduation defense as well as the editing and pressing fees of the school and enterprise joint graduation design.

(4)The proposition enterprise, supporting enterprise and pressing enterprise are responsible for proving the CIID some fund for the school and enterprise joint graduation design activity.

V. Supplementary provisions

(1)The constitution was announced and conducted on January 13, 2013 interpreted by the CIID.

(2) The constitution was fi rst revised in March. 2014; and second revised in April .2015

参加高校毕业设计指导教师等作为评委，参加毕业设计答辩评审、表彰奖励等工作。毕业设计答辩评审与表彰奖励一般安排在春季学期期末（6月上旬）进行。

毕业答辩活动中，每所参加高校优选不超过2个答辩方案组进行陈述与答辩、成果展出；其中陈述不超过15分钟，问答不超过15分钟。此外，每所参加高校可再安排不超过2个自荐方案组进行成果展示，不参加陈述与答辩。

在答辩、成果展示、评审的基础上，学会（或经授权的地方专业委员会）组织评选毕业设计学年奖、毕业设计优秀指导教师、毕业设计优秀组织单位、毕业设计特殊贡献奖等，并给予表彰奖励。毕业设计学年奖按照等级奖（含一、二、三等奖，按照1:2:4比例设置）、优秀奖分别进行评选；其中，等级奖评选仅针对答辩方案设置，优秀奖针对自荐方案设置。

（9）专题展览。学会（或经授权的地方专业委员会）在每届联合毕业设计结束当年学会学术年会上（或经授权的地方专业委员会活动中）安排（1个）毕业设计学年奖作品专题展览（每年10月-11月）；专题展览结束后，毕业设计学年奖作品可在国内相关高校之间巡回展出。

（10）编辑出版。基于每届联合毕业设计活动，各编辑出版（1部）《校企联合毕业设计》（主题卷），作为学会推荐的室内设计、环境设计专业教学参考书。《校企联合毕业设计》（主题卷）编辑工作由学会（或经授权的地方专业委员会）和相应承办高校联合编辑，负责组稿、排版等工作；参加高校作为参编单位，参加高校毕业设计指导教师负责本校排版稿的审稿等工作；出版企业作为责任编辑，负责校审、出版。

（11）对外交流。《校企联合毕业设计》（主题卷）一般在当届联合毕业设计结束当年学会学术年会上举行发行式，并由学会通过联系亚洲室内设计联合会（AIDIA）、国际室内建筑师/设计师联盟（IFI）等开展国际室内设计教育成果交流，打开（1扇）宣传中国室内设计教育、开展国际交流的新大门。

四、校企联合毕业设计活动经费

（1）学会（或经授权的地方专业委员会）负责筹措评选毕业设计学年奖、毕业设计优秀指导教师、毕业设计优秀组织单位、毕业设计特殊贡献奖等等表彰奖励经费，以及学会年会专题展览经费等。

（2）参加高校自筹参加校企联合毕业设计活动相关人员的差旅费，以及毕业设计学年奖作品在本校巡展的场地及相关经费等。

（3）承办高校自筹校企联合毕业设计开题仪式与现场踏勘、中期检查、答辩评审与表彰奖励等相关活动的场地、会议费、组织费，以及《校企联合毕业设计》（主题卷）编辑出版经费等。

（4）命题企业、支持企业、出版企业等负责为向学会提供一定的对校企联合毕业设计活动资助经费等。

五、附则

（1）本章程2013年1曰13日通过并公布施行，由中国建筑学会室内设计分会负责解释。

（2）本章程2014年3月第一次修订；2015年4月第二次修订。

The Framework Of the Task

The 2016 (the 4th) CIID "Interior Design 6+1" University-Business Graduation Projects Competition Framework Specification is sponsored by Beijing Urban Construction Design & Development Group Co., Ltd. Architecture Institute upon the authorization of CIID, and formulated by the advisors taking part in the competition. Universities participating in the competition are required to, based on their existing graduation project programs and this Specification, formulate more detailed specifications for the implementation of the joint graduation project.

I. Project topic

Healthy Aging – Beijing Yaoyang International Retirement Community Environmental Redesign Project

II. Project locaiton

The project is located 10 km to the north of Miyun County, neighboring the Miyun Reservoir on the north, the head of the Beijing-Miyun Diversion Canal on the east, the Xiejia Mountain on the west and Gongzhuangzi Village in Gezhuang Town, Xitian on the south. With a site area of 137,461 m2, the project has a gross building area of 67,230 m2 (including 5,390 m2 underground area), and erects 18 m above the ground.

The project is one of the few large-scale real estate projects themed on aged care in the outskirts of Beijing.

Located in the outskirts of Beijing, the retirement community is surrounded by mountains on three sides, with flat farmland extending southward and the Miyun Reservoir lying 3 kilometers away on the north. Boasting advantaged geographical location, fresh air, clean water, as well as diverse vegetation and year-round mild weather, it is a great summer resort and also an idea place for retirement life.

III. Design scope

C-1: Built in 2010, the building adopts a three-story or four-story reinforced concrete frame structure, with a gross building area of 7,705 m2. It originally acted as a serviced department building, with standard rooms being arranged on one side of the interior corridor. The dimensions of the building is L x W x H = 77.35 m × 81.0 m × 14.45 m.

C-2: Built in 2010, the building adopts a two-story reinforced concrete frame structure, with a gross building area of 1,252.87 m2. It originally served as a care center and was designed as a hospital clinic. The dimensions of the building is L x W x H = 48.75 m × 21.15 m × 8.45 m.

Southern landscape areas: These areas include the south entrances, and the external areas and spaces of buildings in District A, B, C and D.

V. Design topic

According to the overall objectives of the Beijing Yaoyang International Retirement Community Environmental Redesign Project, universities participating in the competition are required to redesign the buildings in District C, and the external areas and spaces of buildings in District A, B, C and D. The competitors can select a design topic from the following options according to their majors and actual situations.

1. Aged care building renovation and interior redesign

Discuss the incorporation of local culture into interior design according to the principle of senior-oriented design. Transform District C into a functional space for entertainment and cultural communication. Functional space re-division, traffic flow design, interior design, and environmental protection design should be conducted according to the psychological and behavior characteristics of the seniors, for the purpose of creating a functional space ideal for the living and rehabilitation of the seniors.

2. Aged care district environmental and landscape design

Discuss the incorporation of seniors' psychological needs into landscape design according to the principle of senior-oriented design. Landscape the outdoor spaces of District A, B, C and D according to the psychological and behavior characteristics of the seniors, meeting their needs on outdoor activities and rehabilitation.

3. Aged care facility and exhibition design

Discuss the incorporation of seniors' physiological and psychological needs into artistic design according to the principle of senior-oriented design. Functional space re-division, facility design, sign system design, exhibition design and public art creation should be conducted in District C according to the psychological and behavior characteristics of the seniors, creating an ideal environment for the seniors to live, communicate and appreciate artistic works.

V. Design documents & materials at various stages

1. Thesis proposal documents: Design orientation analysis, etc.
2. Mid-term reporting documents: Preliminary design scheme
3. Final defense & reporting documents: Detailed design scheme, display boards, and manuscript

VI. Design achievements

(I) Design specificaiton

The design specification shall include general design philosophy description, design orientation, design derivation process analysis, function analysis, economic & technical indexes, graphical representations, diagrams, etc.

(II) Drawings

1. Site plan, landscape plan
2. Floor plan, roof plan, section, typical detailed floor plan
3. Interior and exterior environmental facility arrangement plans, furniture arrangement plan, sign system arrangement plan
4. Exhibition plan, detailed tool and facility arrangement plan
5. Detailed spatial node arrangment plan
6. Lists of main materials and furniture

(III) Color renderings

1. Interior design redenrings
2. Landscape renderings
3. Renderings of environmental facilities, exhibition tools, and public art works

(VI) Achievements submission

1. Thesis detrmination communication.Each university is required to submit and present one thesis proposal in 20 minutes (10 minutes for presentation, and 10 minutes for Q&A), and submit a periodical project report (PPT).
2. Mid-term inspection.Each university is required to select and present three preliminary design schemes in 20 minutes (10 minutes for presentation, and 10 minutes for Q&A), and submit a periodical project report (PPT).
3. Defense

(1) Each university is required to select and present two final design schemes in 30 minutes (15 minutes for presentation, and 15 minutes for Q&A), and submit the graduation design final report (PPT), display boards, and typeset design schemes.

(2) Each design team is required to prepare three A0 display boards (900 × 1,200 mm) with a resolution ration not less than 100 dpi. Annual exhibition display boards are used as the template, and will be provided by the organizer. Design work exhibition will be undertaken by the organizer and arranged by the universities participating in the event.

4. Materials for publication.To edit and publish the Healthy Aging Volume – Beijing Yaoyang International Retirement Community Environmental Redesign - 2016 (the 4th) CIID "Interior Design 6+1" University-Business Graduation Projects Competition, all the organizations and individuals participating in the competition are required to provide electronic versions of corresponding regulations, teaching documents, papers (one paper from one university by the advisors in joint names; within 2,500 characters), typeset design schemes (6 pages for one scheme), experts' comments (a comment of less than 200 characters to a scheme by every advisors), experts' notes (a note of less than 500 characters by every expert), students' thoughts (less than 200 characters by every student), ID photos (one by every expert, advisor and student), award certificates (various types and grades), sidelights (main steps), and candidate introduction (within 1,000 characters).

VII. Site plan and architectural drawing & photos

Site plan, architectural drawings and photos of each building

2016届框架任务书

《CIID“室内设计6+1”2016（第四届）校企联合毕业设计框架任务书》由中国建筑学会室内设计分会委托北京城建设计发展集团股份有限公司建筑院牵头提出，经2016（第四届）校企联合毕业设计指导教师联合编制形成。活动参加高校应结合本校毕业设计教学工作实际，据此进一步编制相应的毕业设计详细任务书，指导毕业设计教学工作，开展联合毕业设计活动。

一、项目名称

鹤发医养——北京曜阳国际老年公寓环境改造设计

二、项目地点

北京曜阳国际老年公寓建设用地位于密云县城北10km处，北临密云水库，东临京密引水渠渠首，西侧为卸甲山，南侧为西田各庄镇龚庄子村。总用地面积为137461㎡，总建筑面积为67230㎡（含地下5390㎡），建设高度18m。

本项目是北京市城郊区县范围内为数不多的以老年人居住和养生为主题的大型开发项目。

本项目园区三面环山，山北侧3km外为密云水库，南侧为平原农地，地形地貌、水源大气和地下水等自然条件独特。由于园区处于京郊山脚下，园区内冬暖夏凉、植被丰富、空气清新，成为京郊的避暑养生胜地。

三、设计范围

C-1：建于2010年，建筑为三层或四层钢筋混凝土框架结构，总建筑面积为7705㎡，原为服务型公寓，内部相通，按单侧走廊的客房标准间式公寓设计。建筑物长×宽×高为77.35m×81.0m×14.45m。

C-2：建于2010年，建筑为二层钢筋混凝土框架结构，总建筑面积为1252.87㎡，原为健康护理中心，按医院门诊部设计。建筑物长×宽×高为48.75m×21.15m×8.45m。

南区景观：园区南半部分的出入口以及A区、B区、C区、D区建筑外部场地与周边空间环境。

四、设计专题

参加高校基于北京曜阳国际老年公寓环境改造设计的总体目标，在给定的北京密云北京曜阳国际老年公寓园区C区中的建筑及南区A、B、C、D区域周边空间环境中，结合本校参加毕业设计的专业特色和实际，在以下设计专题中选择确定毕业设计的侧重点。

1. 医养建筑改造与室内设计

遵从老年人居住养老的设计原则，开展将本地建筑的地方文化气息与室内设计相结合的探讨。将C区改造成适宜老年人生活娱乐和文化交流的功能空间；基于老年人的生理心理和行为特点，分别进行更新功能空间区划、动线设计、室内设计、绿色环保设计等适合老年人生活及康复的功能空间。

2. 医养园区环境与景观设计

遵从老年人居住养老的设计原则，开展将老年人生理心理与景观设计相结合的探讨。基于老年人的心理和行为特点，分别对南区景观即A、B、C、D等四个不同区域的外部空间环境做景观设计等配套，满足老年人室外生活及康复需求。

3. 医养建筑设施与展示设计

遵从老年人居住养老的设计原则，开展将老年人生理心理与艺术设计相结合的探讨。基于老年人的生理心理和行为特点，借助此次C区更新功能空间环境进行及设施设计、导识系统设计、展示设计、公共艺术等满足老年人生活、交流及参观的需求。

五、设计深度

（1）开题汇报文件：设计定位分析等内容。
（2）中期检查汇报文件：设计初步方案。
（3）答辩汇报文件：设计深化方案、展板及书稿。

六、设计成果

1 设计说明

设计说明内容主要包含整体设计思想的阐述、设计定位、设计衍生过程分析、功能分析、经济技术指标等，可作图示及图表。

2 图纸

（1）总平面图、场地设计图、景观设计图。
（2）建筑室内平面图、顶面图、剖立面图、典型房间平面详图。
（3）建筑室内外环境设施布置图、家具陈设布置图、导识系统布置图。
（4）展陈设计布置图、道具设施详图。
（5）重要节点详图。
（6）主要材料表、家具陈设清单等图表。

3 彩色效果图

（1）室内效果图。
（2）景观效果图。
（3）环境设施、展示道具、公共艺术等效果图。

4 成果提交

（1）开题交流。每个参加高校汇总1个汇报文件进行陈述，每校答辩时间限20min（含陈述10min、问答10min）；提交阶段成果汇报PPT文件。

（2）中期检查。每个参加高校优选3个方案组进行陈述，每组答辩时间限20min（含陈述10min、问答10min）；提交阶段成果汇报PPT文件等。

（3）毕业答辩。

1）每个参加高校优选2个答辩方案组进行陈述，每组答辩时间限30min（含陈述15 min、问答15 min）；提交毕业设计答辩陈述PPT文件、作品展板、作品排版等。

2）每个方案设计组的作品展板限3张，展板幅面A0（900mm×1200mm），分辨率不小于100dpi。展板模板由学会按照年会展板要求统一提供。展览由毕业设计答辩活动承办高校负责布置。

（4）出版素材。为编辑出版《鹤发医养卷：北京曜阳国际老年公寓环境改造设计——CIID“室内设计6+1”2016（第四届）校企联合毕业设计》，活动相关参加单位和个人等需积极响应学会要求，负责提供相应的活动章程、教学文件、专家演讲提要（每位专家中文3000字左右）、教研论文（每所高校指导教师联名写1篇，中文3000字以内）、排版作品（每个方案在一个专业设计专题占4P或6P）、专家点评（每位专家、导师对每个方案点评中文200字以内）、专家寄语（每位专家中文500字左右）、学生感想（每个方案组中文200字以内）、工作照片（每位专家、导师、学生1张）、奖励证书（各级各类）、活动花絮（活动各主要环节）、单位简介（中文1000字以内）等出版素材的电子文档。

七、附件：建筑与场地图

养老机构总平面图、单体建筑图等。

附件：建筑与场地图

C1 首层平面图

C1 二层平面图

C1 三层平面图

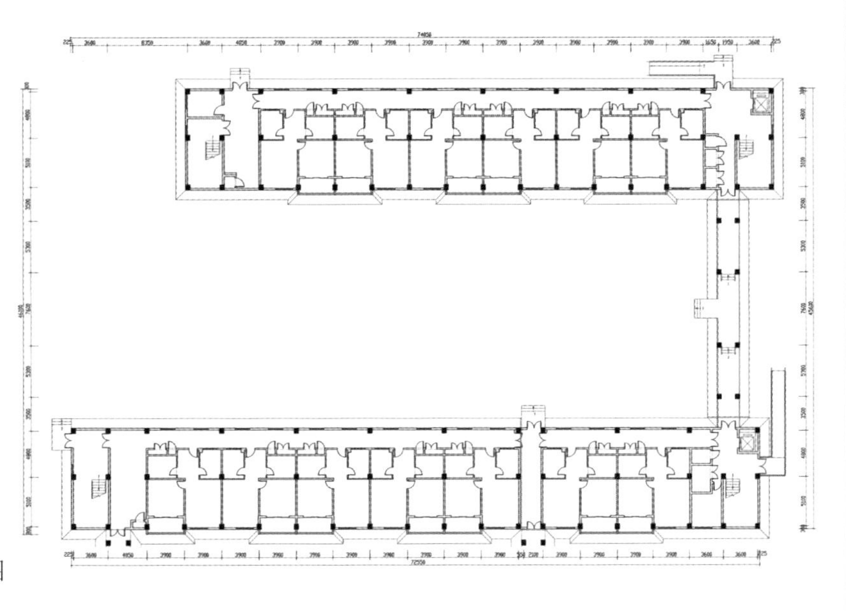

C2 首层平面图

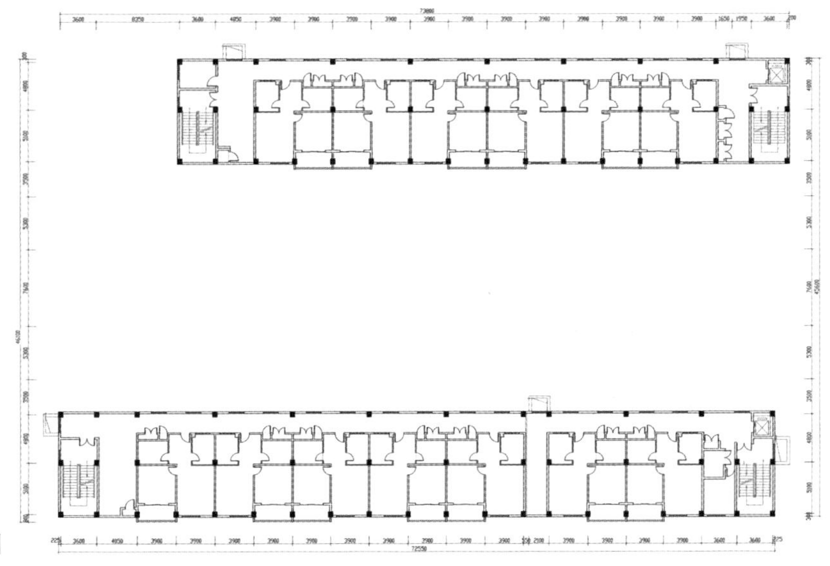

C2 标准层平面图

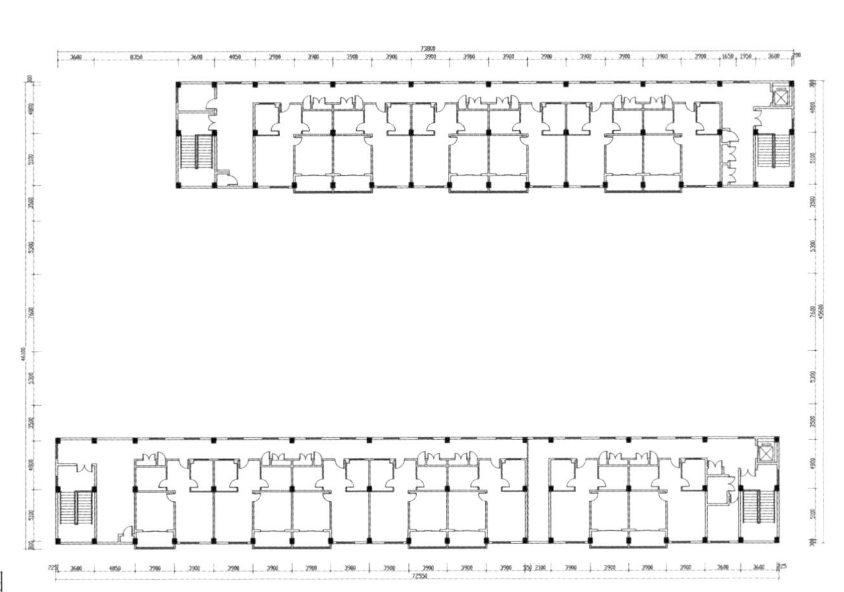

C2 四层平面图

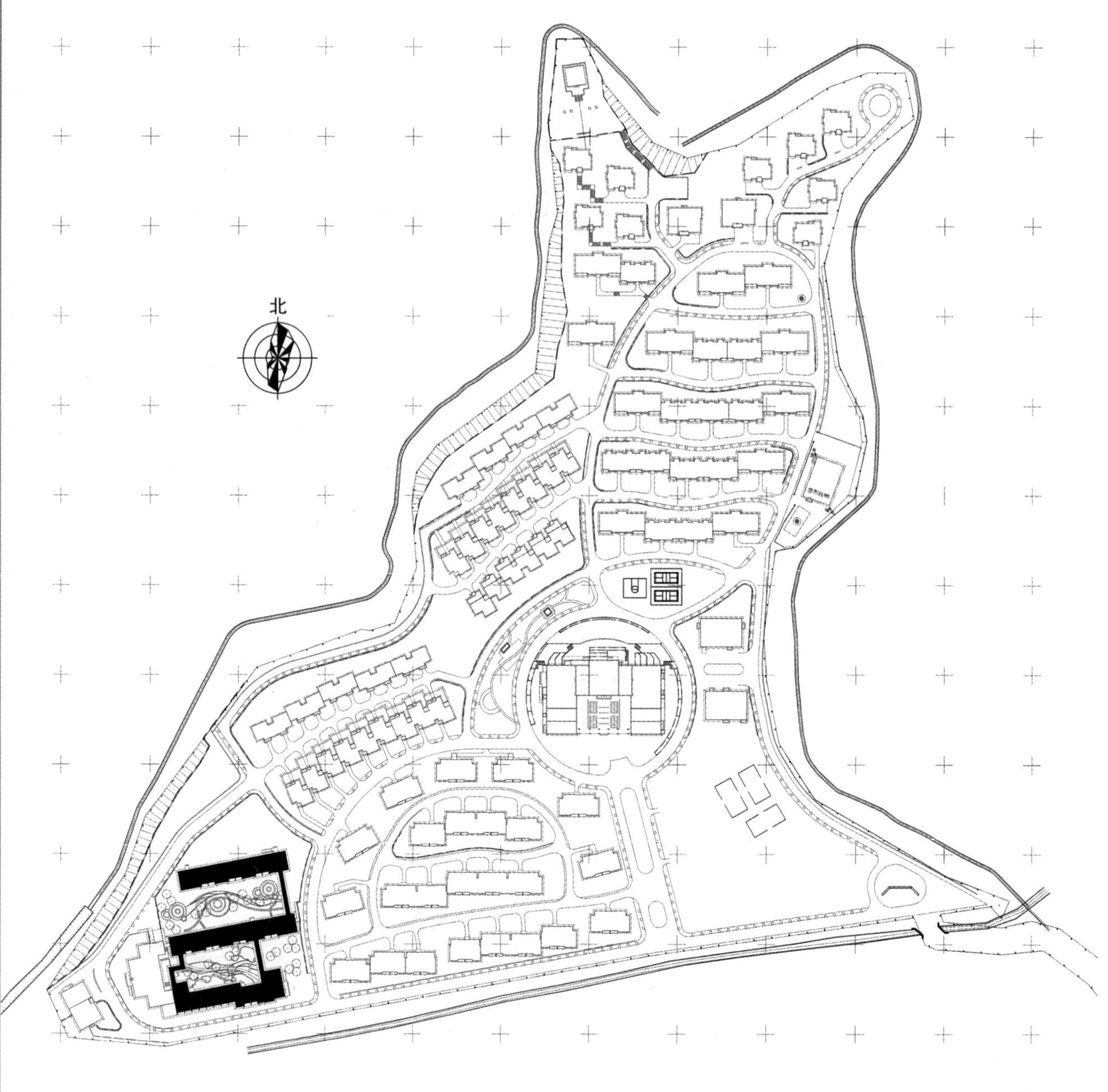

总平面图

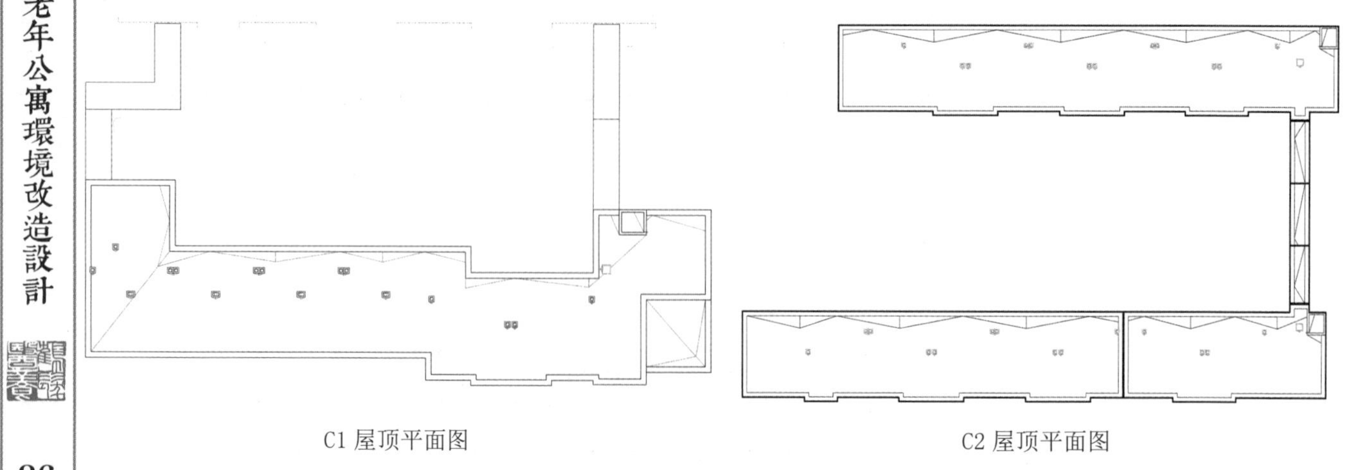

C1 屋顶平面图

C2 屋顶平面图

2016届答辩规则

一、答辩准备

（1）参加高校每校到场指导教师不超过2名，到场学生总人数限6～8名。

（2）参加高校每校安排2个毕业设计答辩方案组，参加毕业设计答辩活动、“毕业设计等级奖”评选和成果展出；此外，最多可再报送2个自荐方案，不参加毕业设计答辩和“毕业设计等级奖”评选，但参加“毕业设计佳作奖”评定和成果展出。

（3）每个毕业设计答辩方案组提前准备毕业设计答辩陈述PPT等电子文档，于毕业设计答辩活动报到时提交活动组委会。

（4）每个毕业设计答辩方案和自荐方案需提前准备成果展板3张；展板幅面为A0加长：900mm×1800mm（与学会当年年会展览要求一致），分辨率不小于300dpi，使用学会发布统一模板编辑。展板电子版须于答辩活动前1周发送到答辩活动承办高校指定的工作邮箱；由承办高校负责汇总打印、布展等。

（5）参加高校按《CIID“室内设计 6+1”校企联合毕业设计联合毕业设计书稿排版要求》编辑书稿，于毕业设计答辩活动现场提交活动组委会。

（6）毕业设计专题方向选报。“室内设计6+1”校企联合毕业设计的参加高校，结合本校毕业设计教学实际，分别按照当届《“室内设计6+1”校企联合毕业设计框架任务书》设置的毕业设计专题，对2个毕业设计答辩方案设计方向进行选报（可多选），并组织毕业生进行答辩。

二、答辩与评奖

1. 答辩组评委由特邀评委和高校评委组成

(1) 特邀评委由命题企业、学会（含地方专业委员会）专家、活动观察员、支持企业等在内的5～7位专家担任；命题企业专家担任评委组长。

(2) 高校评委由参加高校各推选1位毕业生指导教师担任。

2.“校企联合毕业设计等级奖”评选

(1)“校企联合毕业设计等级奖”按当届《“室内设计6+1”校企联合毕业设计框架任务书》设置的毕业设计专题分别设置一、二、三等奖，奖项设置相应比例一般为1:2:4。

(2) 首先，参加高校每个毕业设计答辩方案组按选报设计专题进行答辩，每组答辩陈述时间不超过15分钟，问答不超过15分钟。由特邀评委、高校评委共同填写选票，进行排序评选（如，1为建议排序第一，2为建议排序第二，依次类推）。活动组委会负责排序选票统计，形成相应设计专题“校企联合毕业设计等级奖”评选建议排序。

(3) 最后，由特邀评委以“校企联合毕业设计等级奖”评选建议排序为基础，对照答辩方案展板进行审议，特邀评委再次填写选票（高校评委须回避），进行排序评选。活动组委会负责排序选票统计，形成相应设计专题“校企联合毕业设计等级奖”评选排序结果，经特邀评委审议确定当届“校企联合毕业设计等级奖”获奖方案。“校企联合毕业设计等级奖”可以空缺。

3.“校企联合毕业设计佳作奖”评选

特邀评委和高校评委共同对参加成果展示的高校自荐方案展板进行评议，对是否认定为“校企联合毕业设计佳作奖”方案进行投票；同意票数超过两类评委总人数1/2（含）的高校自荐方案，确定获得佳作奖。

4.“校企联合优秀毕业设计导师”

获得“毕业设计等级奖”、“毕业设计佳作奖”作品的命题企业导师、参加高校导师成为“校企联合优秀毕业设计导师”。

5.“校企联合毕业设计最佳组织单位”

承办当届校企联合毕业设计活动开题踏勘、中期检查、毕业答辩、编辑出版的高校成为“校企联合毕业设计最佳组织单位”。

6.“校企联合毕业设计突出贡献单位”

负责当届校企联合毕业设计活动的命题企业、重大支持企业等成为“校企联合毕业设计突出贡献单位”。

三、表彰奖励

（1）在活动颁奖典礼上，由学会分别向“校企联合毕业设计等级奖”“校企联合毕业设计佳作奖”“校企联合优秀毕业设计导师”“校企联合毕业设计最佳组织单位”“校企联合毕业设计突出贡献单位”的获得者或代表颁发证书。

（2）获奖证书由学会盖章有效，“毕业设计等级奖”“优秀毕业佳作奖”证书印有活动特邀评委和高校评委签名，以示纪念。

四、附则

（1）本规则2013年6月8日公布施行，由中国建筑学会室内设计分会负责解释。

（2）本规则2015年5月第一次修订，2016年5月第二次修订。

零貳 02

調研踏勘

Reconnaissance Survey

CIID“室内设计 6+1”2016（第四届）校企联合毕业设计

CIID"Interior Design 6+1"2016(Fourth Session)University and Enterprise Joint in Graduation Design

鹤发医养卷——北京曜阳国际老年公寓环境改造设计

White hair volume medical support

—Beijing Yao Yang International Apartments for the elderly environmental reconstruction design

同济大学

Tongji University

CIID“室内设计 6+1”2016（第四届）校企联合毕业设计

CIID“Interior Design 6+1”2016(Fourth Session)University and Enterprise Joint in Graduation Design

小组成员：

胡　楠 Hu Nan	邹天格 Zhou Tiange	丁思岑 Ding Sicen
王雨林 Wang Yulin	王舟童 Wang Zhoutong	张晗婧 Zhang Hanjing

基地概况 || Site profile

密云县历史悠久，有大量的人文和自然旅游资源，密云有独特的乡村文化、习俗、文化遗产、特色美食和商品。基地周边的黑龙寺、云佛山大溶洞和黑龙潭风景区等为老年人的户外活动提供了条件。

Known for a long history, Miyun County is home to rich cultural and natural tourism resources, boasting unique local culture, customs, cultural heritages, foods and commodities. Around the project site, there are a number of scenic spots such as Black Dragon Temple, Yunfo Mountain Cave and Black Dragon Pool which are ideal places of outdoor activities for seniors.

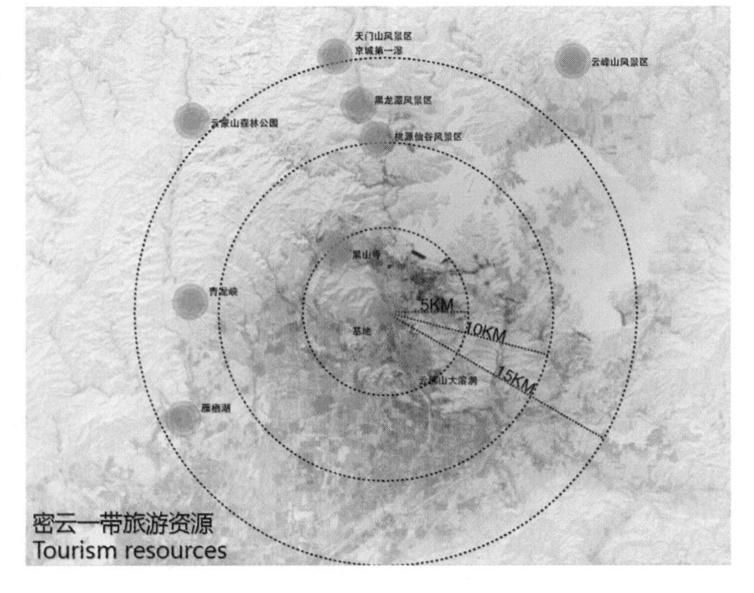

密云一带旅游资源
Tourism resources

社区服务中心位于小区中心，其余建筑均为居住功能

The community service center is located at the center of the retirement community. Other buildings are all for living.

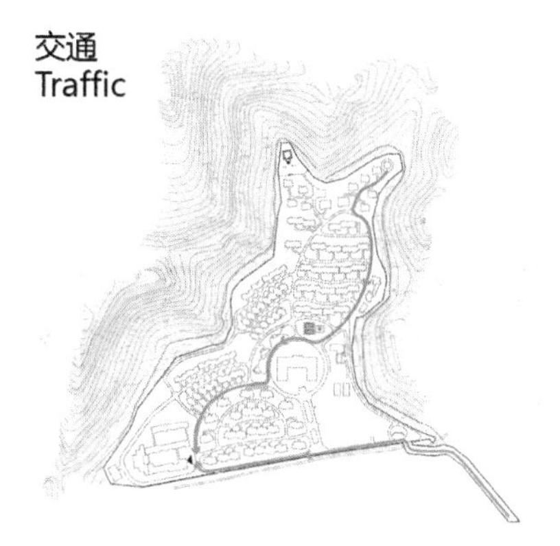

待改造建筑位于小区西北角

The buildings to be renovated stand in the northwest corner of the retirement community.

待改造建筑北立面在小区多个视点可见

The north façades of the buildings to be renovated can be seen from several viewpoints in the community.

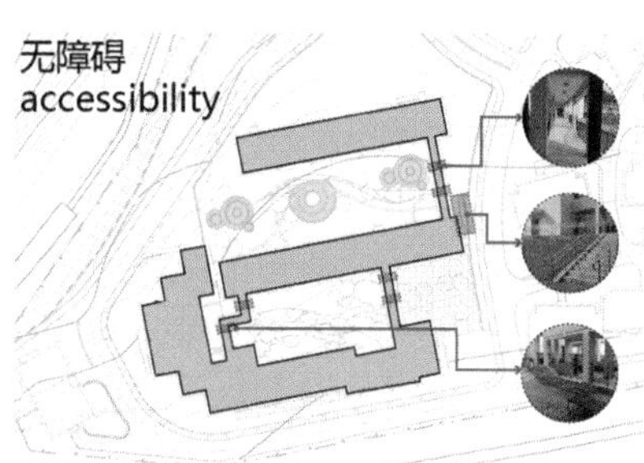

公共空间
public space

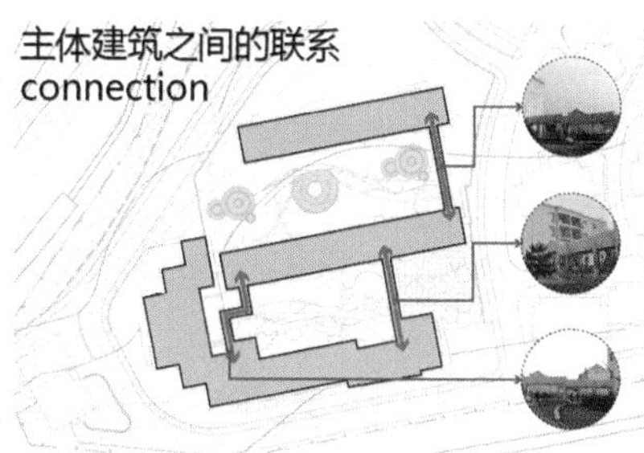

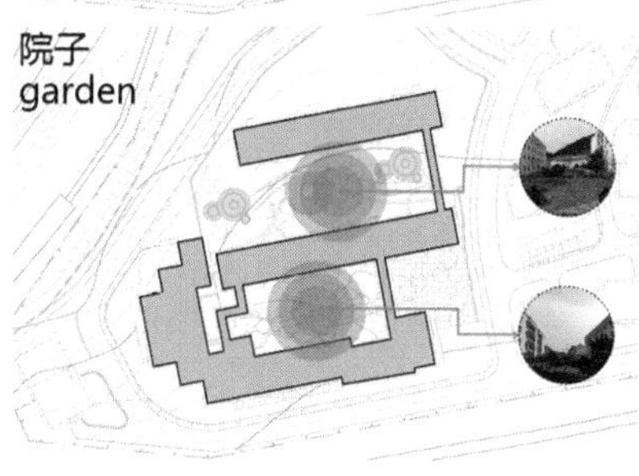

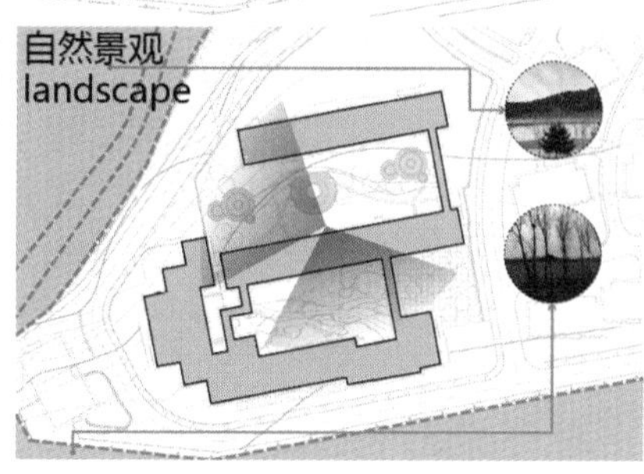

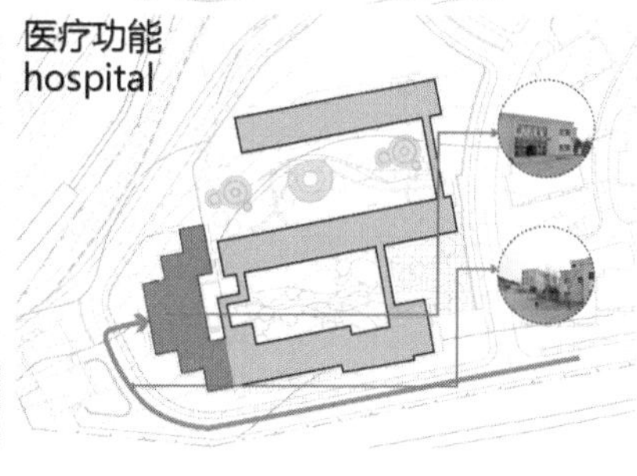

- 原有建筑不满足无障碍的要求
- 原建筑缺少完整的公共活动空间
- 原有建筑的三栋主体建筑之间仅通过约两米宽的廊道相连
- 原建筑室外空间只有两个较为封闭的院子，冬天的光照不充足
- 基地周边的挡土墙和栅栏将很好的景观与建筑隔绝开来
- 原医疗功能可达性不好，整体造型与三栋主体建筑不和谐

- The original buildings are not wheelchair friendly.
- The original buildings lack complete public spaces.
- The three main buildings are only connected by 2-meter-wide corridors
- In the project site, there are only two yards with insufficient exposure to sunlight in winter as outdoor spaces.
- The retaining walls and fences around the community separate the buildings from beautiful landscape.
- The original medical facility offers poor accessibility to its functions, and its exterior design is not in harmony with that of the three main buildings.

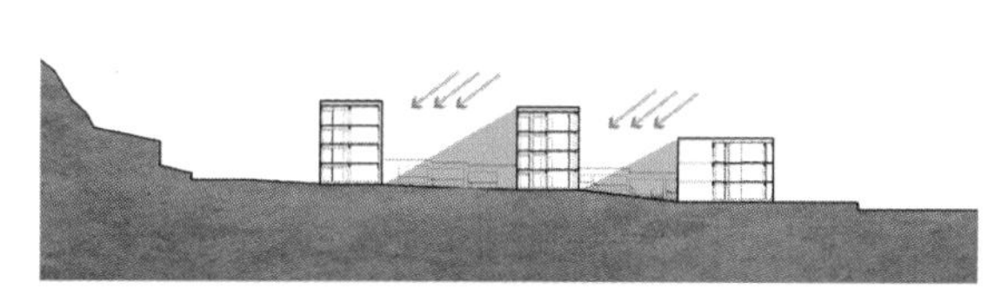

只有最北面的一层有加建的可能性

Only the northernmost building can be raised.

基地北边是山地，中间有两个庭院，南面是平原

There are two yards in the community. To the north of the community are mountains and to the north a plain.

华南理工大学

South China University of Technology

CIID“室内设计 6+1”2016（ 第四届 ）校企联合毕业设计

CIID“Interior Design 6+1”2016(Fourth Session)University and Enterprise Joint in Graduation Design

小组成员：

郑潇童 Zheng Xiaotong　　　　陈谢炜 Chen Xiewei

前期调研分析 || Early-stage survey and analysis

- 调研途径

  场地现场调研，养老院走访，网上资料查阅和书籍查阅。
- Survey methods

  Survey methods include site survey, visiting retirement homes, and collecting information on line and from books.

- 调研目的

  了解现场情况，分析场地所针对人群，以及了解人群活动、行为。
- Survey objectives

  To know the conditions of the project site, and to analyze the target population, including their activities and behaviors.

- 调研结果
- Survey results

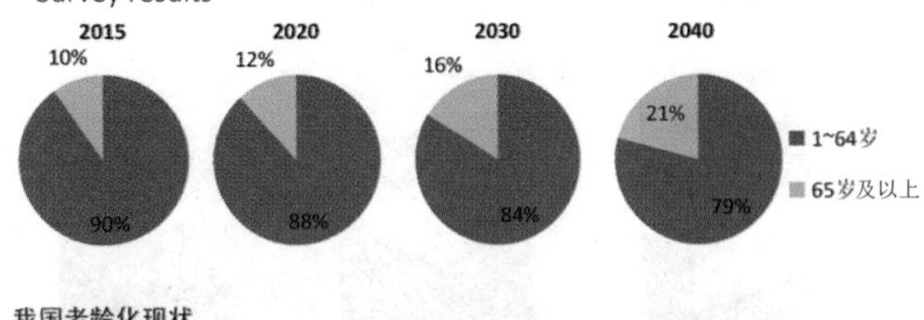

我国老龄化现状

2000年第五次人口普查显示，我国已进入老龄化社会，并呈现出老年人口基数大、增速快的态势。再加上我国未富先老的国情和家庭小型化的结构叠加在一起，养老问题异常严峻。

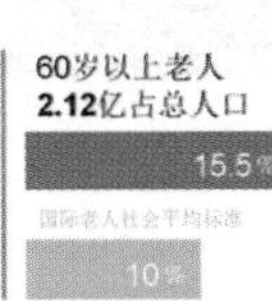

调研发现问题 || Problems discovered

通过走访周边养老院发现以下问题。

（1）老人没有丰富的娱乐活动，生活节奏单一，令老人们感到不适应和孤单。从而导致情绪不稳定，消极以及生活饮食差，缺少寄托。

（2）园区内公共场所的休息场所太过暴露，没有隐私和舒适感，利用率不高。

（3）建筑内部没有美感，只有功能性没有美观性的设计会感觉索然无味。空旷的房间会让人感觉萧条。

Problem discovered during the survey include:

1. The seniors live a monotonous and lonely life due to the lack of entertainment activities and spiritual sustenance, presenting unstable and negative emotions, and poor appetite.
2. The public rest areas are underutilized due to their lack of privacy and comfort.
3. The interior design only focuses on functions and ignores aesthetics which makes the rooms empty and depressing.

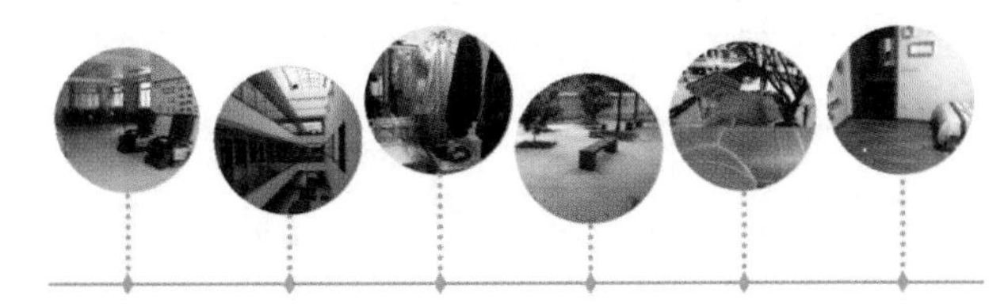

园区分析 || Analysis of the project site

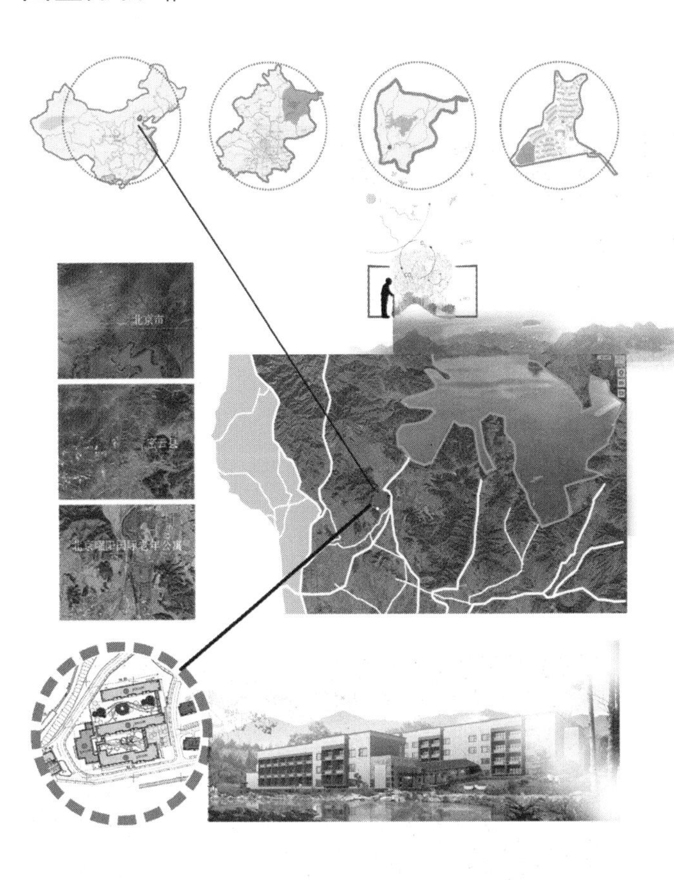

本项目园区三面环山，山北侧 3km 外为密云水库，南侧为平原农地，地形地貌、水源大气和地下水等自然条件独特。由于园区处于京郊山脚下，园区内冬暖夏凉、植被丰富、空气清新，成为京郊的避暑养生胜地。

Located in the outskirts of Beijing, the retirement community is surrounded by mountains on three sides, with flat farmland extending southward and the Miyun Reservoir lying 3 kilometers away to the north. Boasting advantaged geographical location, fresh air, clean water, as well as diverse vegetation and year-round mild weather, it is a great summer resort and also an idea place for retirement life.

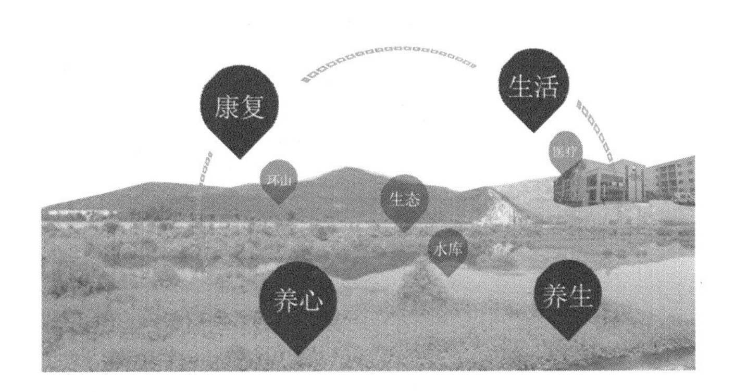

现存问题 || Current problems

- 两个庭院之间缺少联系
- 庭院利用率低
- 无公共活动空间
- 交通形式单一

- Lack of communications between the two yards
- Underutilization of the yards
- Absence of spaces for public activities
- Very limited means of transportation

使用人群分析 || Target population analysis

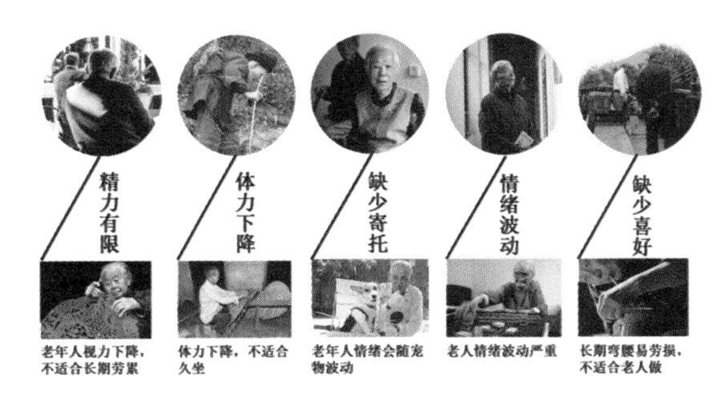

他们能做什么？
如何让他们生活的更好？

各个组团依山而建的建筑布局，充分利用山坡台地的特色，建筑物迭次升高，高低错落，空间变化丰富。建筑环境自然、和谐、素雅、幽静，组团间高低错落，为老年人营造温馨、舒适的休闲情调。基地周围环山，有水库，并设有医疗功能，所以有很好的契机成为生活、康复、养生、养心的地方。

On the bench terraces, tiers of houses rise up, taking on expressive edges against the mountains. All the things here incorporate into nature tranquilly and harmoniously, creating a comfortable and cozy environment for the seniors. Aside from a pleasant natural environment, well-established medical facilities are another highlight of the retirement community, meeting the seniors' need for healthcare and rehabilitation.

哈尔滨工业大学

Harbin Institute of Technology

CIID“室内设计 6+1”2016（第四届）校企联合毕业设计

CIID“Interior Design 6+1”2016(Fourth Session)University and Enterprise Joint in Graduation Design

小组成员：

张相禹 Zhang Xiangyu　　张玲芝 Zhang Lingzhi　　朱梦影 Zhu Mengying

袁思佳 Yuan Sijia　　李佳楠 Li Jianan　　巴美慧 Ba Meihui

李　皓 Li Hao　　李曼园 Li Manyuan

开题立意—外婆桥 || Inspiration–Grandma's Bridge

原人群定位：中产阶级及其以上——少数人 / 不需要

原市场定位：山地别墅的文化精神价值——缺少吸引力

原建筑定位：养生住宅——可在此基础上增添其他功能与趣味增添吸引力

区位分析：依山而建——偏僻、缺少配套设施
远离市区——空气清新、适合老人与儿童

老人生活模式
Life patterns of the seniors

设计过程 Design Routes

由外及内

由功能到形式

小时候　外婆家是我们最喜欢的地方

长大后　我们渴望能够过上与朋友比邻而居的日子

年老后　我们希望需找一个可以依托的地方

……　我们的故事，未完待续事

老人是暖巢
子孙是飞鸟
常有归时。

社区层面——团聚生活

百米长街巷。
何处不相见？
一天一小聚，
两天一大聚。

家庭层面——团团圆圆

景观草图 || Landscape sketch

景观采用园艺治疗的手段。

园艺疗法：日本称为园艺疗法，韩国称为园艺治疗，简单的定义是利用园艺来治疗。
种类：
植物疗法；芳香疗法；花疗法；园艺疗法；药草疗法；艺术疗法之一（插花、押花）组合花园制作等。

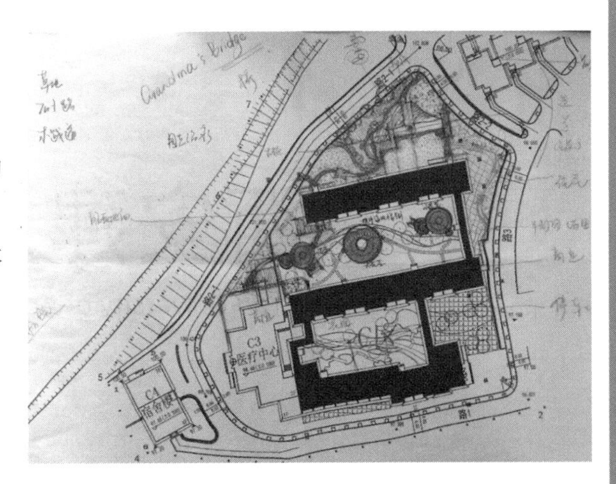

设计初衷

案例调研 || Case study

北京英智康复机构调研

调研空间布置

探究家具尺度

了解设备作用

老年人长期居住——主体
第二代偶尔居住——探亲
第三代短期居住——节假

居住时间
Length of stay

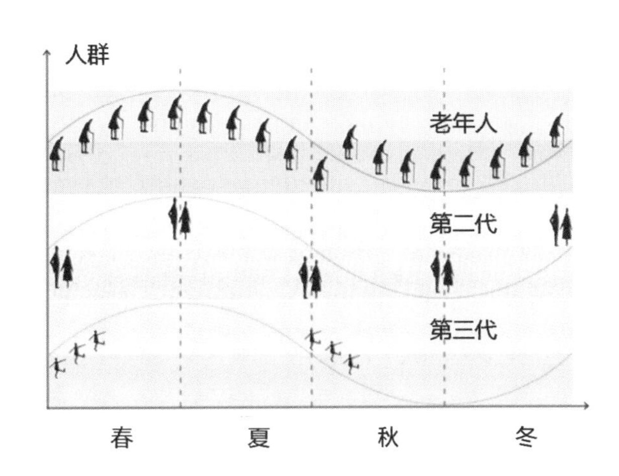

西安建筑科技大学

Xi'an University of Architecture and Technology

CIID“室内设计 6+1”2016（第四届）校企联合毕业设计

CIID"Interior Design 6+1"2016(Fourth Session)University and Enterprise Joint in Graduation Design

小组成员：

赵凯文 Zhao Kaiwen　　张　峰 Zhang Feng　　颜　强 Yan Qiang

李坷欣 Li Kexin　　刘雨鑫 Liu Yuxin　　曾　嘉 Zeng Jia

老龄化现状 || Population aging

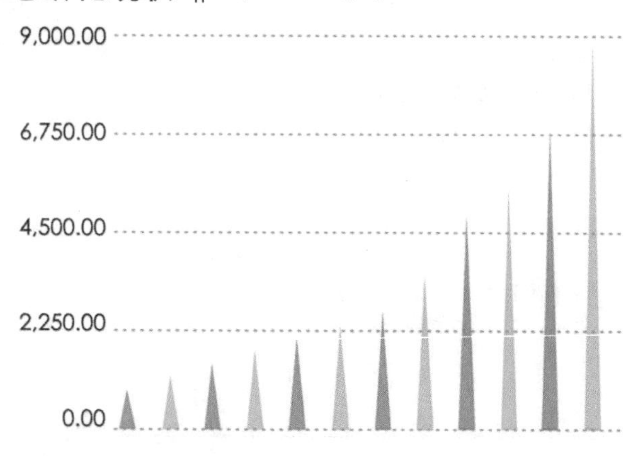

全球进入人口老龄化的国家和地区已达到 72 个，2006 年全球老年人口已达到 6.88 亿，全球 80 岁以上高龄老人占世界老年人口总数的 11%。世界百岁老人约 14.5 万。

Currently, 72 countries and regions worldwide have become aged societies. A survey shows that by 2006, there were totally 688 million old people around the world, among whom, people aged 80 and above accounted for 11%, and people aged 100 and above reached 145,000.

我国老龄化特征及养老选择因素

我国老年人口的绝对数量大。我国老龄人口绝对值为世界之冠，占世界老龄人口总数的 1/5。高龄化趋势显著随着我国经济持续发展和人民生活水平的提高，我国人均预期寿命大大延长。“未富先老”是显著特色。发达国家在进入老龄化社会时，人均国民生产总值基本上在 5000 美元至 1 万美元，而我国人均国民生产总值尚不足 1000 美元。“空巢”老人迅速增加，关于老年人居住情况调查结果显示，“三代同堂”式的传统家庭越来越少，“四二一”的人口结构（一对夫妻同时赡养 4 个老人和 1 个小孩）愈加明显。

China has the largest aging population in the world, with elderly people accounting for 1/5 of the world's aging population. In the context that China is seeing increasingly growing economy and living standard, which will undoubtedly lead to a longer life span, the trend of aging will be more obvious. "Getting old before getting rich" becomes a problem faced by many Chinese people. Compared with developed countries, China is still economically backward, with per capita GNP staying below $1,000, falling far behind the value of $5,000-10,000 of developed countries. Another problem is the rapidly increasing population of seniors living alone. Surveys show that the traditional family model of three generations living together has been disappearing, and the 4-2-1 (a couple supporting four seniors and one child) family structure has become a main trend in today's China.

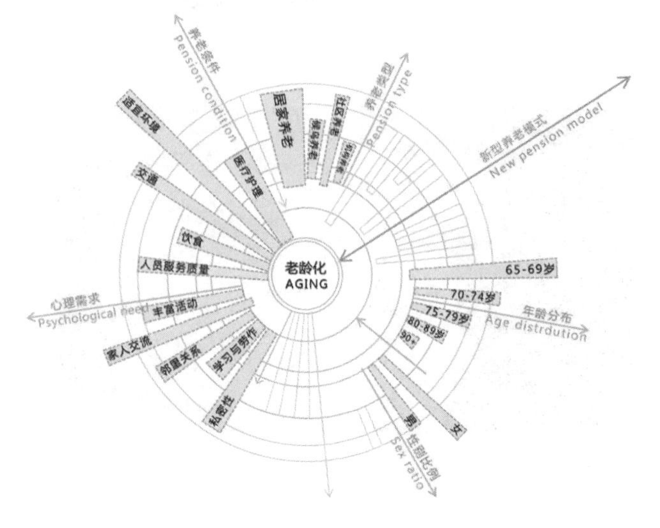

基地现状

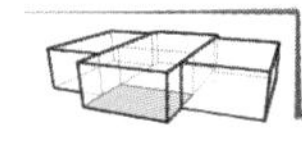

个性化感知

当我们本身置身于场地中，我们的感受...

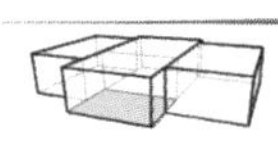

空间理论论证

北京曜阳国际老年公寓位于北京密云区的东北方向。

Beijing Yaoyang International Retirement Community is located in the northeast of Miyun, Beijing.

距离北京市区远，周围交通单一，周边缺少大型医院以及等级相当的一系列医疗设施。

Current problems faced by the community mainly include a long distance from downtown Beijing, limited transportation means, and lack of large hospitals and advanced medical facilities.

交通路线

北京市

怀柔区

顺义区

密云区

北京曜阳国际老年公寓

密云水库

北京建筑大学

Beijing University of Civil Engineering and Architecture

CIID“室内设计 6+1”2016（第四届）校企联合毕业设计

CIID"Interior Design 6+1"2016(Fourth Session)University and Enterprise Joint in Graduation Design

小组成员：

陆　昊 Lu Hao　　魏　卿 Wei Qing　　曾　宁 Zeng Ning

王海月 Wang Haiyue　　李丽阳 Li Liyang　　王逸开 Wang Yikai

张博闻 Zhang Bowen

场地分析

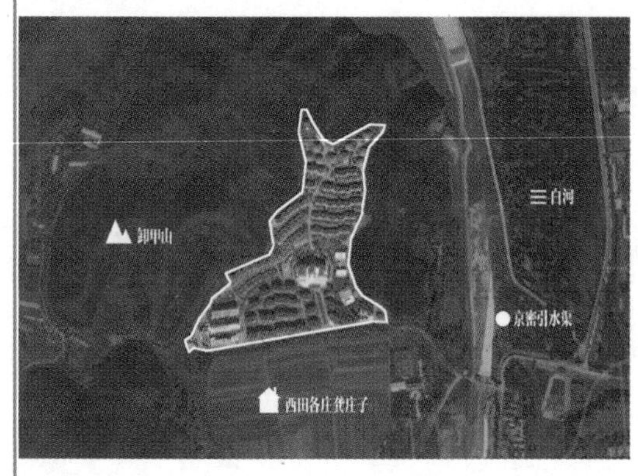

• 北京长青国际老年公寓建设用地位于密云县城北 10km 处，北临密云水库，东临京密引水渠渠首，西侧为卸甲山，南侧为西田各庄镇龚庄子村。总用地面积为 137461 ㎡，总建筑面积为 67230 ㎡（含地下 5390 ㎡），建设高度 18m。

• C1：三层或四层钢筋混凝土框架结构，总建筑面积为 7705 ㎡，原为服务型公寓；

• C2：建筑为二层钢筋混凝土框架结构，总建筑面积为 1252.87 ㎡，原为健康护理中心，按医院门诊部设计。

C区干道

职工之家

C区景观

C区入口

C1楼梯

过廊

C2室内

南侧干道

场地分析

C1：三栋建筑过于分离，在功能上不易形成联系，老人的管理上也存在一定困难，如当老人出现突发情况时，存在从室内到室外再到室内的流线。
针对的老人群体过窄，不能对老人形成多层级、连续的照护模式，如失智老人和介护老人，需要单独照护。
老人的居住模式单一，每户之间的联系较少，老人孤独感仍然很强，比如缺少同层老人的公共活动空间。

C2：功能过于偏向医疗，容易给老人形成压抑的印象。
大厅进深较深，白天较为昏暗。

文献研究

“高龄化；失能化；空巢化。”

——我国当前的老龄背景

• 2014 年国土资源部发布《养老服务设施用地指导意见》，将养老用地供应纳入国有建设用地的供应计划中，明确了养老服务设施供地政策。
• 2015 年初，民政部，发改委等部委发布《关于鼓励民间资本参与养老服务业发展的实施意见》，调动社会资本进入养老领域的积极性，改善养老服务供求关系，逐步使社会力量称为发展养老服务业的主体。
……

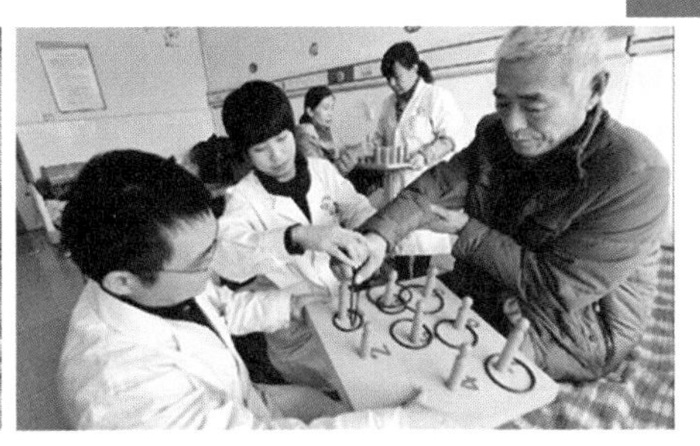

老年宜居项目分析——建设类型

• 全龄社区配建养老产品
在新建居住区时，配建一定比例的养老产品（老年住宅、养老公寓、养老服务）。
• 新建综合性养老社区
专门针对老年人群体建设的养老社区。通常包含老年住宅、养老公寓、养老设施等多种居住产品类型，以满足自理、半自理到不能自理各阶段身体状况老人的居住需求。
• 既有社区中插建、改建养老服务设施
指利用已建成社区中的空闲用地或空闲建筑，插建、改建或扩建能够提供多元化服务的养老设施。该建设类型主要见于在城市中建成年代较早，周边配套设施发展已较为成熟的社区中。
• 将闲置建筑改造为养老服务设施
将闲置建筑改造为养老服务设施是指将闲置的厂房、办公楼、会所、招待所等建筑通过重新安排空间格局、改变内部装饰等方式改造为养老服务设施。

老年宜居项目分析——开发模式

• 与医疗资源结合　• 与养老保险产品结合　• 与旅游资源结合
• 主要问题
高端扎堆，忽略主流需求；规模过大，与老人需求不符；定位不清，产品缺乏细分；水土不服，盲目照搬国外模式；适老性公共环境远未形成。
• 对策
场地类型多样：不同室外活动场地的规划与设计应与居住区的规模、构成相协调，基于差异化的社区条件，合理配置不同规模活动场地的数量，并根据场地类型配备相应的附属设施（包括卫生间、休憩空间等）。
场地易感知、便捷可达：居住区户外场地要使用易于识别的基本组织模式，在规划布局中，将活动和服务设施适当地集中布局，不同活动场地之间要有明显的联系，满足视线上的通透性，且活动场地平面模式应当易于被识别确定，特别是较大的场地。同时也要满足交通的易达性，室外活动场地应与交通系统有良好的衔接。此外，所有的户外活动场地都要满足无障碍设计要求，例如有舒适的扶手和栏杆，以鼓励老年人的使用。

南京艺术学院
Nanjing University of the Arts

CIID“室内设计 6+1”2016（第四届）校企联合毕业设计
CIID“Interior Design 6+1”2016(Fourth Session)University and Enterprise Joint in Graduation Design

小组成员：

龚月茜 Gong Yueqian　华彬彬 Hua Binbin　戚晓文 Qi Xiaowen
魏　飞 Wei Fei　冼佩茹 Xian Peiru　张梦迪 Zhang Mengdi

前期调研 || Early-stage survey

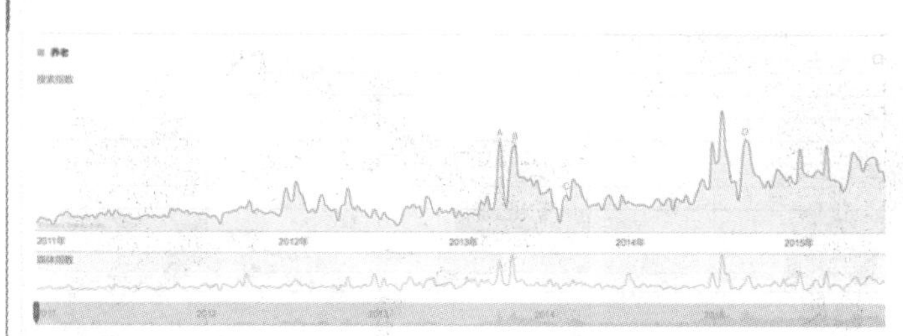

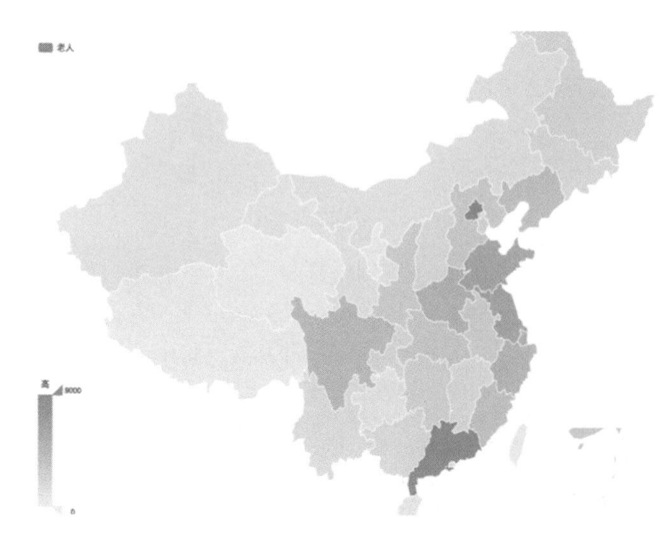

问卷调查

1. 您的性别
A 男　　B 女
2. 您的年龄是多少
A 40～50　B 51～60　C 61～70　D 70～80　E 80 以上
3. 您从事的职业
A 服务业　B 教育业　C 金融业　D IT 业　E 工人　F 农民　G 其他
4. 您目前的婚姻状况是
A 有老伴　B 老伴已过世　C 离异　D 无婚姻经历
5. 您的文化程度
A 初中及以下 B 高中 C 大专 D 本科　E 研究生及以上
6. 您的年收入
A 5 万以下　B 5 万～10 万　C 10 万以上
7. 您现在的居住状况
A 自己单住　B 夫妻同住　C 和子女或者孙辈一起住　D 和未成年孙辈一起住　E 其他
8. 您的养老费用由谁支付
A 自己　B 子女　C 政府补贴　D 其他
9. 您的闲暇时间最喜欢做什么
A 公园散步　B 旅游　C 聊天　D 读书看报　E 上网　F 钓鱼　G 书法绘画　H 下棋　I 打牌，麻将　J 其他
10. 您期望的养老模式
A 家庭养老（主要由家庭成员提供日常照料）　B 机构养老（老年公寓，敬老院等）　C 社区居家养老（居住在家里，社区可以提供生活照料、家政服务、医疗保健等服务　D 其他
11. 您期待的养老环境应该有哪些必备的条件
A 生活照料（洗衣做饭、打扫卫生、买菜购物、洗澡穿衣、陪同外出）　B 医疗保健　C 日托服务　D 紧急救助 E 休闲娱乐活动　F 老年人学习培训　G 参与社会活动　H 心里护理（聊天解闷、心理开导）
12. 您期待的环境需要提供的社区精神文化生活服务
A 举办兴趣班、培训班等　B 专设老年工作室（老年交流活动、老年互学活动等）　C 不清楚　D 其他
13. 您认为养老机构最大的缺点是
A 收费高　B 没有家庭温暖　C 伙食差　D 卫生差　E 老弱病残集中，产生压抑感　F 其他
14. 您觉得老人需要的居住条件是什么
A 安静整洁的环境　B 与人沟通方便　C 活动场所宽阔　D 医疗设施齐备　E 家政服务到位　F 交通、购物方便　G 其他
15. 您认为老年人的幸福是怎样的
A 身体健康　B 儿女孝顺　C 有稳定的经济来源　D 受到社会和家庭的尊重　E 丰富的文化娱乐活动　F 能够继续发挥余热　G 其他

	居家养老	社区养老	机构养老
特点：	老年人居住在自己或血缘亲属的家庭中，由其他家庭成员提供养老服务。服务提供者主要有：居家养老服务机构、老年社区、老年公寓、托老所、志愿者等。服务内容包括基生活照料（饮食起居照顾、打扫卫生、代为购物等）、物质支援（提供食物、安全措施等）、心理支持（治病、护理等）、休闲娱乐设施等。	老年人居住在自己家中，由社会提供商业化的养老服务。老年人居住在自己熟悉的环境里，既可以得到适当的照顾，也随时欢迎子女的探望。	老年人集中居住在特别的养老机构中。养老机构提供专业的医疗及养老服务。根据老年人身体健康状态、生活自理程度及社会交往能力，实行分级管理，一般分为自理型、半自理型和完全不能自理型三级。不同级别的老人入住不同类型的养老机构。但是一般探望不便，容易造成老年人和子女的隔阂。
现状：	受传统文化的影响，是东亚国家普遍的养老模式。	目前在欧美发达国家比较普遍。	目前在欧美发达国家比较普遍。
经营模式：	美国	日本	欧洲
	建筑规模大，有各种各样的俱乐部，开设的课程和组织的活动超过80%以上。代表楼盘：80%以上。代表楼盘：太阳城中心	日本老龄人的生活品质是在良好的社会保险的基础上实现的。提供无障碍设施的老龄人住宅产品、具有看护性质的老龄人住宅产品、能和家人共同生活（二代居）的住宅产品。代表楼盘：港北新城	根据国家政策倾向于让老年人居住在独立的公寓中，建筑将三种元素结合在一起：城市意味、社区功能和生态目标。代表楼盘：荷兰弗莱德利克斯堡老年人公寓

南京各大楼盘调研

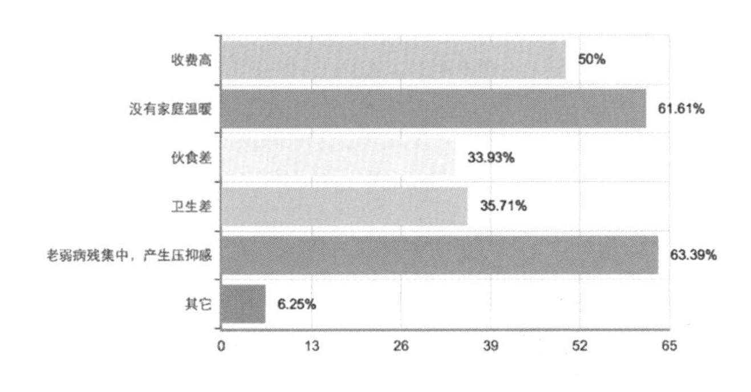

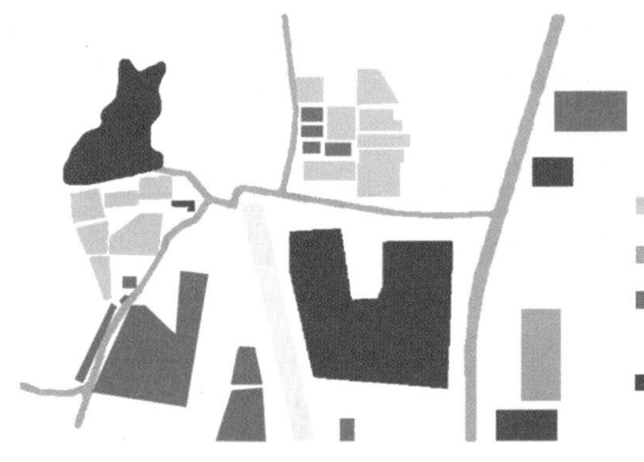
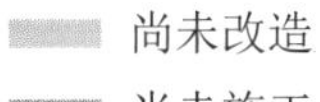
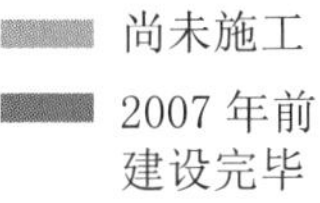
尚未改造
尚未施工
2007 年前建设完毕
2016 年前建设完毕

南京鼓楼医院

浙江工业大学

Zhejiang University Of Technology

CIID“室内设计 6+1”2016（第四届）校企联合毕业设计

CIID“Interior Design 6+1”2016(Fourth Session)University and Enterprise Joint in Graduation Design

小组成员：

张　怡 Zhang Yi　　罗　忆 Luo Yi　　卢倩文 Lu Qianwen

邱丽珉 Qiu Limin

调研基本情况 || Survey profile

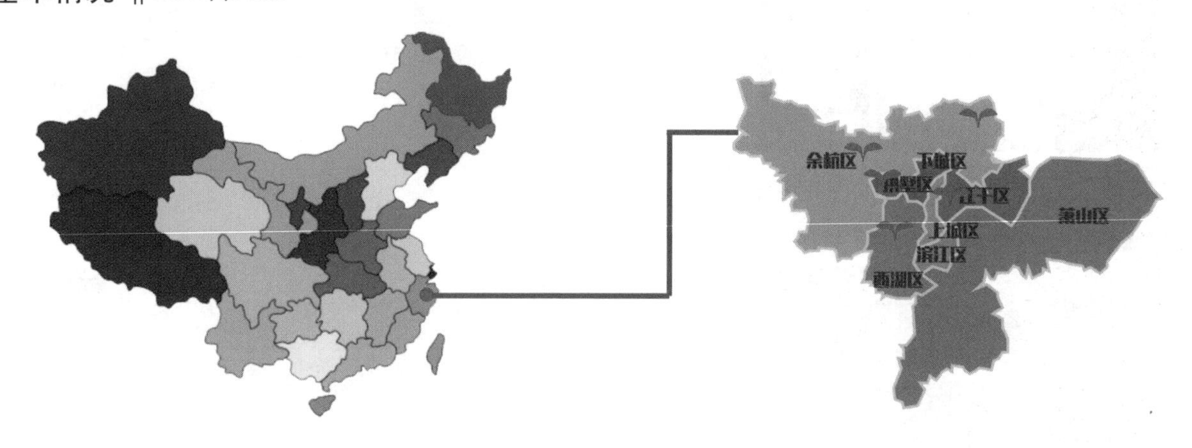

实地调研 || Site survey

1、海华幸福公寓（宾馆式养老机构）

2、山缘公寓（复合式养老社区）

3、绿城蓝庭颐养公寓（酒店养老公寓）

4、爱康温馨家园（非营利性综合养老机构）

5、万科良渚随园嘉树（大型中国式邻里养老社区）

1、海华幸福公寓（拱墅区）
2、山缘公寓（西湖区）
3、绿城蓝庭颐养公寓（余杭区）
4、爱康温馨家园（江干区）
5、万科良渚随园嘉树（余杭区）

调研数据分析 || Survey data analysis

基础分析
Basic analysis

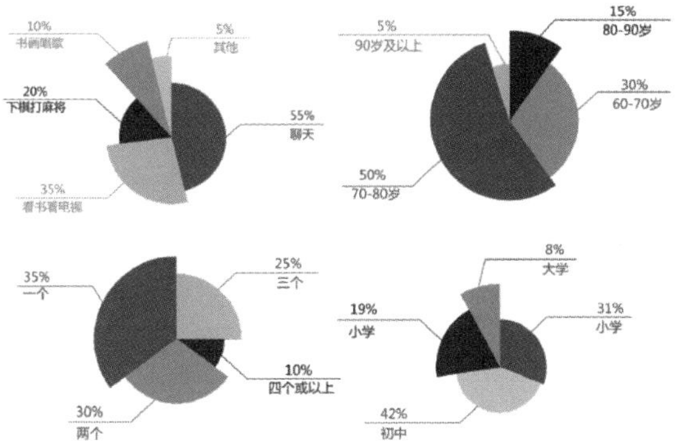

10%
书画唱歌
5%
其他
20%
下棋打麻将
55%
聊天
35%
看书看电视

5%
90岁及以上
15%
80-90岁
30%
60-70岁
50%
70-80岁

35%
一个
25%
三个
10%
四个或以上
30%
两个

8%
大学
19%
小学
31%
小学
42%
初中

行为分析
Behavior analysis

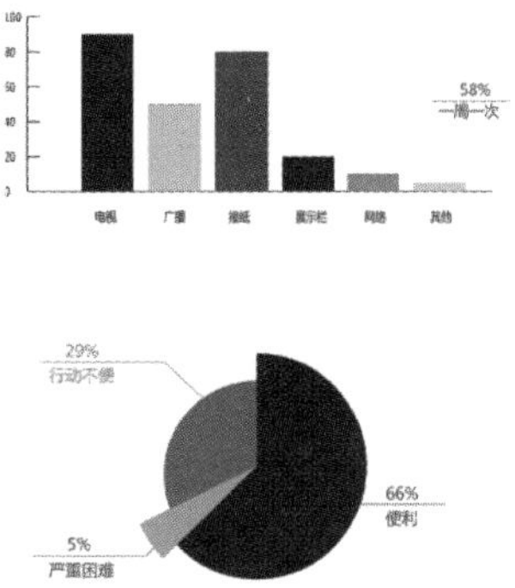

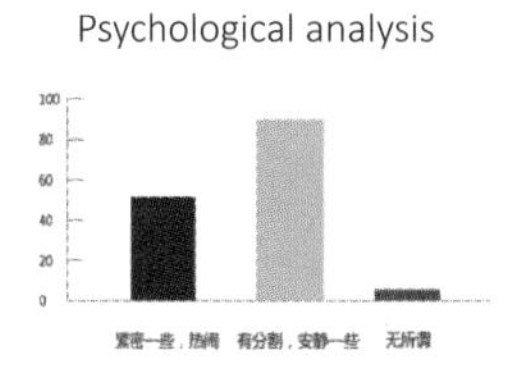

29%
行动不便
66%
便利
5%
严重困难

心理分析
Psychological analysis

晴天 阴天 雨天 雪天

老年人常见疾病分析 || Analysis of common diseases among seniors

手术后恶性肿瘤化学治疗 9.52%
Cancer chemotherapy after surgery 9.52%

脑梗死 3.99%
Infarction 3.99%

冠状动脉粥样硬化性心脏病 3.76%
Coronary heart disease 3.76%

肺部感染 1.5%
Lung infection 1.5%

慢性肾脏病 1.45%
chronic kidney disease 1.45%

肺炎 1.38%
Pneumonia 1.38%

样本容量：5350人
Sample size: 5,350

压抑的生存环境
养老院普遍像医院，
冰冷而充满距离感。
来自环境的暗示。

社会
老人是家庭和社会的
负担，但必须尽到赡
养的义务。

自身
身体机能下降，记忆力
衰退；觉得不中用，认
为自己失去了价值。

子女
想要尽到孝心，但苦
于没有时间没有精力，
只能委托机构养老。

未来
老龄人口逐爆炸式
增长，机构养老的
比例还会继续提升。

场地分析
Site analysis

功能分析
Functional analysis

别墅区
Villas

高档区
Premium apartments

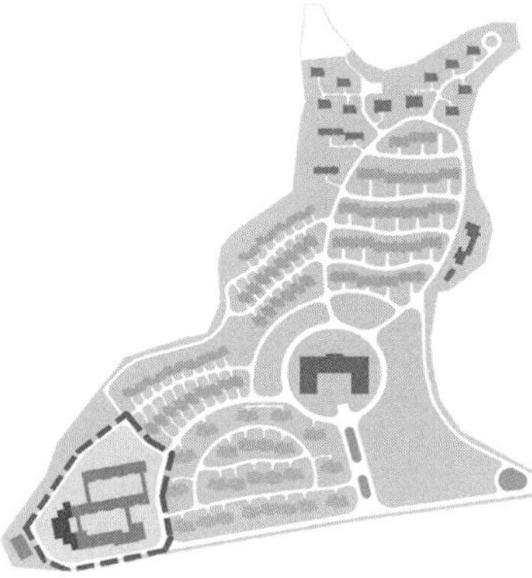

普通区
Common apartments

中档区
Mid-range apartments

公共建筑
Public buildings

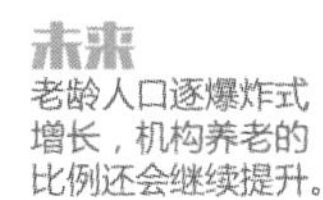

交通分析
Transport analysis

风向分析
Wind direction analysis

夏季：东南风
冬季：西北风
Summer: Southeast wind
Winter: Northwest wind

老年人群分析 || Analysis of the elderly

A 人群
性格乐观，外向；
喜欢打牌，热情
好客

B 人群
性格沉闷，内向；
喜欢看书，不爱
说话

C 人群
性格活泼，外向；
喜欢活动，乐于
交际

03

過程方案

progress

CIID "室内设计 6+1" 2016（第四届）
校企联合毕业设计

CIID"Interior Design 6+1"2016(Fourth Session)University
and Enterprise Joint in Graduation Design

鹤发医养卷
——北京曜阳国际老年公寓环境改造设计

White hair volume medical support
—Beijing Yao Yang International Apartments
for the elderly environmental reconstruction design

同济大学

Tongji University

CIID“室内设计 6+1”2016（第四届）校企联合毕业设计

CIID“Interior Design 6+1”2016(Forth Scssion)University and Enterprise Joint in Graduation Design

小组成员：

胡　楠 Hu Nan　　邹天格 Zhou Tiange　　丁思岑 Ding Sicen

王雨林 Wang Yulin　　王舟童 Wang Zhoutong　　张晗婧 Zhang Hanjing

项目定位 || Project positioning

仚

居

行

音仙，山居长往也。古通仙，入山长生曰仙。

凥处也。从尸，得几而止也。引孝经，仲尼凥，凥谓闲居，如此会意。

行，从彳从亍，人之步趋也。彳，小步也；亍，步止也。

Fairyland, as it literally indicates, is where immortals live in.

Undoubtedly, a fairyland is a great place to live, and to ramble around.

一类患病人群 Patient of some category

55-70 岁	Aged 55-70
中等收入	Middle income
骨科疾病、腿脚不便者	Orthopedic patients such as people with disabled or weak legs

二套护理服务 Two sets of services

短期康复：针对患病较轻、有自理能力或骨病术后康复的老年人提供服务。

长期护理：针对长期患病或患病严重、生活需要照料、需要轮椅的老年人提供服务。

Short-term rehabilitation: The service is provided to seniors who suffer minor injuries and have the ability to take care of themselves, and seniors needing postoperative rehabilitation.

Long-term nursing: The service is provided to patients needing care and wheelchairs due to long-term sickness or serious diseases.

三种居住模式 Three kinds of living mode

护理单人间：针对患病严重，需一级护理，单独居住的老年人

护理双人间：针对患病严重，需一级护理，双人居住的老年人

家庭房：针对患病较轻，需康复指导，家庭居住的老年人

Single nursing room: It's designed for seniors who need Grade 1 nursing due to serious diseases, and choose single room accommodation.

Double nursing room: It's designed for seniors who need Grade 1 nursing due to serious diseases, and choose double room accommodation.

Family suite: It's designed for seniors who need rehabilitation guidance due to minor diseases, and choose family suite accommodation.

四大核心功能 Four main functions

医疗	Medical center
志愿者服务中心	Volunteer service center
居住	Living area
公共	Public area

文献调研 || Literature survey

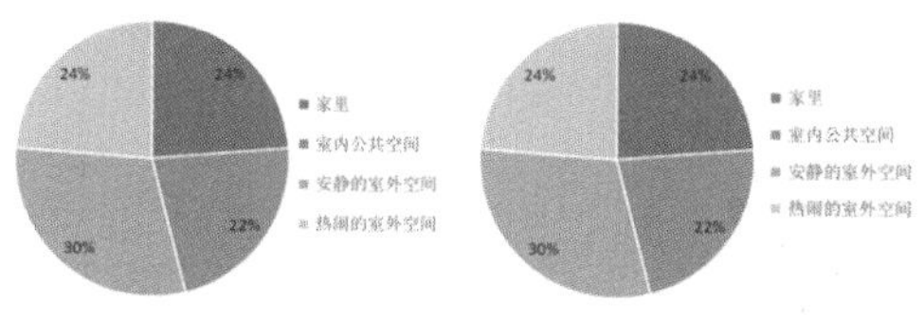

50-70岁的人群，最喜欢的交流空间是热闹的室外空间和室内公共空间。

患有骨科疾病的人群，最喜欢的交流空间是热闹的室外空间和室内公共空间。

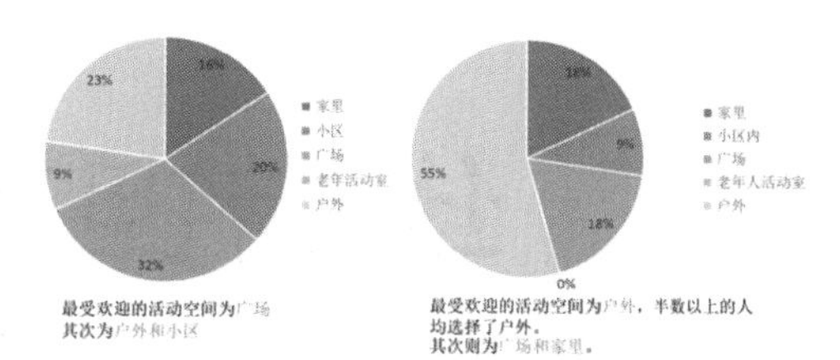

最受欢迎的活动空间为广场 其次为户外和小区

最受欢迎的活动空间为户外，半数以上的人均选择了户外。其次则为广场和家里。

总体人群社区服务需求

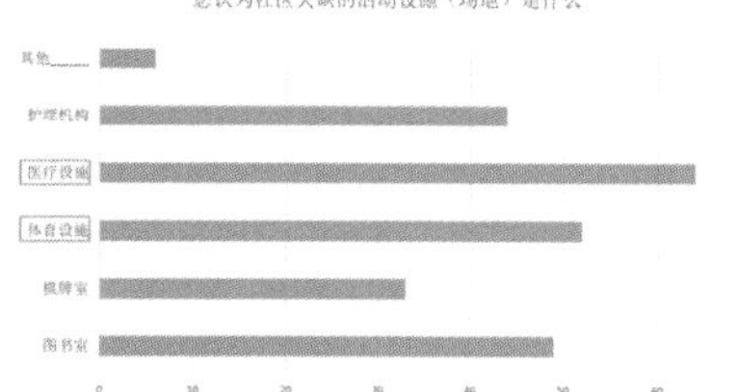

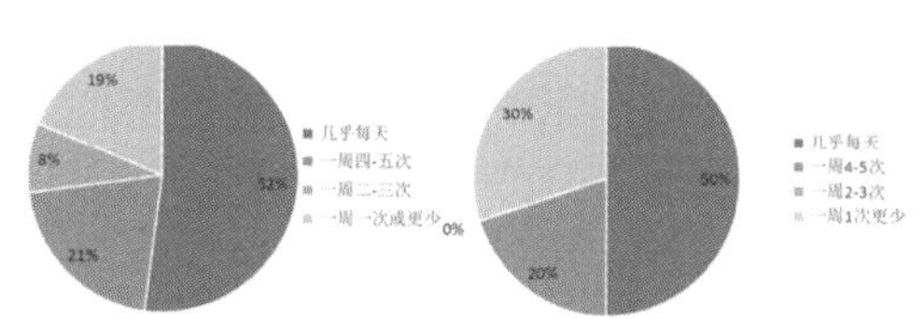

半数以上的老年人活动频率为几乎每天 19%的老年人则由于不便外出或不愿外出，活动频率为一周一次或更少。

半数老年人活动频率为几乎每天 而30%的老年人则由于行动不便，只能多待在家中，活动频率为一周一次或更少。

活动、服务、交流

老人偏好户外活动，且活动频率较大
在行动不便的情况下，待在家较多

1. 医疗服务　2. 体育设施
3. 周边环境质量　4. 餐饮问题

与年轻人交流的意愿强烈
交流发生的场所：热闹的室外公共空间，室内公共空间

设计要求

户外设计
无障碍设计
医疗功能　户型设置厨房
公共餐饮　景观设计
吸引年轻人的场所
气氛热闹的公共空间

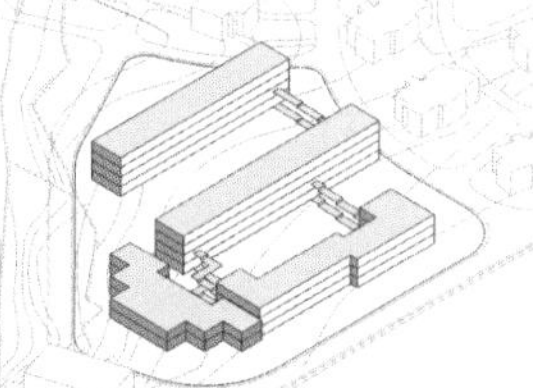

原有建筑
Original building

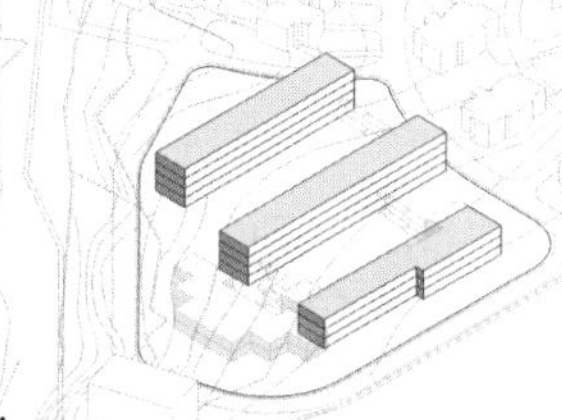

拆除
Dismantle

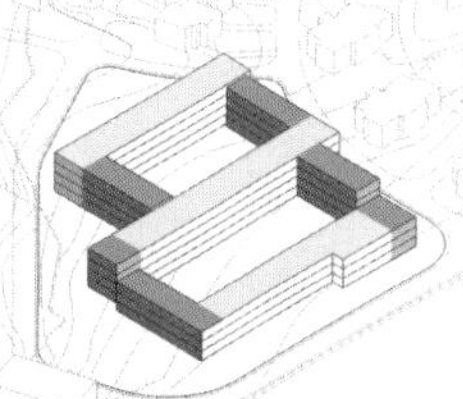

加建实体
Build entity

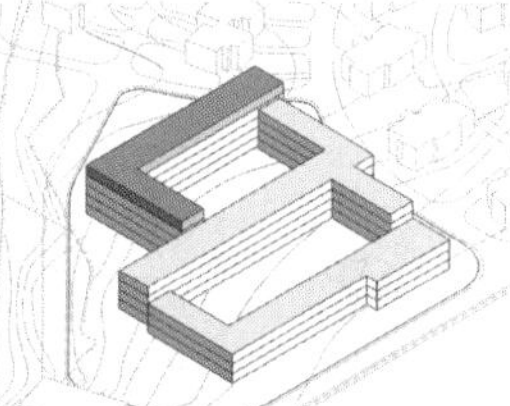

北侧加建一层轻钢结构，平衡房间户数
Build a storey，balance room number

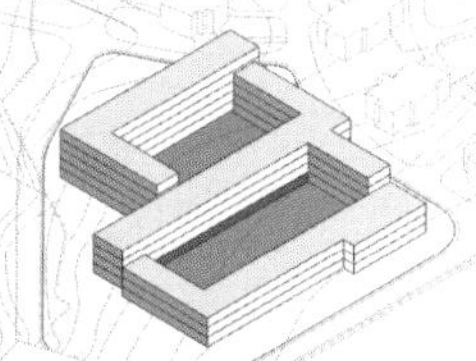
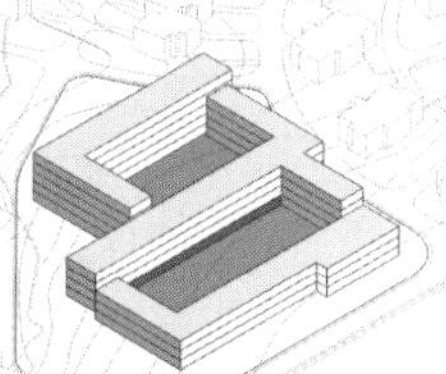

打通底层，联系内院
Open bottom, contact inner court

加建大空间
Build bigger space

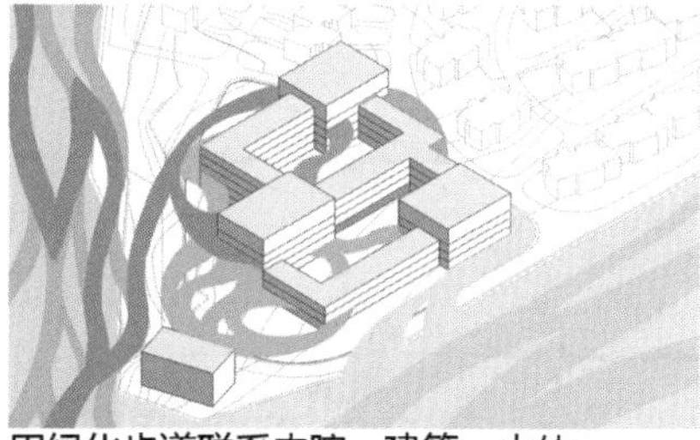

用绿化步道联系内院，建筑，山体
Build greenline to connect court building and mountain

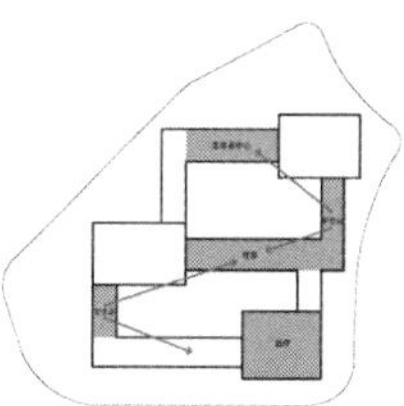

无障碍设计
Barrier free design

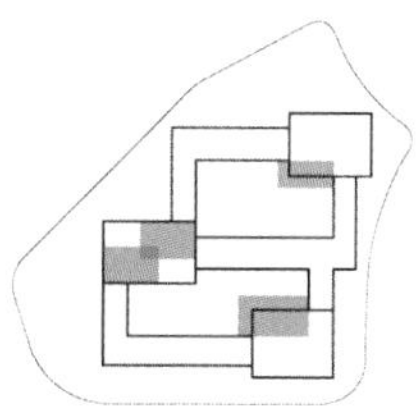

功能设置
Function setting

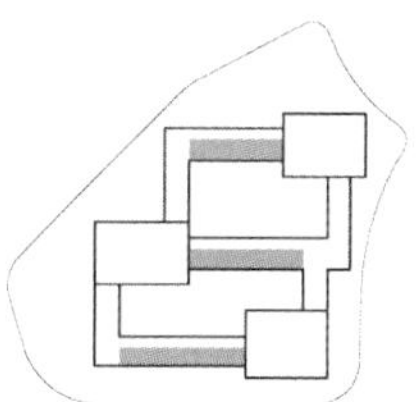

户型设置
Apartment setting

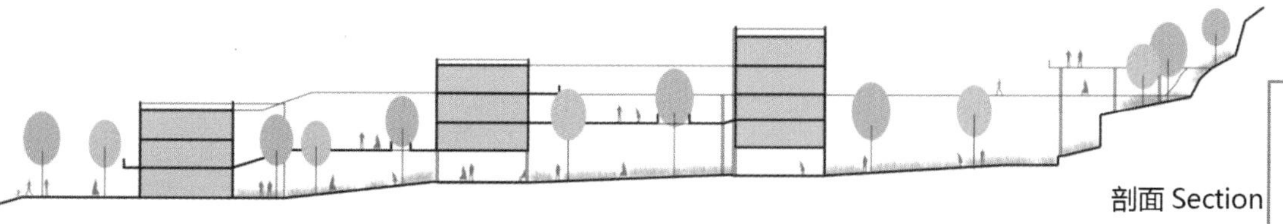

剖面 Section

过程方案

华南理工大学

South ChinaUniversity Of Technology

CIID“室内设计 6+1”2016（第四届）校企联合毕业设计

CIID“Interior Design 6+1”2016(Forth Scssion)University and Enterprise Joint in Graduation Design

小组成员：

郑潇童 Zheng Xiaotong　　陈谢炜 Chen Xiewei

设计愿景 || Versions

- 创建一个活力四射的庭院空间
- 旧建筑与新景观的融合
- 营造出不同的社区空间
- To create vibrant yards
- To integrate existing buildings with newly added landscape
- To create different community spaces

北京曜阳国际老年公寓环境改造设计的目标是努力建设成社区农场，更专注打造一种田园氛围，形成公园化的农场。在农场的养老产业中，老年人可亲自参与或监督整个农场的生产、生活环境的建设工作，从吃到住，完全可以满足农场老年社会中所需的一切，老年人组成的团队，甚至可以走向外面的社会，从而产生更大的价值和影响力。

Beijing Yaoyang International Retirement Community Landscape Renovation Project proposes the idea of building vegetable gardens in the retirement community as a part of its landscape. Seniors can take part in or supervise the construction and production of the gardens, and in return, the yields can satisfy the needs of the seniors for vegetables. In the process of group activities, the seniors can have more contact with the outside society, creating more value and influence.

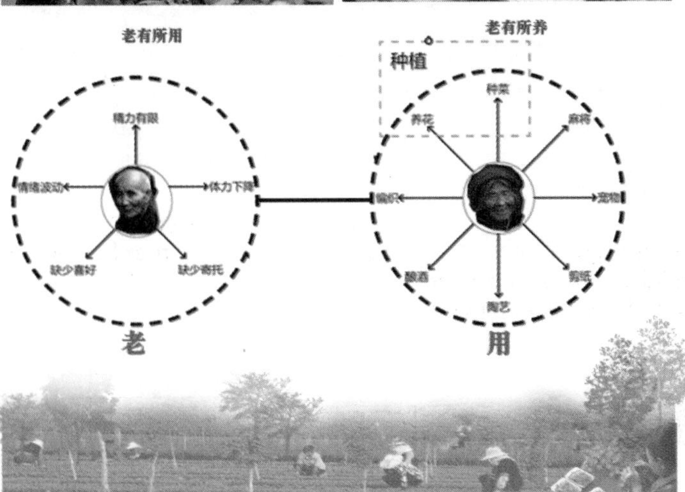

园区分析 ‖ Site analysis

分析得到种植是比较适合老年人这个群体的，一方面让老年人从心理上有个寄托，另一方面通过种植，收获，提高老年人自身价值。所以我们采用了种植的方法来实现“老有所用”的目标。

Site analysis results show that growing vegetables is suitable for seniors living in the retirement community. On the one hand, it gives a kind of emotional sustenance to the seniors, and on the other hand, the seniors can prove their value during the process of growing and harvesting.

种植点分析 ‖ Locations of vegetable gardens

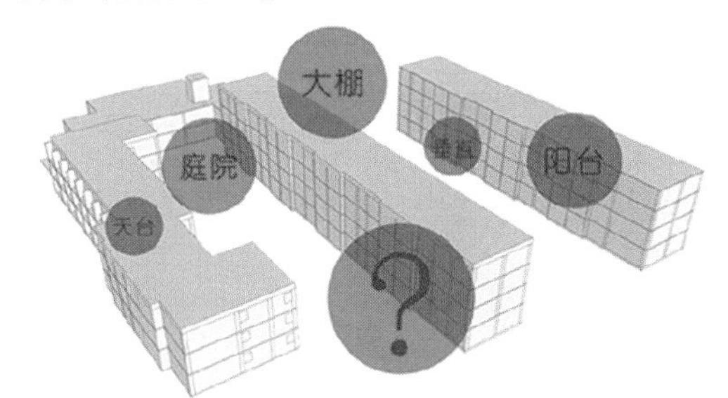

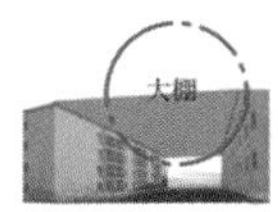

优点：植物生长速度快　便于预防自然灾害

缺点：温度较高　通风较差　维护成本较高

优点：方便老人种植

缺点：阳台使用率降低　互动性差

优点：光照雨水充足　面积充沛　屋面平坦

缺点：老人参与度低　互动性低　存在安全隐患

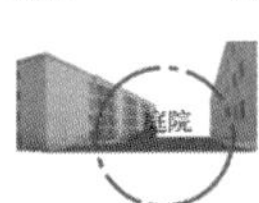

优点：维护成本低　互动性高　参与度高　观赏价值得以发挥

缺点：光照不足　二层以上老人参与度低

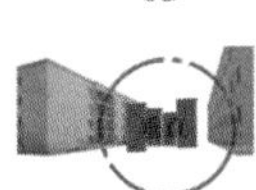

优点：参与度高　互动性强　光照较充足　观赏性强

缺点：维护成本较高

那么在园区哪里进行种植也是值得关注的问题之一，我们运用了 5 种种植方法：

（1）庭院种植：在原有的庭院进行种植。

（2）垂直绿化墙：在庭院中组团状分布垂直绿化墙，可控制旋转，吸取阳光。

（3）温室种植：在北方寒冷的冬天也可以满足蔬菜的需求。

（4）阳台种植：简单的阳台种植可以方便老年人种植。

（5）屋顶种植：有足够的面积，阳光充足，雨水充沛，易于浇灌。

Where to build vegetable gardens is one of the key issues in the project. We provide five solutions to the issue:

1. Yard gardens: Grow vegetable in the existing yards.
2. Vertical gardens: Build groups of rotatable vertical green walls which ensure sufficient sunlight exposure on both sides in the yards.
3. Greenhouse gardens: Vegetables can be grown even in the cold winter in North China.
4. Balcony gardens: The seniors can grow vegetables conveniently.
5. Roof gardens: Sufficient area, sunshine and rainwater, and convenient irrigation can be realized.

功能置换 ‖ Function replacement

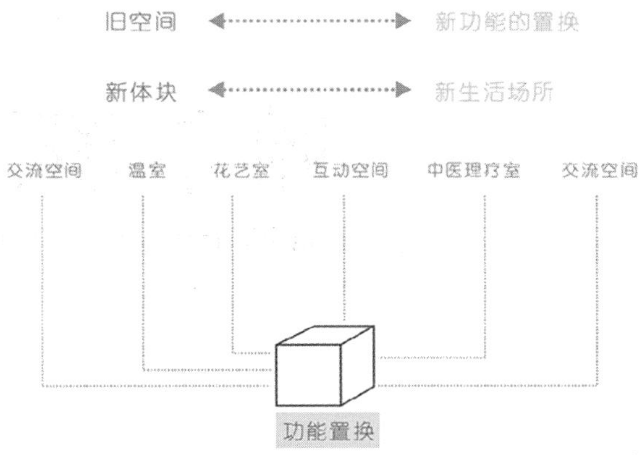

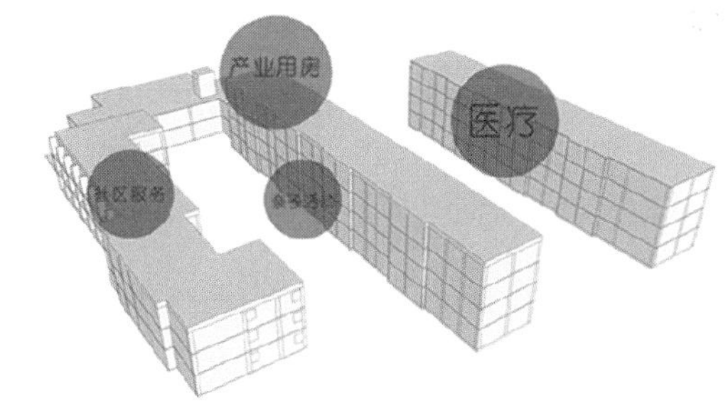

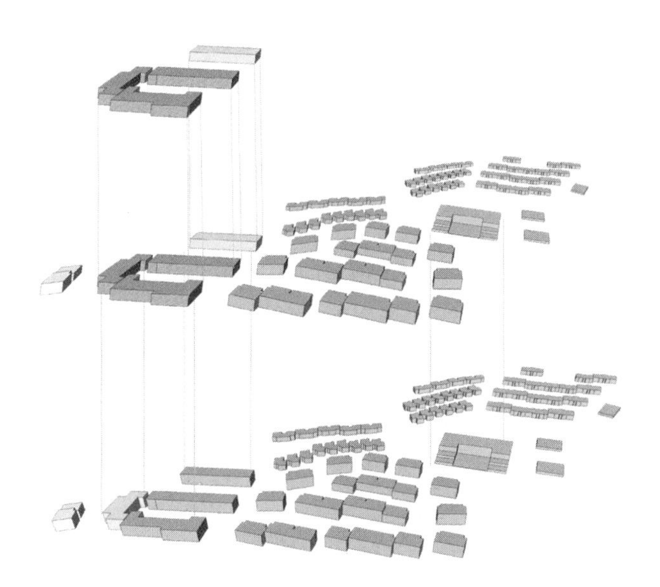

方案构想 ‖ Conceptual framework

哈尔滨工业大学

Harbin Institute of Technology

CIID“室内设计 6+1”2016（第四届）校企联合毕业设计

CIID"Interior Design 6+1"2016(Forth Scssion)University and Enterprise Joint in Graduation Design

小组成员：

张相禹 Zhang Xiangyu　张玲芝 Zhang Lingzhi　朱梦影 Zhu Mengying

袁思佳 Yuan Sijia　李佳楠 Li Jianan　巴美慧 Ba Meihui

李　皓 Li Hao　李曼园 Li Manyuan

生成理念 || Philosophy

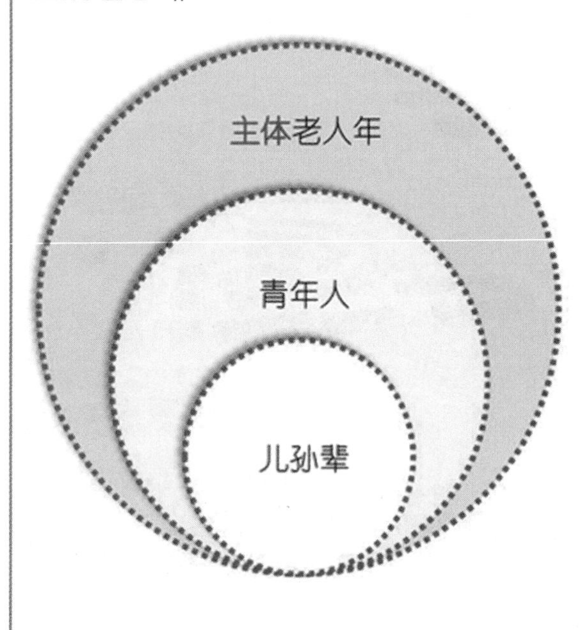

打造家庭度假式养老居住空间，以老年人为居住主体，家人可在假期时间居住陪伴。同时打造邻里交互式居住空间，模仿旧时集体居住模式，消除老年人孤独感。

A resort-style retirement community providing a home-like living environment is built for seniors. Its resort-style design allows the company of the occupants' families during holidays, and its spatial design imitating the old-fashioned communal living space promotes the communication among the seniors, leaving no room for loneliness.

形成以老年人为主，年轻人时常光顾的服务综合体。

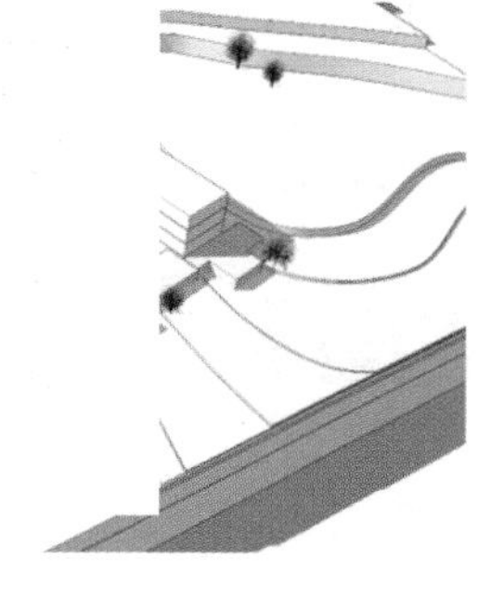

空间改造过程 || Space redesign process

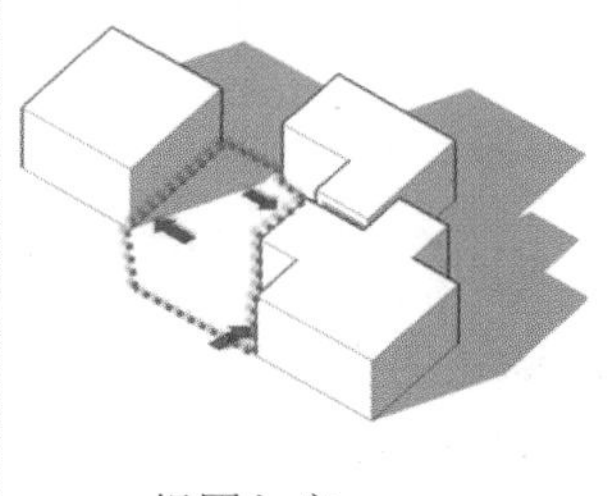

组团入户

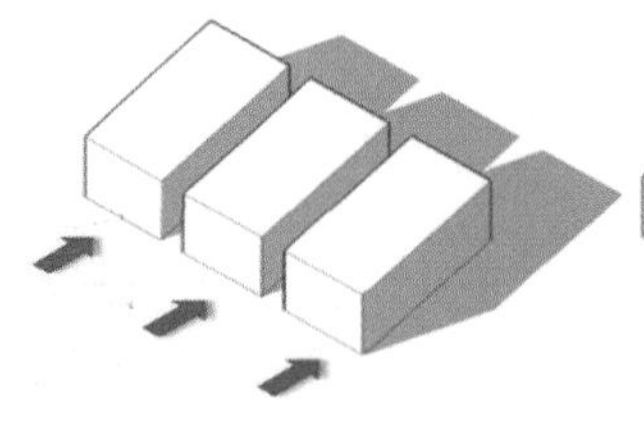

并列入户

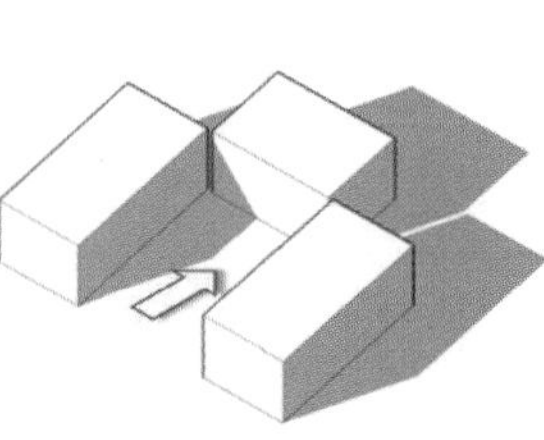

形成公共空间

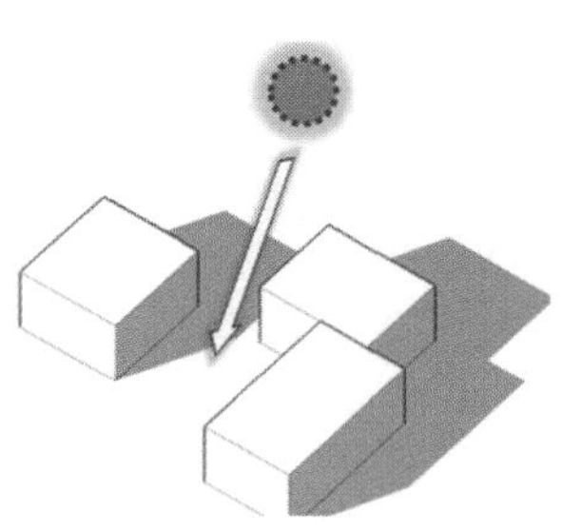

解决采光问题

空间生成过程 || Space formation process

休闲娱乐空间 · 娱乐养老理念

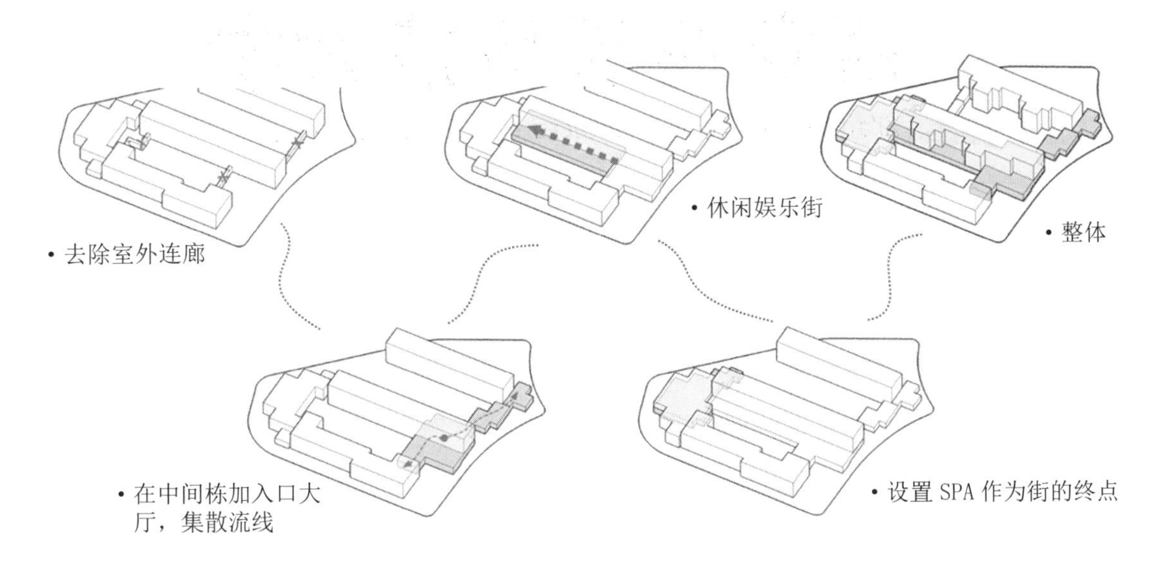

医疗空间 · 养生养老理念

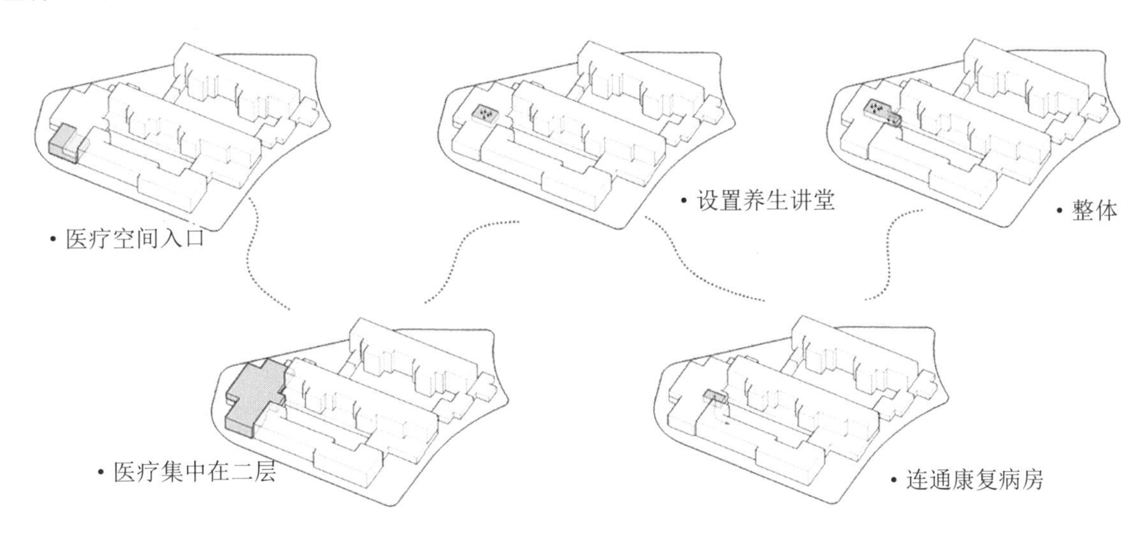

改造分析 || Redesign analysis

保留原有柱网、管径与垂直交通空间，通过添加楼体突凸出结构，重新打造组合空间，形成"团"式的居住模式。

On the basis of retaining the original column grids, piping and vertical circulation spaces, some protruding structures are added to reshape the buildings, forming a communal living space.

原平面

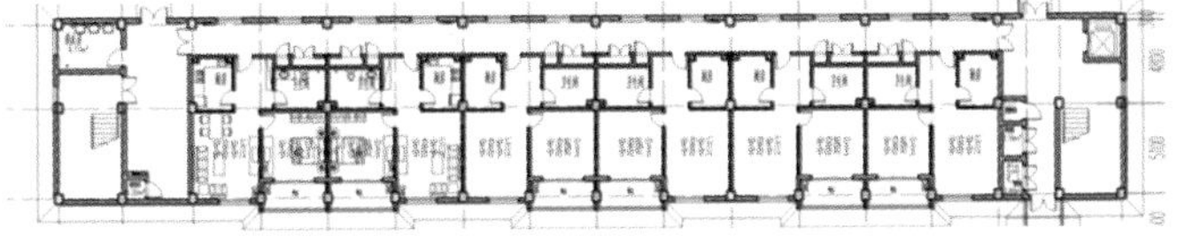

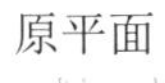

改造后平面

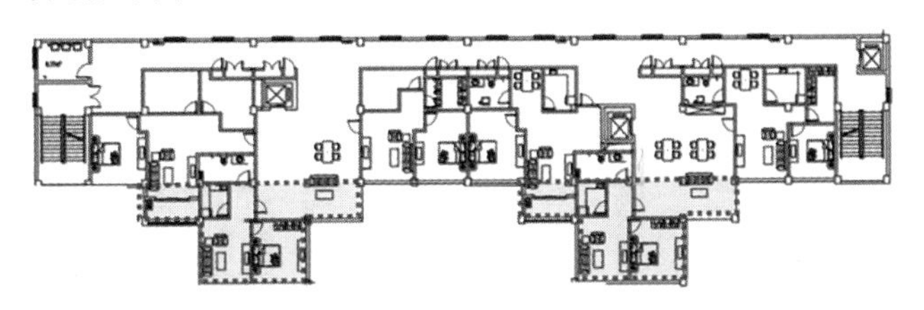

西安建筑科技大学

Xi'an University of Architecture and Technology

CIID“室内设计 6+1”2016（第四届）校企联合毕业设计

CIID“Interior Design 6+1”2016(Forth Scssion)University and Enterprise Joint in Graduation Design

小组成员：

赵凯文 Zhao Kaiwen　　张　峰 Zhang Feng　　颜　强 Yan Qiang

李坷欣 Li Kexin　　刘雨鑫 Liu Yuxin　　曾　嘉 Zeng Jia

深入调研 || In-depth survey

北京人愿意去密云养老吗？

Do people living in downtown Beijing want to spend their retirement years in Miyun?

北京市养老机构由首都功能核心区向城市拓展区和城市发展新区，远郊区的生态涵区递减。

In Beijing, core functional areas have the largest number of retirement homes, followed by urban expansion areas, urban new development areas, and suburbs with good natural environment.

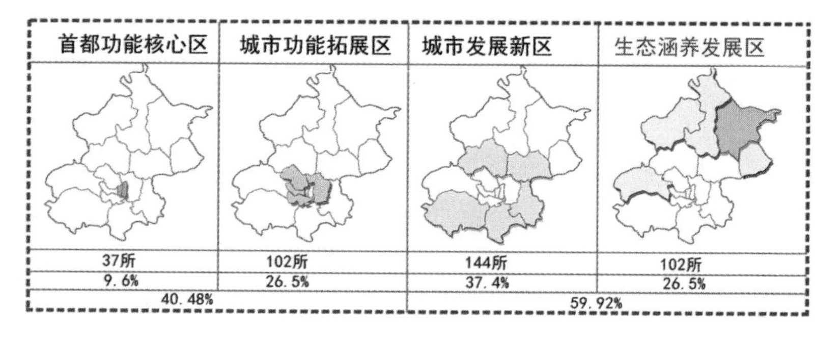

首都功能核心区	城市功能拓展区	城市发展新区	生态涵养发展区
37所	102所	144所	102所
9.6%	26.5%	37.4%	26.5%
40.48%		59.92%	

设计者：对空间的新鲜感，基地原有空间的好处

Designers: The retirement community features spatial advantages.

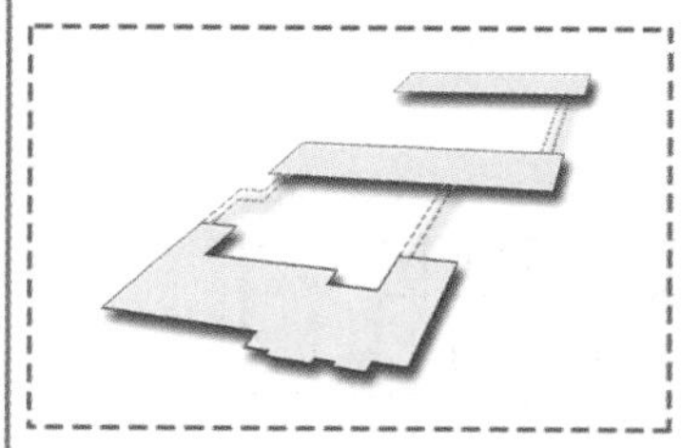

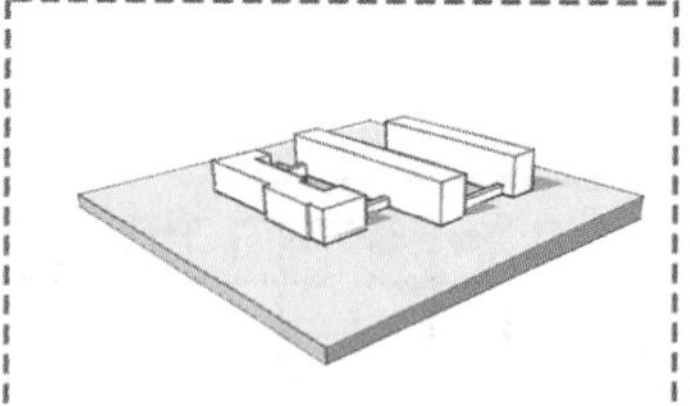

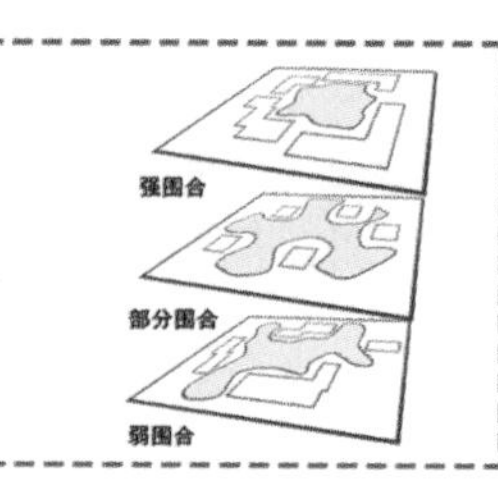

使用者：老年人是特殊群体

Target group: Seniors are a special group

提出问题 || Raise problems

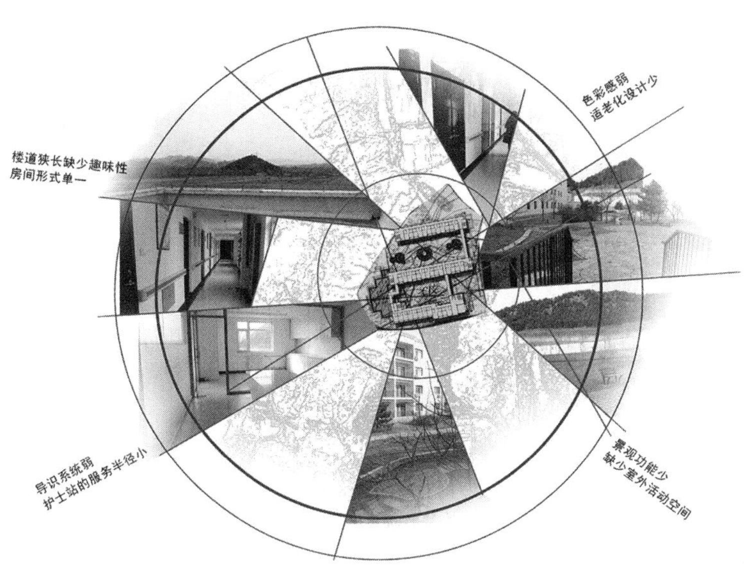

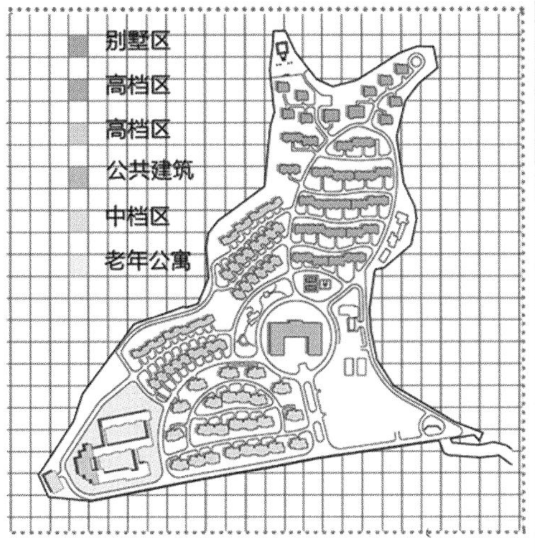

基地虽距离北京市区较远，但是基地周围环境好，景区多，可以吸引老年人来参观。首先我们要提供给老年人并没有脱离城市没有脱离家的感觉，可以吸引他们的子孙让他们更加舒适在这里生活。

Although the retirement community is far away from downtown Beijing, it features good natural environment and is close to many scenic spots which are attractive to seniors. What we need to do is creating a home-life environment with modern facilities, so that seniors and their younger generations will be attracted to and live comfortably here.

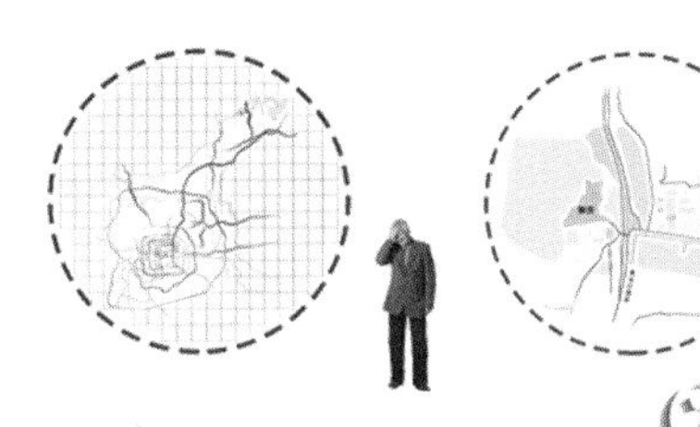

怎么样吸引老年人来？

首先，要让老年人爱上这里……

老年人喜欢？……

基地位于京郊山脚下，三面环山，园内冬暖夏凉，植被丰富，空气新鲜成为京郊的避暑养生胜地。老年公寓西侧有医疗空间，可以提供给老年人基本的医疗保障。

Located at the foot of a mountain in the suburb of Beijing, the retirement community is surrounded by mountains on three sides which give it comfortable temperatures all year round. Good weather and bush vegetation make it an ideal place to spend hot summers. Standing to the west of the retirement apartments is a medical center which can offer basic medical service to the seniors.

人群定位：

中高档老年公寓
有一定经济基础的老年人
生活可以自理或者半自理
丁克家庭或独居老人

解决问题 || Solve problems

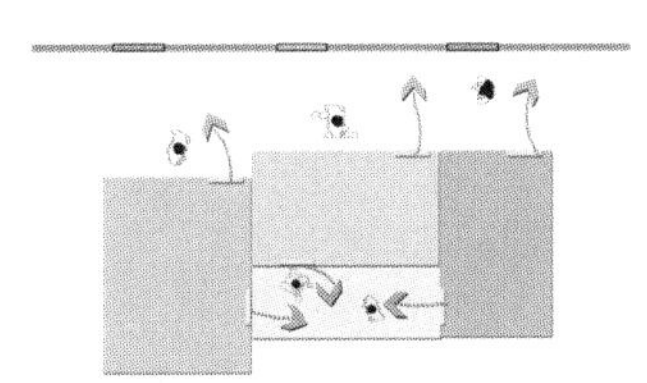

组团概念

提供给老年更多的交流空间，满足他们日常生活的娱乐需求。

Create more spaces for communication and entertainment.

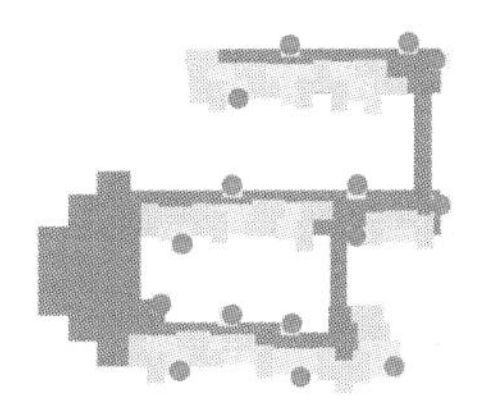

概念总平

点状分布的标志物，让老年人更好的找到方向，不易迷路。

Dot signs across the project site so that the seniors can easily find their way, avoiding getting lost.

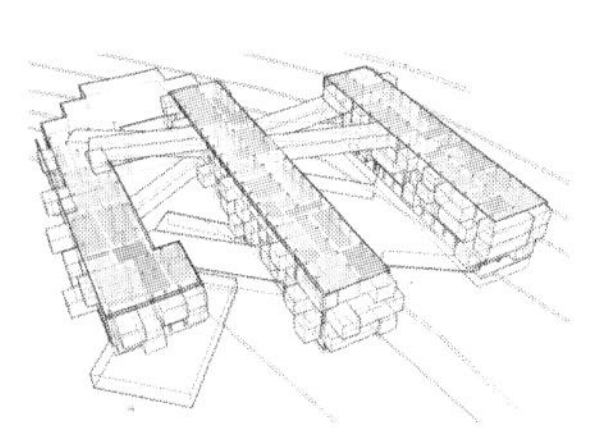

屋顶绿化

屋顶的绿化及种植池，增加老年人与空间的参与性。

Build roof gardens and planting beds, encouraging the seniors to participate in relevant activities.

区位分析 ‖ Location analysis

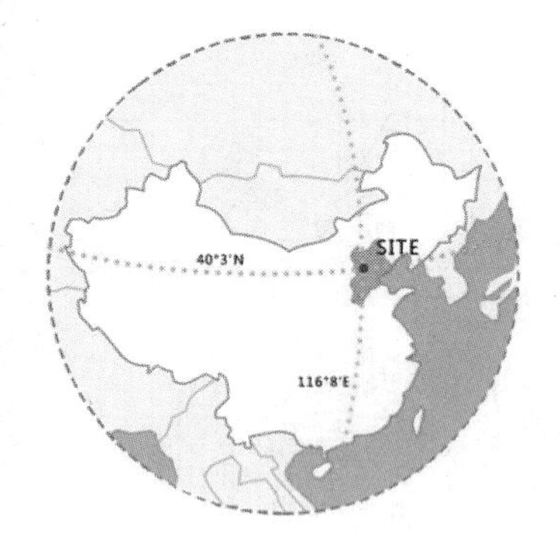

位于北纬 40° 东经 116°
Location: 116° E, 40° N

距离北京主城区约 80km，3 小时的车程，交通不够便利
The retirement community is about three hours' drive (80 km) from downtown Beijing. The transport is not convenient.

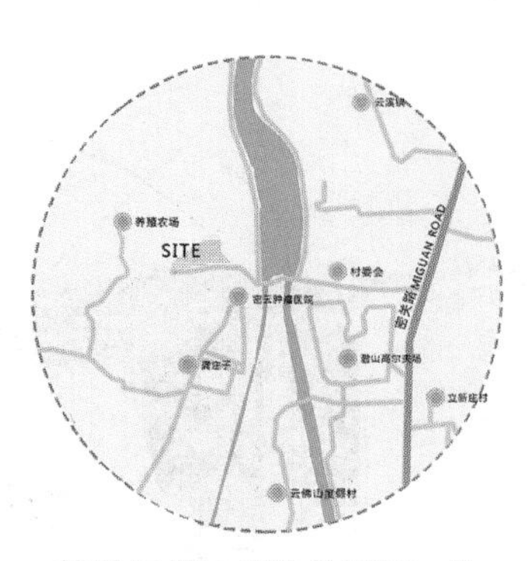

较城区相比周边的医疗，生活，娱乐等等服务资源少
Fewer medical, living and entertainment facilities and resources

场地内外资源现状与潜力 ‖ Current and potential resources in and outside of the project site

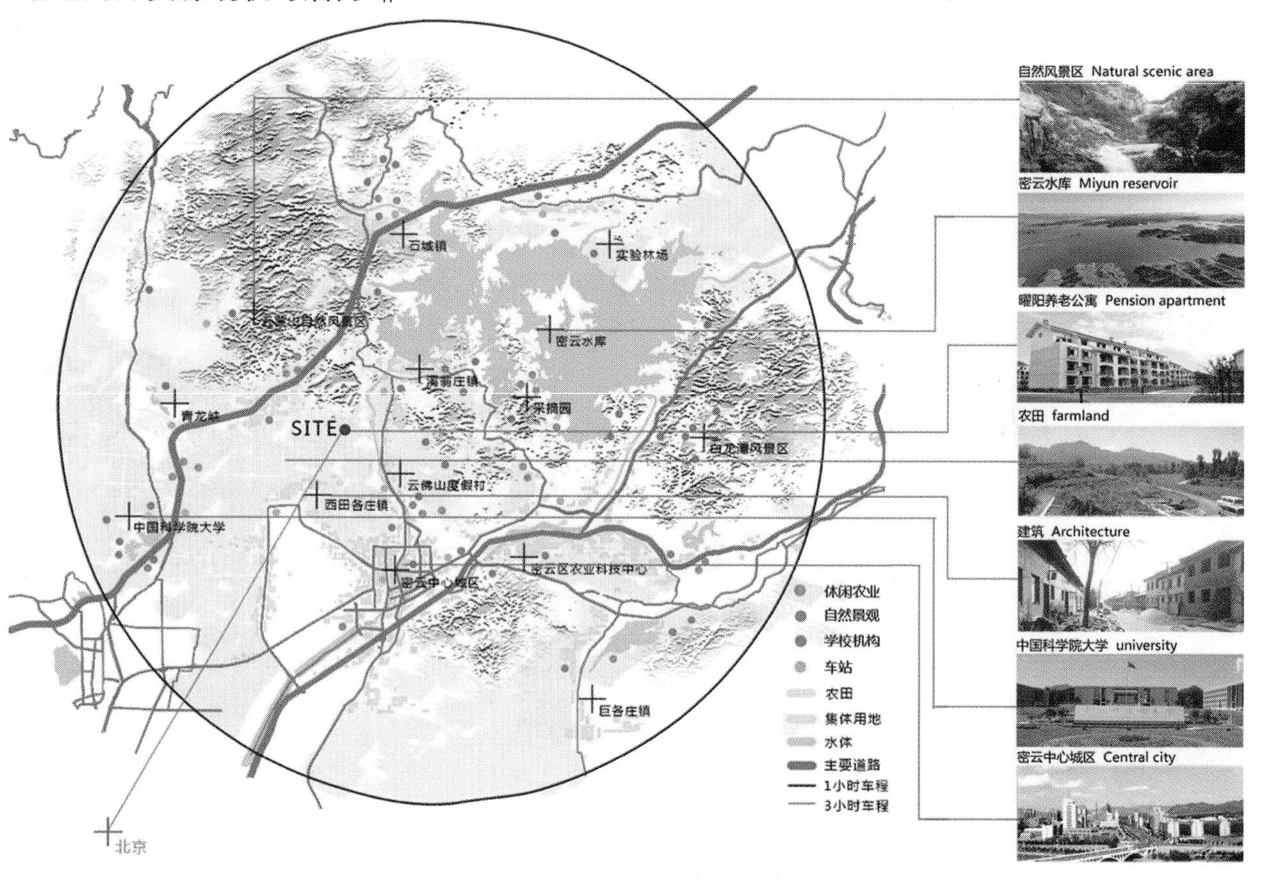

场地现存问题分析 ‖ Analysis of current problems faced by the project site

长廊式布局，功能空间不足
Rooms standing along long corridors, and insufficient functional spaces

标识系统颜色大小不明显
Small signs with inconspicuous color

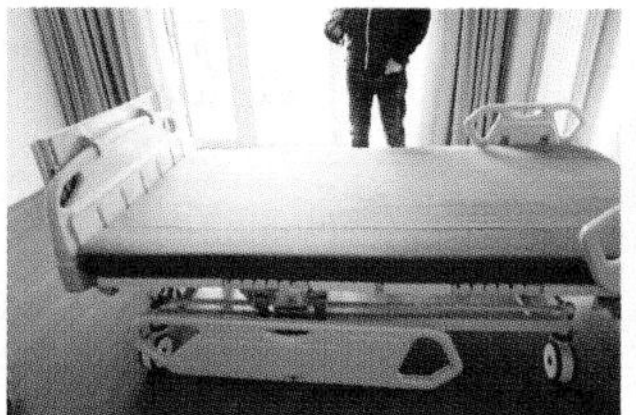

居室户型病房化，空间空旷
Ward-like and empty apartments

无障碍设计不完整
Incomplete wheelchair friendly facilities

過程方案

主题介绍 || Theme introduction

居住组团，区域自足远离城市背景下老年公寓的重构

redesign of retirement apartments advocating self-sufficient life in groups in the suburbs

居住组团 Group living

人员组团	居住抱团	活动社团

尝试一种“朋友圈”的养老模式。合理调配入住区域和楼层，方便老人结伴，相互照应。

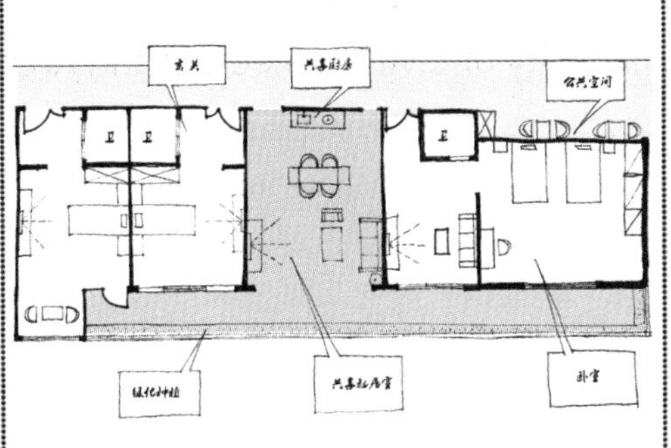

以共享空间为中心，单元围合合成一个庭院聚落。

公共空间节点形成活动空间开展，联谊，老年课程等活动。

区域自足 Regional self-sufficiency

精神自足	管理自足	食物自足

组团的形式加强老人交流。共享空间提供更多可能性如冥想。通过共同生活，劳动价值达到一定满足感。

在活动空间，老人可以主动承担活动的发起者、主持者、参与者以及知识文化技能的传播者，提供老人办公场所达到自足。

通过建立小型农场自给自足的形式来提高老人的兴趣和活力。在劳作不劳累的情况下通过收获达到一定自足。

建筑改造构思 Building renovation concepts

①

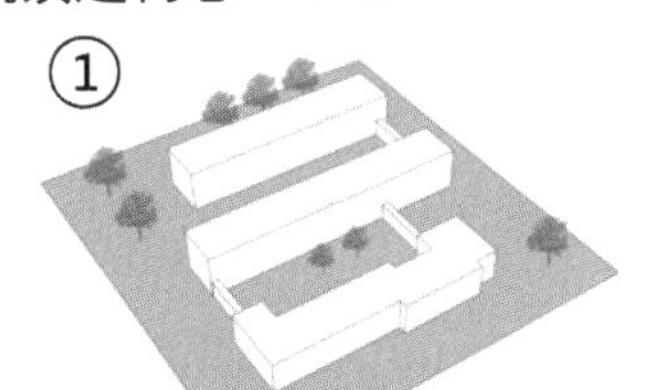

原有三栋建筑相对独立，简单的用交通廊道连接。

②

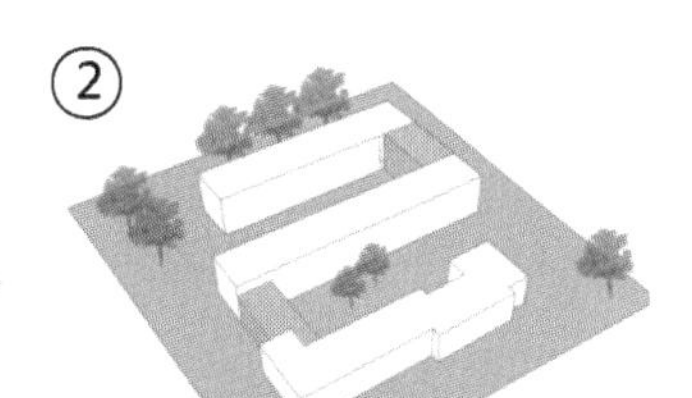

拆除东北向通廊，保留并扩大剩下的交通廊道。

③

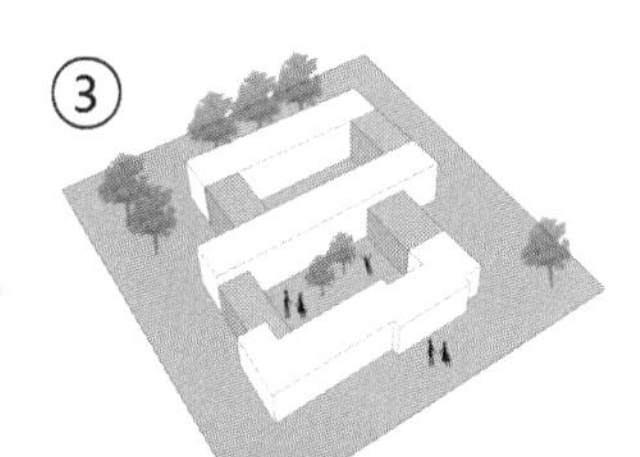

新增体块，围合成两个庭院并形成完整的交通环线。

④

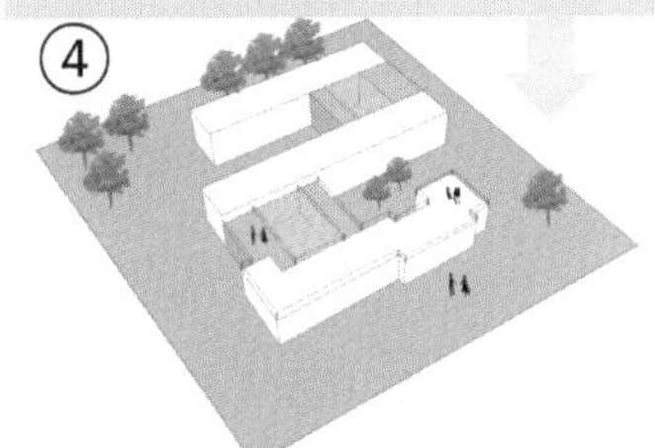

考虑到四周过于封闭和北京天气，将体块往里推得到两个内庭。

⑤

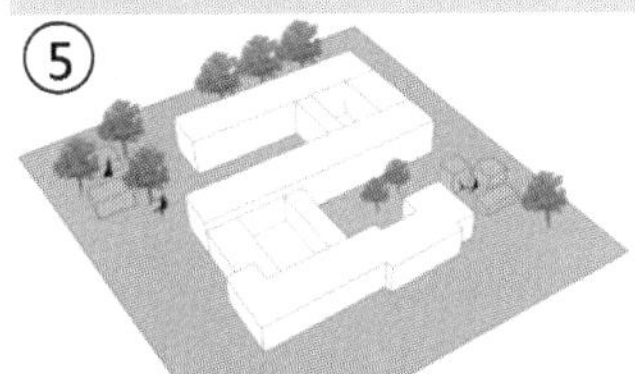

然后在建筑外侧加入若干构筑物供老人自由使用。

⑥

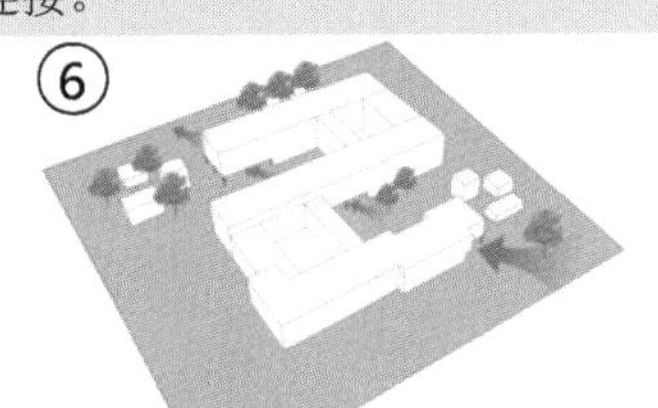

考虑到通风和交通便捷性将一层部分架空，将室内外环境连为一体得到更多活动空间。

北京建筑大学

Beijing University of Civil Engineering and Architecture

CIID“室内设计 6+1”2016（第四届）校企联合毕业设计

CIID“Interior Design 6+1”2016(Forth Scssion)University and Enterprise Joint in Graduation Design

小组成员：

陆　昊 Lu Hao　　魏　卿 Wei Qing　　曾　宁 Zeng Ning

王海月 Wang Haiyue　　李丽阳 Li Liyang　　王逸开 Wang Yikai

张博闻 Zhang Bowen

思维过程分析

我国养老模式下的养老机构的主要问题
- 高端扎堆，忽略主流需求
- 规模过大，与老人需求不符
- 定位不清，产品缺乏细分
- 水土不服，盲目照搬国外模式

曜阳国际老年公寓的建筑设计问题
- 建筑分散，联系困难
- 不能对老人形成多层级、连续的照护模式
- 居住模式单一，每户之间的联系较少
- C2 过于偏医疗，容易给老人形成压抑的印象
- 大厅进深较深，白天较为昏暗
- 形式单调乏味，不能吸引老人
- 植物过于低矮，老人无法触摸
- 植物种类缺乏四季变化

老年人排斥养老机构的问题
- 目前国内养老院服务条件不能满足老年人需求
- 传统家庭养老模式的优势
- 中国传统孝道思想束缚
- 未富先老（经济问题）

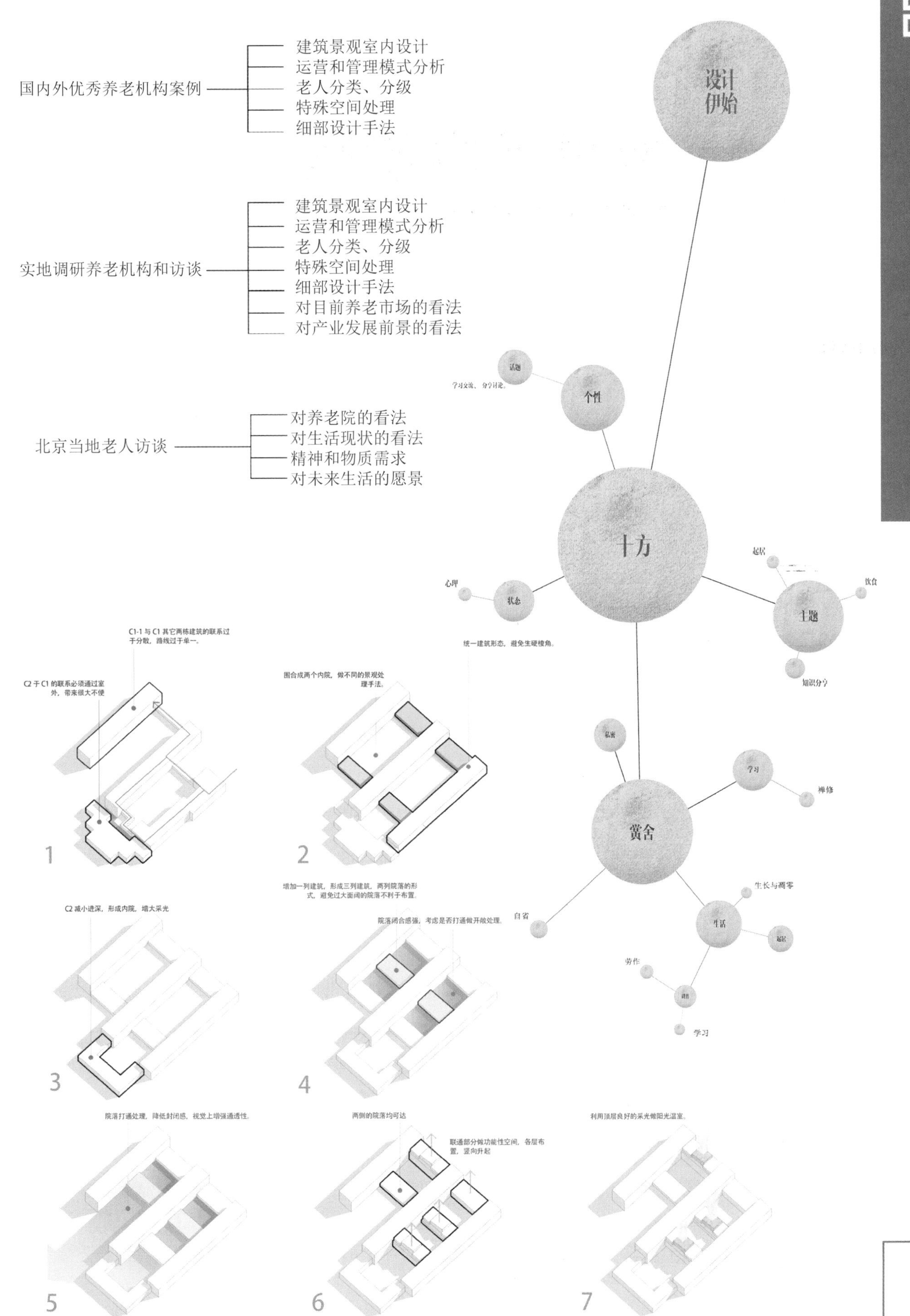
国内外优秀养老机构案例
建筑景观室内设计
运营和管理模式分析
老人分类、分级
特殊空间处理
细部设计手法
实地调研养老机构和访谈
建筑景观室内设计
运营和管理模式分析
老人分类、分级
特殊空间处理
细部设计手法
对目前养老市场的看法
对产业发展前景的看法
北京当地老人访谈
对养老院的看法
对生活现状的看法
精神和物质需求
对未来生活的愿景
设计伊始
个性
话题
学习交流、分享讨论。
十方
状态
心理
主题
起居
饮食
知识分享
统一建筑形态，避免生硬棱角。
黉舍
私密
学习
禅修
自省
生活
生长与凋零
劳作
学习
C1-1 与 C1 其它两栋建筑的联系过于分散，路线过于单一。
C2 于 C1 的联系必须通过室外，带来很大不便
1
围合成两个内院，做不同的景观处理手法。
2
C2 减小进深，形成内院，增大采光
3
增加一列建筑，形成三列建筑，两列院落的形式，避免过大面阔的院落不利于布置。
院落闭合感强，考虑是否打通做开敞处理。
4
院落打通处理，降低封闭感，视觉上增强通透性。
5
两侧的院落均可达
联通部分做功能性空间，各层布置，竖向升起
6
利用顶层良好的采光做阳光温室。
7

南京艺术学院

Nanjing University of the Arts

CIID“室内设计 6+1”2016（第四届）校企联合毕业设计

CIID“Interior Design 6+1”2016(Forth Scssion)University and Enterprise Joint in Graduation Design

小组成员：

龚月茜 Gong Yueqian　　华彬彬 Hua Binbin　　戚晓文 Qi Xiaowen

魏　飞 Wei Fei　　冼佩茹 Xian Peiru　　张梦迪 Zhang Mengdi

方案过程 || Scheme process

太极阴阳元素表现形式

Incorporation of the concept of Taiji and the Yin-Yang theory into design

养生　中医养生(道家养生)　养气　顺气一日分为四时　气和而神形　阴阳平衡

元素提取：太极
元素运用：形式装饰，空间分布，功能划分，材质表现，主题颜色，情感表现
元素拓展：金，木，水，火，土，道家，阴阳平衡
元素提取：中医养生，道家养生，气场养生
涉及相关：中国水墨艺术，装饰艺术，室外庭院

Design element: Taiji
Element application: formative decoration, spatial division, function setting, material texture, theme color, emotion expression
Element expansion: Metal, wood, water, fire, earth, Taoism, Yin-Yang balance
Design elements: TCM-based health maintenance, Taoism-based health maintenance, aura-based health maintenance
Relevant fields: Chinese ink painting art, decorative art, outdoor yards

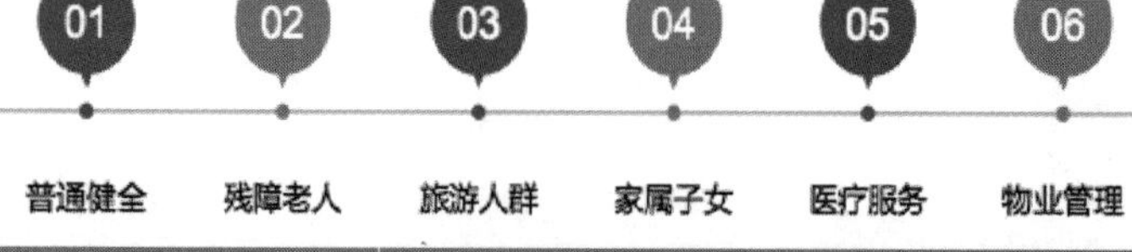

类型	研究分析
展示型和观赏型老人	现代老年人都有一定的特长，能歌善舞、能画会写，有一定的艺术表现能力，他们往往喜欢参加展示活动，如老年合唱团、老年时装模特表演、老年书画展、老年舞蹈队、乐队、街头广告宣传队等活动。而没有表现能力的老年人则喜欢“看热闹”，他们成群结队地看表演、听故事、看展览，相互交流观看心得，不亦乐乎。展示型和观赏型的两个群体在老年人中形成了良好的互补互动。
根据受众，搭建展示公共平台方便老年人交流与互动	人各有所长，很多老年人往往有很多艺术才能，生活技能等，可因为居住空间，居住群体，而没有使其得到良好的展示与交流互动，因此其兴趣也随之减弱，从而影响身心发展，设计者的我们如果能将空间构造成一个展示平台，会促进老人们的交流与互动，增进其各方面的兴趣，从而老有所爱，老有所乐。

老有所养，是中国人自古即寄语的美好愿望，随着我国老龄化人口的不断加剧，老年人的医养空间的建设吸引了人们的眼球，也同时进一步改善了目前老年人的居住环境，六合宓馆的设计是依据现代中老年公寓体系中比较完善的机制和设施，以人为本，倡导中医与道家的传统养生法则，围绕太极主题进行规划的方案。

Being properly cared for after getting old is a traditional Chinese concept. In the context that population aging is increasingly becoming an issue for concern, the construction of aged care facilities has attracted a lot of attention, and consequently, the living environment of seniors has been improved. Themed on Taiji, Six Harmonies is a human-oriented design scheme which adopts modern retirement home operation system and facilities, and advocates TCM and Taoist health maintenance principles.

元素规划 || Element application

山

关于山的元素，结合太极五行中的土元素，从形状，颜色，含义等丰富空间，在庭院设计中，放入假山石，同时根据山峰层层叠叠的外观，在绿植外观的设计上我们也会依势而造：庭院地面设计中，根据原先的地势，结合沙石等材质，设计庭院道路。在室内空间中，置入山石小景同时根据山的颜色将部分空间的主色调设计成满足老年人视觉的深色。

Mountains

Mountains stand for earth in the Five Elements. The shape, color and meaning of mountains are used in the design. Rockeries are arranged and landscaped in the yards to create vivid "rolling mountains with lush vegetation". Besides, materials like sand and stones are used to pave paths running through the yards based on its original terrains. Small-sized rockeries are arranged in the buildings, and dark color similar to the color of rockeries is adopted as the dominant color of some interior areas.

水

关于水的元素，我们会从墙面，地面等，根据水的颜色，状态，利用纱，香，石，木等营造气氛。在空间中，利用一小部分具态的水结合植物，自然。使室内与室外串联起来，具有一种流动感与整体感。

Water

Materials like translucent fabric, incenses, stones and wood are used on walls and ground to imitate the colors and shapes of water. In addition, real water is also used, together with plants, to link the interior and exterior environments, integrating the buildings with nature through flowing water.

材料规划 || Materials

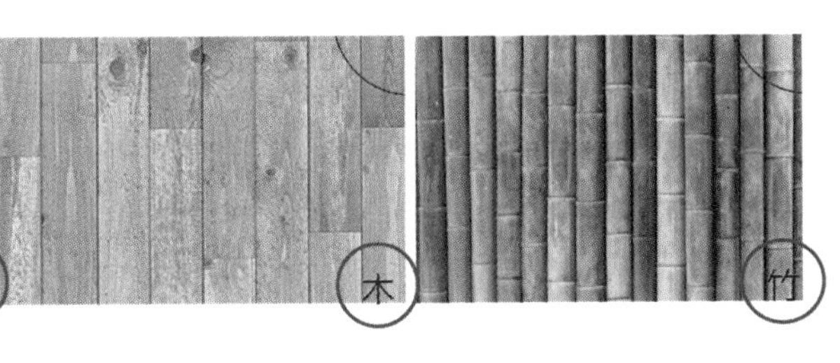

内容策划 || Scheme content

序号	内容	主要展品	展示方式	展示等级	目标
1-1	《黄帝内经》	照壁	主题性篆刻文字	★★	向受众传达养生概念，作为过渡区。
1-2	楼层功能分布	立体图片文字展板	LOGO灯、单色LED	★★	介绍各个楼层的大概功能和走访顺序。
1-3	天圆地方	数字影像	数字投影、电脑灯	★★	结合大堂中的部分空间,向受众传达天圆地方,天人合一的养生概念。 运用数字投影的方式引起受众共鸣，并传播太极养生的理念。
1-4	《步出太极》	投影机、图像采集设备、控制主机	地面互动感应装置、数字投影、电脑灯、声控装置	★★	此展项核心为互动感应装置，是通过视觉识别系统判断经过该区域受众的脚步，从而实现相应的预先制作好的互动效果。地面投影会随着人的脚步变化，变幻出不同的水墨太极拳步伐。不同的太极拳步伐会伴随着不同的太极声乐。
1-5	六合 恋馆简介	立体图片文字展板，数字影像	展板，数字投影	★★	向受众介绍六合 恋馆的大体概况，设施人员配备，人文理念，服务目标与主题等。
2-1	六合山水	假山，庭院，沙石等	场景复原	★★	利用人工造景的形式，建造中国传统园林，同时考虑到老年人安全问题，利用沙石排布，石块堆叠，经营出假山假水的意蕴，表现出山水的元素。
2-2	移步太极	健身设施器材	文字说明，器具陈列	★★	
2-3	气生太极	绿植	绿植设计	★★	借助其原有的地势高低，同时当地适宜的绿植进行修剪，规律排布，营造出由多变少，集聚分散的感觉，给人以流动的感觉。
2-3	《太极拳》	雕塑小品	文字说明，小品展示	★★	利用几何石头雕像，陈列出太极拳的招式，同时配以文字说明，使受众可以了解太极拳的内涵，并引领受众进行正确的健身练习。

浙江工业大学

Zhejiang University Of Technology

CIID“室内设计 6+1”2016（第四届）校企联合毕业设计

CIID“Interior Design 6+1”2016(Forth Scssion)University and Enterprise Joint in Graduation Design

小组成员：

张　怡 Zhang Yi　　罗　忆 Luo Yi　　卢倩文 Lu Qianwen

邱丽珉 Qiu Limin

项目背景 ‖ Project background

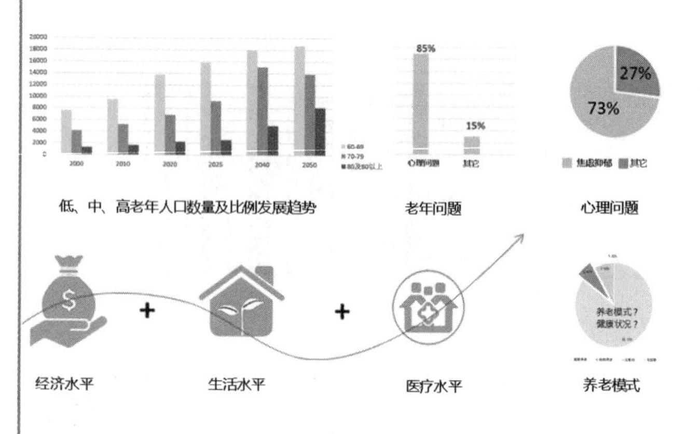

随着老龄化社会的到来，老年人的健康状况与养老的模式成为了社会热切关注的一个问题。老年人的平均寿命随经济、生活水平和医疗水平的提高而提高。但在全国，仍有85% 的老年人或多或少存在着心理问题，每年至少有 10 万老年人自杀死亡。

In the context that China has become an aged society, aged care, including the health care of seniors, has attracted a lot of attention. Although old people are living longer thank to increasingly growing economy, living standard and medical level, 85% of them have psychological problems more or less, and at least 100,000 seniors suicide every year.

区位分析 ‖ Location analysis

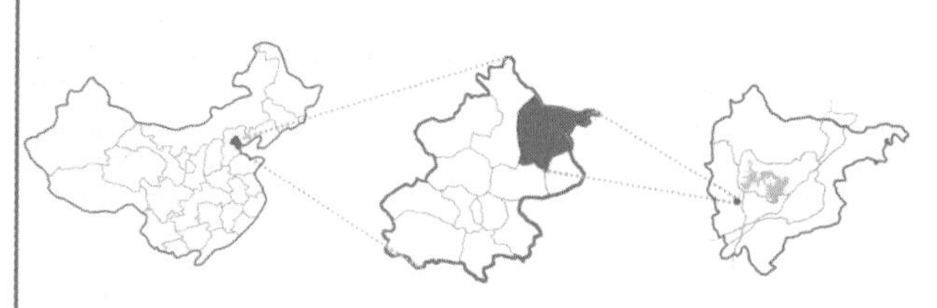

北京曜阳国际老年公寓建设用地位于北京市密云县城北 10km 处，西侧为卸甲山，南侧为西田各庄镇龚庄子村。远离市区，交通较为不便。

Beijing Yaoyang International Retirement Community is 10 km to the north of Miyun County, Beijing, with Xiejia Mountain to the west, and Gongzhuangzi Village in Gezhuang Town, Xitian to the south. Since it's far away from downtown Beijing, the transportation is not very convenient.

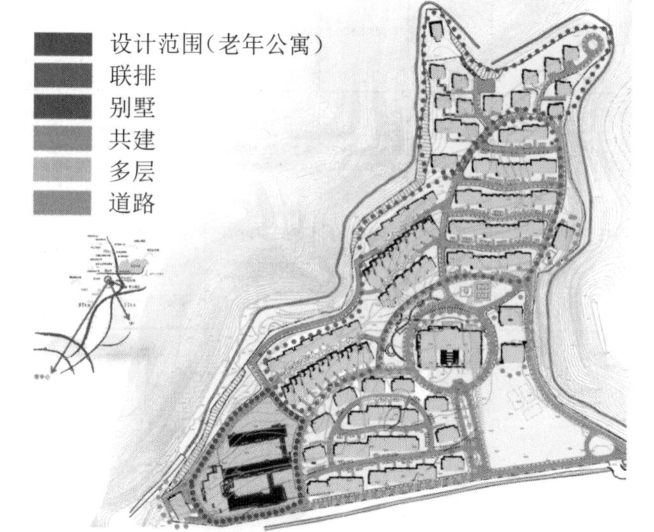

场地分析 || Site analysis

北京曜阳国际老年公寓北临密云水库，东临京密引水渠渠首，自然资源得天独厚，是北京郊区县范围内为数不多的以老年人居住和养生为主题的大型开发项目。

Neighboring the Miyun Reservoir on the north, and the head of Beijing-Miyun Diversion Canal on the east, Beijing Yaoyang International Retirement Community, which boasts advantageous natural resources, is one of the few large-scale real estate projects themed on aged care in the outskirts of Beijing.

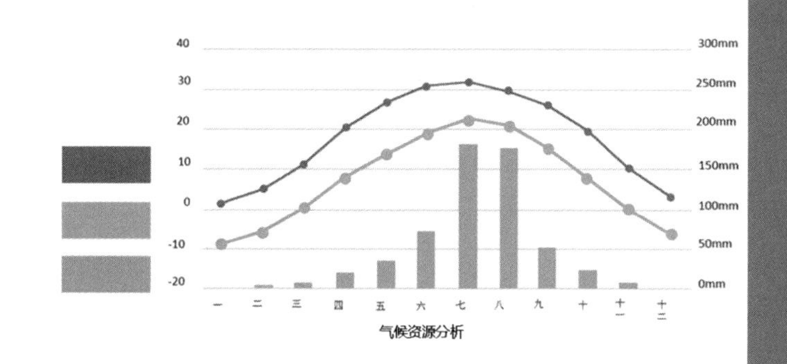

现状分析 || Current situation analysis

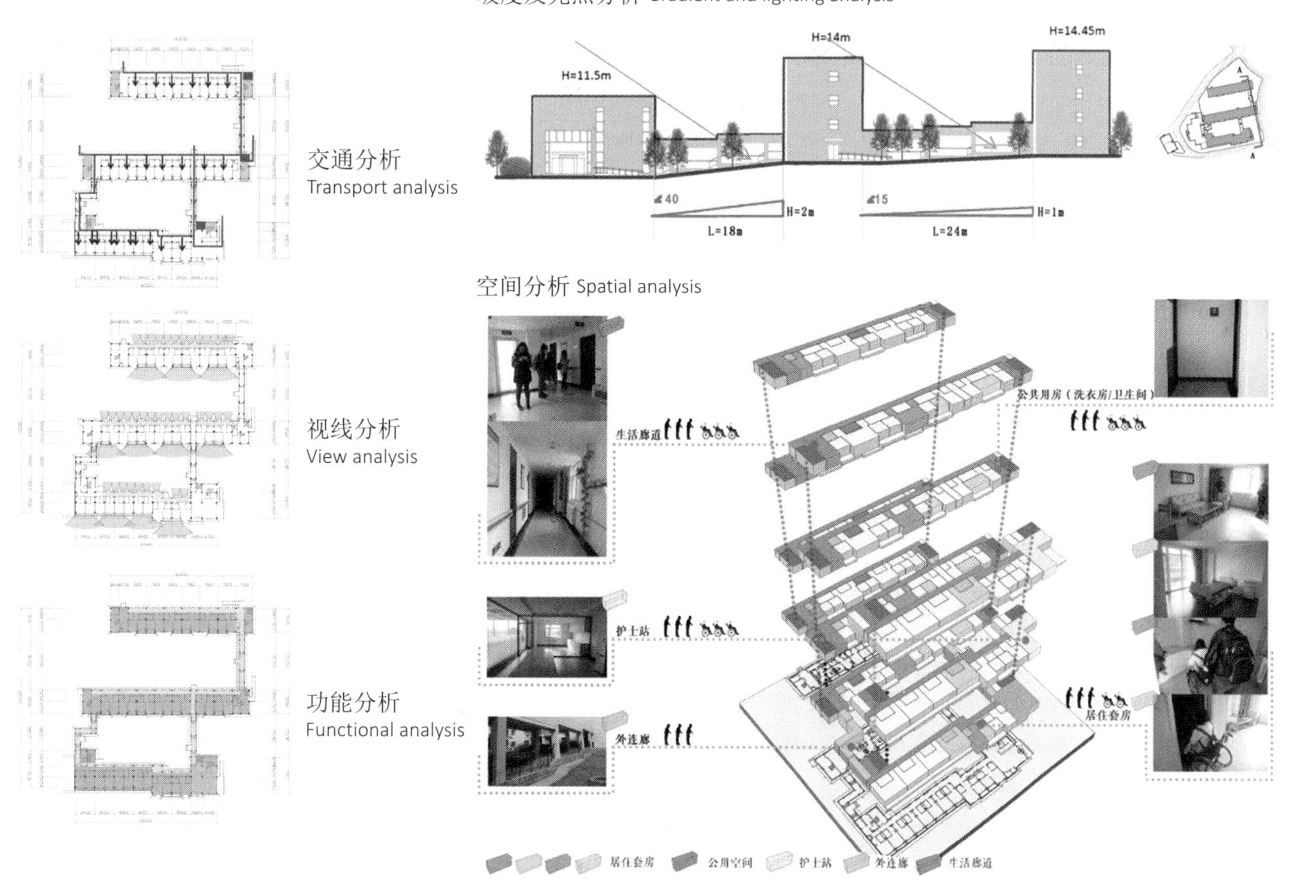

大部分空间为居住空间，公共空间功能少，利用率低。空间功能单一，过于私密化，缺少互动。老年人最需要的交流消遣、娱乐活动、健身养心的空间缺乏。

Most spaces are for living, and the public spaces, with very limited functions, are seldom used. Limited functions, too much privacy and lack of communication are also problems faced by the retirement community. Therefore, spaces for communication, entertainment, and physical & spiritual fitness are needed by the seniors.

人群定位 ‖ Target group

基地现状　　　　社会事件

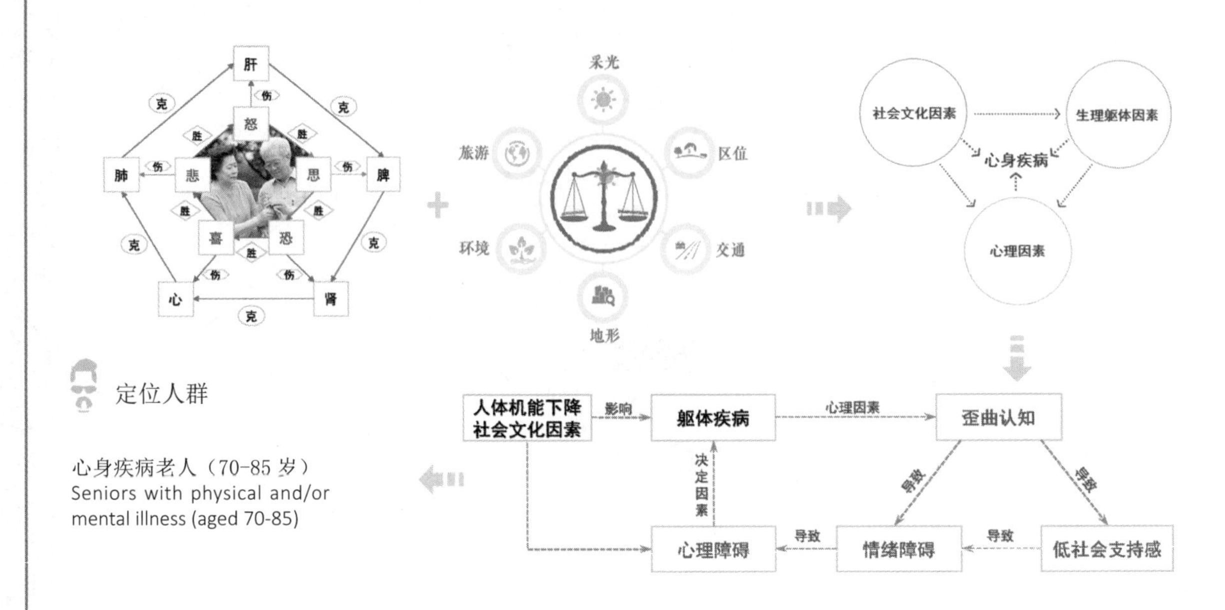

定位人群

心身疾病老人（70-85 岁）
Seniors with physical and/or mental illness (aged 70-85)

生活事件	生活改变值	生活事件	生活改变值	生活事件	生活改变值	生活事件	生活改变值
配偶去世	100	家人健康的转变	45	丧失贷款抵押品的续取权	30	工作时数或工作条件的改变	20
离婚	73	怀孕	44	工作职责的转变	29	搬家	20
分居	65	性功能障碍	39	子女离家	29	转校	19
亲密家人去世	63	新生儿诞生	39	吃官司	29	教堂活动的改变	19
入狱	63	工作变动	39	个人杰出的成就	28	娱乐的改变	19
自己受伤或生病	53	经济状况改变	38	配偶开始工作或停止工作	26	社交活动的改变	18
结婚	53	好友去世	37	学业的开始或结束	26	贷款（少于一万美元）	17
被老板解雇	50	从事不同性质工作	36	生活水平的改变	25	睡眠习惯的改变	16
婚姻调和	47	与配偶吵架次数的改变	35	个人习惯上的修正	24	饮食习惯的改变	15

老年人的心理疾病造成晚年生活质量低下，改善老人的心理健康成为提高老人晚年生活幸福指数的重点。北京曜阳老年公寓环境优美，宜养宜居，适宜健康出行，但远离市区，医疗设施不齐全，不宜介护老人的生活。因此，选择 70~85 岁心身疾病老人作为服务对象。

Mental illness lowers the quality of life of seniors; therefore, improving their mental health is important for their happiness. Embraced by the beauty of nature, Beijing Yaoyang International Retirement Community is an ideal destination for rehabilitation and health maintenance. However, since it's far away from downtown Beijing with insufficient medical resources, it's not proper for seniors needing full-time care. In this project, our target group is set as seniors aged 70-85 with physical and/or mental illness.

人群特点 ‖ Target group's characteristics

	人群需求 Target group's demands		功能需求	空间需求
生理特点 Physiological characteristics	健康需求 Demand on health		体育锻炼 娱乐活动 动态活动	公共空间 可活动 安全性
心理特点 Psychological characteristics	安全需求 Demand on safety 价值需求 Demand on value		亲近自然 静态活动	半公共空间 采光通风好 较封闭
行为特点 Behavior characteristics	参与需求 Demand on participation		多人集会交流 动静皆宜活动	可驻足空间 尺度感 围合感

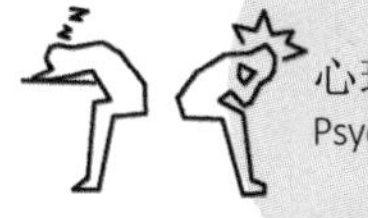

设计理念 || Design philosophy

问题提出 Raise problems

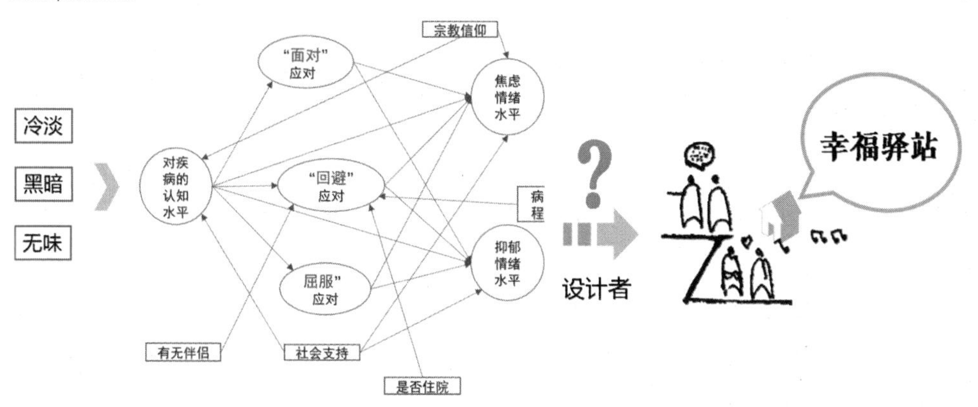

应对方式（前）Before interaction design　　应对方式（后）After interaction design

如何根据老年人自身需求和场地的特点，以心理为出发点，结合生理及行为特点，利用交互，打造一个幸福驿站的主题空间，使老年人在生命的最后一个站点老有所想，老有所乐，老有所用？

How to, according to site conditions, the needs of the seniors and their physical, physiological and behavioral characteristics, create a "happy station" where they can enjoy their life and create their value at the last station of their life using interaction design?

理念提出 Concepts　　交互形式 Forms of interaction

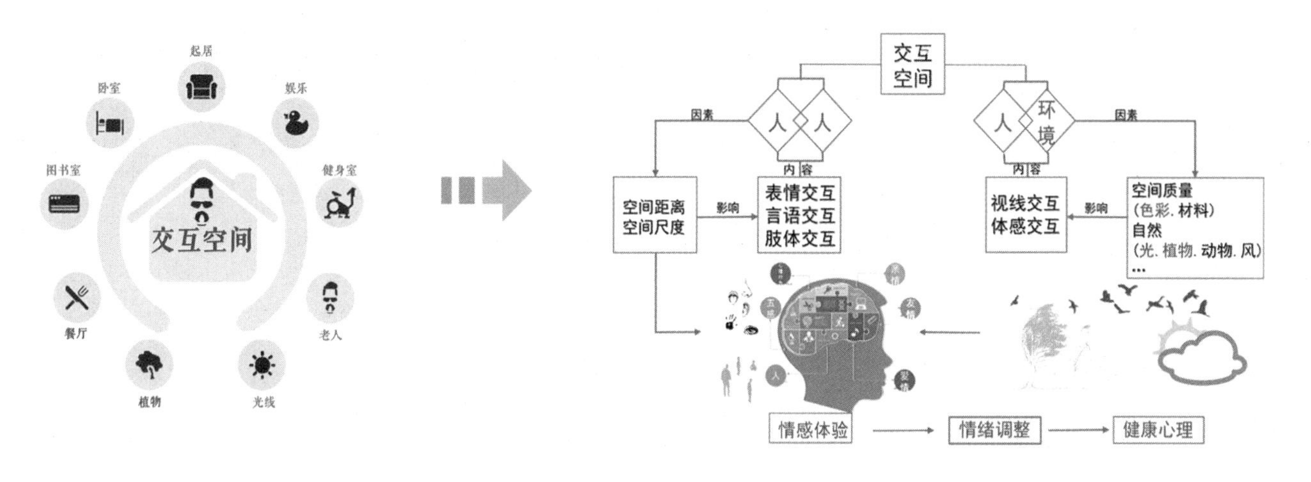

交互目标 Objectives of interaction

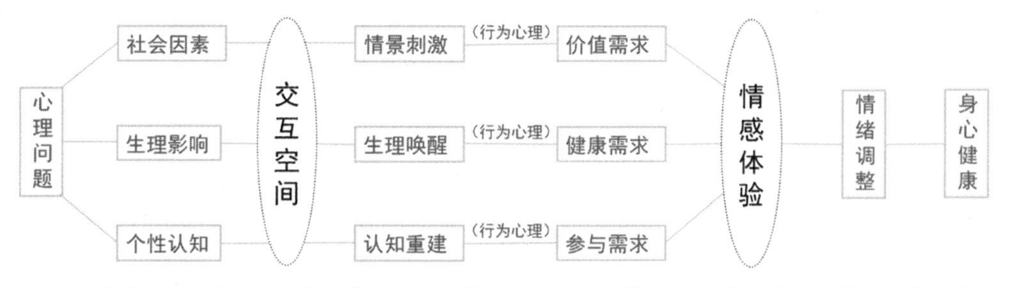

将交互设计定义为情感体验空间。老年人通过不同交互形式与空间元素的交流互动，产生情景刺激、生理唤醒、认知重建。交互空间满足其心理生理需求，使其获得情感体验，得到情绪调整，实现心身健康。

In this project, interaction design is defined as emotional experience. Through various forms of interaction, the seniors communicate and interact with spatial elements to obtain emotional stimulation, physiological activation and cognitive reconstruction. In this process of interaction, their psychological and physiological needs are satisfied, and consequently, their emotions are adjusted, leading to physical and mental health.

遵從老年人居住養老的設計原則，開展將老年人生理心理與景觀設計相結合的探討。基於老年人的心理和行爲特點，分別對南區景觀即A、B、C、D等四個不同區域的外部空間環境做景觀設計等配套，滿足老年人室外生活及康復需求。

零肆

04

醫養園區環境與景觀設計

Environment and landscape design of medical and support Park

CIID“室内设计 6+1”2016（第四届）

校企联合毕业设计

CIID"Interior Design 6+1"2016(Fourth Session)University and Enterprise Joint in Graduation Design

鹤发医养卷

——北京曜阳国际老年公寓环境改造设计

White hair volume medical support

—Beijing Yao Yang International Apartments for the elderly environmental reconstruction design

CIID“室内设计 6+1”2016（第四届）校企联合毕业设计

CIID“Interior Design 6+1”2016(Fourth Session)University and Enterprise Joint in Graduation Design

高　　校：　哈尔滨工业大学
College：　Harbin Institute of technology
学　　生：　张相禹　张玲芝　朱梦影　袁思佳
Students：　Zhang Xiangyu　Zhang Lingzhi　Zhu Mengying　Yuan Sijia
指导教师：　马辉　兆翚　周立军
Instructors：　Ma hui　Zhao hui　Zhou Lijun
参赛成绩：　医养园区环境与景观设计组一等奖
Achievement：　Frist Prize for Aged Care District Environmental and Landscape Design

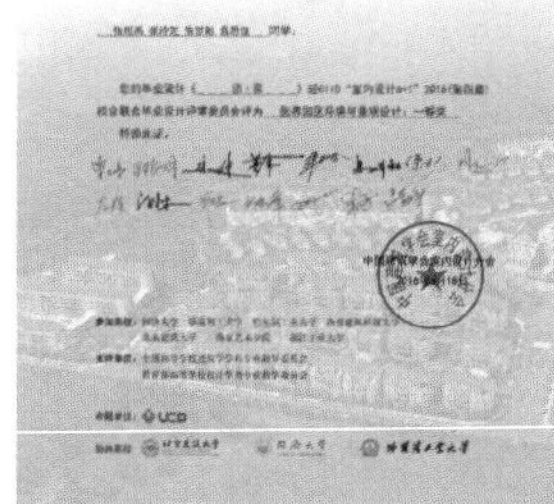

张相禹
Zhang Xiangyu

张玲芝
Zhang Lingzhi

朱梦影
Zhu Mengying

袁思佳
Yuan Sijia

学生感悟：

经过对原项目定位及策划的分析，我们决定把该项目改造成主要针对半自理老人的、引入休闲度假理念的娱乐养老型老年公寓。北京四合院式的邻里关系非常珍贵，家庭式的入住模式也非常适合老人，基于这两个层面，我们提出了“团聚”主题。“团”代表家庭入住择邻而居的生活模式，“聚”则代表有朋自远方来不亦乐乎的生活态度。这个地方不再是冰冷的病房，老人们有权力择邻而居，有权力逛街做 SPA，可以有尊严地在这个温馨的地方安度晚年。

Students' Thoughts:

Based on an analysis of original project positioning and scheme, we decided to transform the project into a resort-style retirement community mainly aiming at partially disabled seniors. Given that precious neighborhood in courtyard houses and home-like living environment are what the seniors in Beijing really need, we determined the theme of “Communal Living” – to live with favorite neighbors and enjoy a happy life together with friends. We want to turn lonely and depressing awards into homes where seniors living in have the right to choose their neighbors, to go shopping, to enjoy SPA, and most importantly, to spend their remaining years with dignity and the company of love.

团·聚

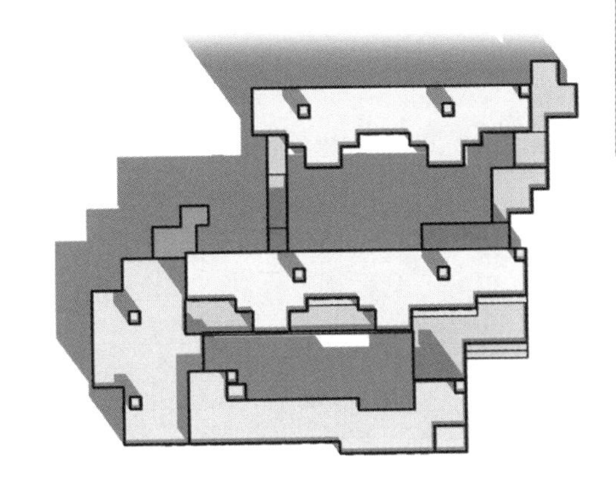

概念分析 || Conceptual analysis

1 社区层面：一种生活方式
大面积的公共服务空间；
休闲SPA、养生医疗、拜神祭灵、棋牌娱乐促进社区社区范围内的交往让老人们相聚在一起，共度时光

2 邻里层面：一种居住模式
形成“大杂院”形式的邻里组团；
抱团居住，互通有无；
爱亲睦邻，互帮互助

3 家庭层面：一种养老理念
将密云作为老人的家；
每逢假期家人将在这里相聚；
是祖孙三代相聚的地方；
在密云便会阖家团聚

百米长街巷。何处不相见？
一天一小聚，两天一大聚。

“大杂院”多年前的记忆，
怀旧复古。

老人是暖巢，子孙是飞鸟，
常有归时。

主题立意 || Conceptual framework

《 当你老了 》

——当你老了 头发白了 睡意昏沉

——当你老了 走不动了

SPA 区所在位置
Location of SPA Area

SPA 室外区效果图
Exterior SPA Area rendering

SPA 室内区效果图
Interior SPA Area rendering

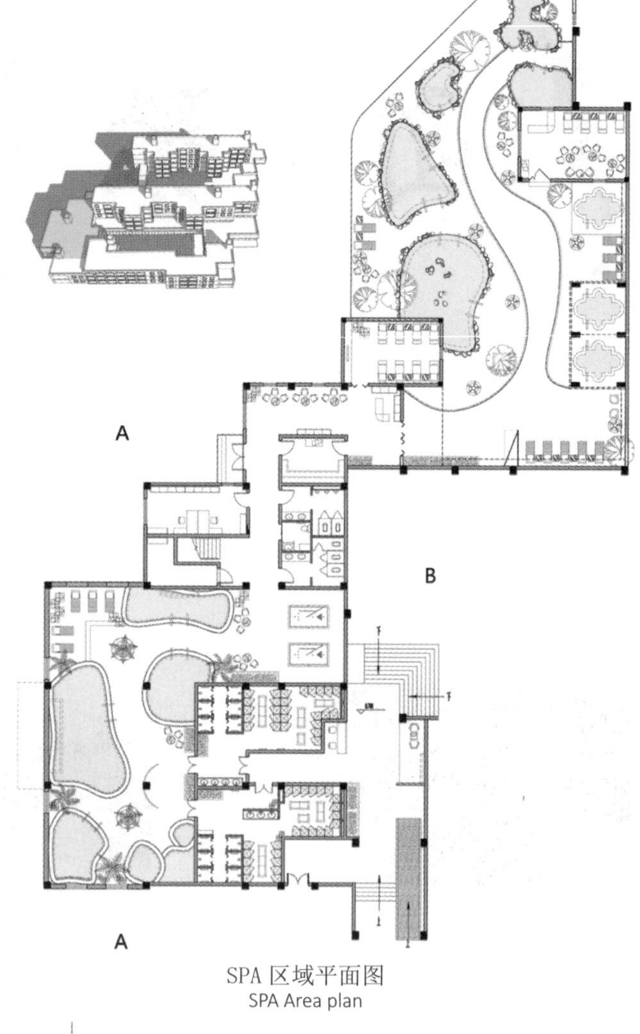

SPA 区域平面图
SPA Area plan

B-B 剖面图

B-B 剖面图

门厅 || Hallway

原园区的三栋楼共用一个在场地东南角的小门厅，门厅的联系作用很弱，从而也阻隔每栋楼老人之间的交往。改造后的门厅选址在中间栋的端头，联系行加强，同时将门厅扩大，布置更丰富的功能，提供一个足够的吸引点，吸引老人走出房门，互相陪伴。

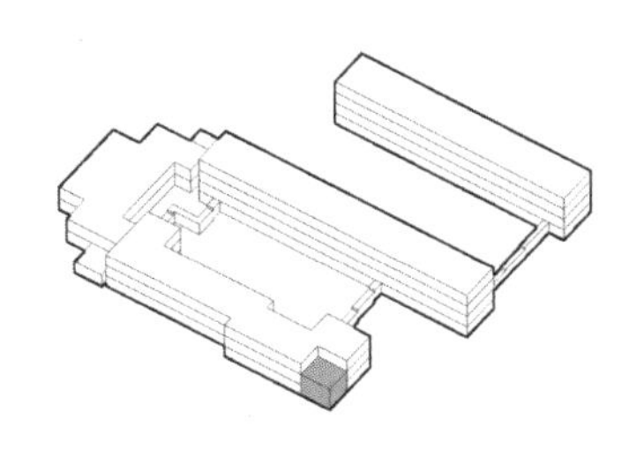

改造前门厅位置

C-1 楼：半自理老人康复病房

Building C-1: Rehabilitation wards for partially disabled seniors

原建筑病房平面布置有诸多不合理之处，针对这些问题对症下药，进行建筑改造和室内设计。

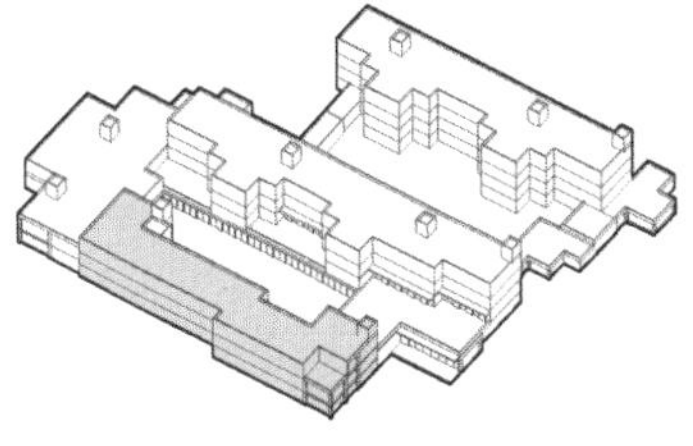

位置示意图

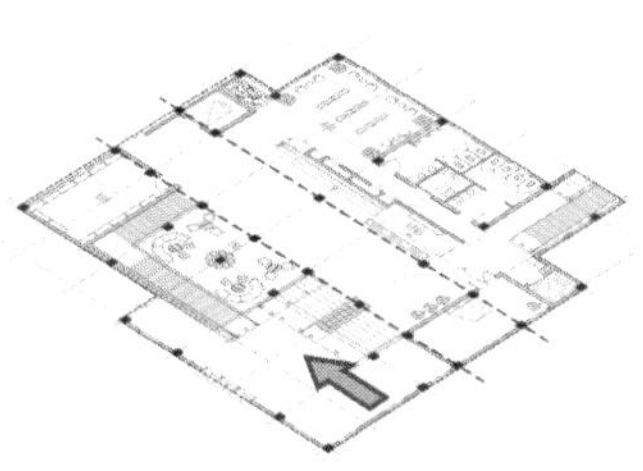

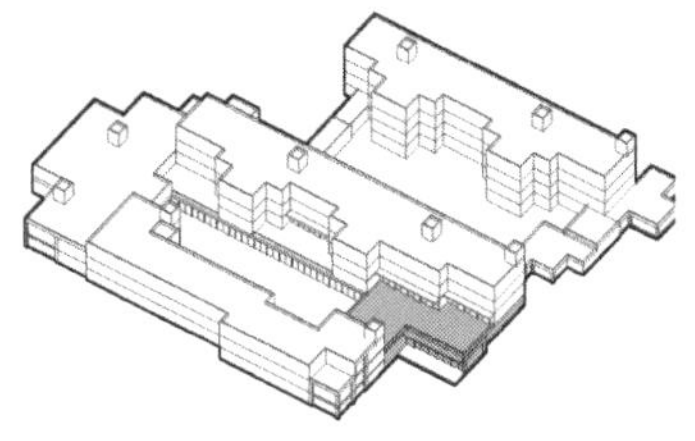

将门厅大致分为三个区域：依次是动区－导向区－静区。

改造后门厅位置

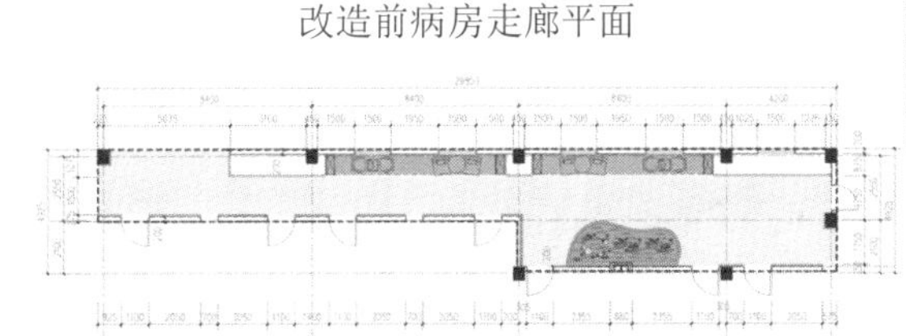

改造前病房走廊平面

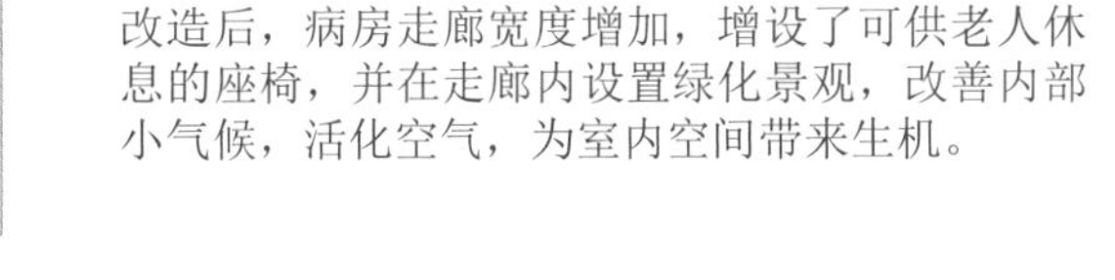

改造后病房走廊平面

改造后，病房走廊宽度增加，增设了可供老人休息的座椅，并在走廊内设置绿化景观，改善内部小气候，活化空气，为室内空间带来生机。

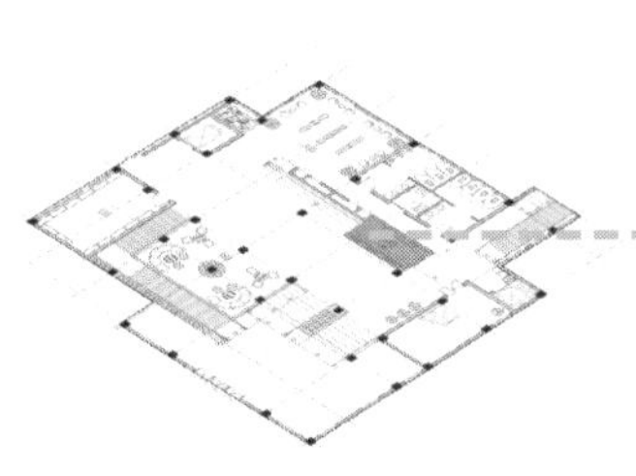

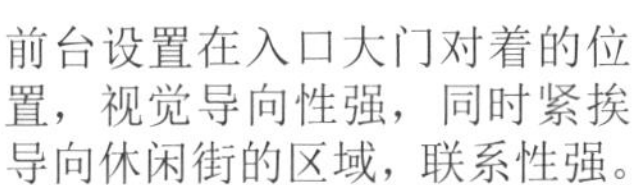

前台设置在入口大门对着的位置，视觉导向性强，同时紧挨导向休闲街的区域，联系性强。

前台桌子选择外部凹进的形式，适宜坐轮椅的老人腿伸进；背景墙以“树林”为意向，聚木为林，聚人为朋。

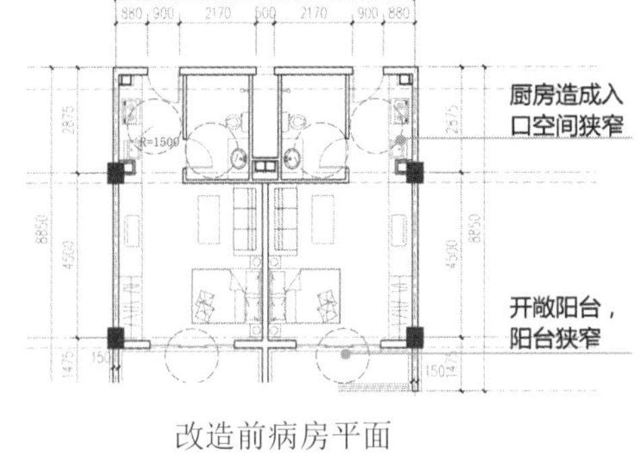

改造前病房平面

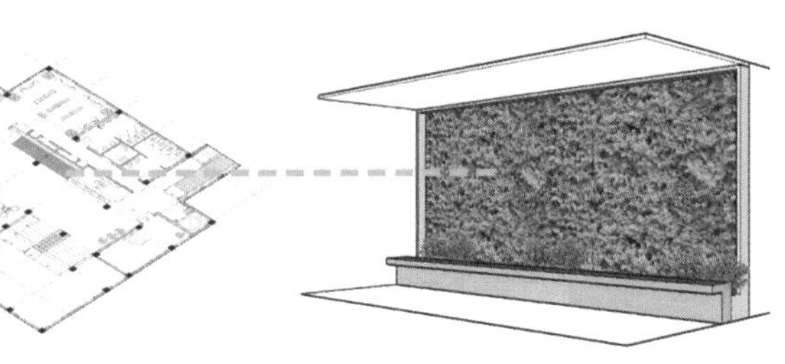

原建筑管井对空间的限制，化弊为景，设计成室内墙面绿化及休息区。

墙面绿化的形式为壁挂式绿化装饰，设置不锈钢种植槽，有利于调节室内小气候，净化空气。

经过内部景观绿化后的室内效果图

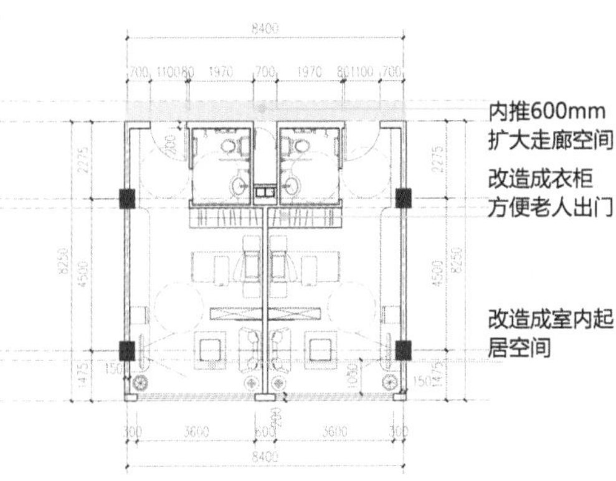

改造后病房平面

门厅细部设计 || Detailed design of the hallway

原建筑管井对空间的限制，化弊为景，设计成室内墙面绿化及休息区。墙面绿化的形式为壁挂式绿化装饰，设置不锈钢种植槽，有利于调节室内小气候，净化空气。

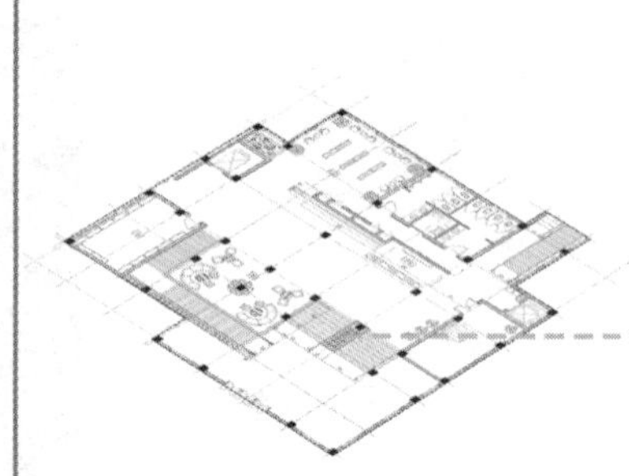

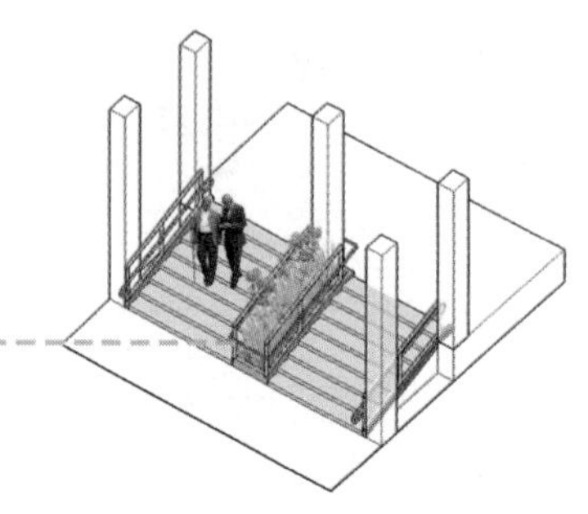

由于原建筑结构的限制，层高为3.3m，为了提高门厅的净高，将门厅入口新建部分地基下降1m。

缓步台阶导向门厅的第二区域；台阶为适合老年人的缓步台阶；结合原建筑结构柱设置花坛。

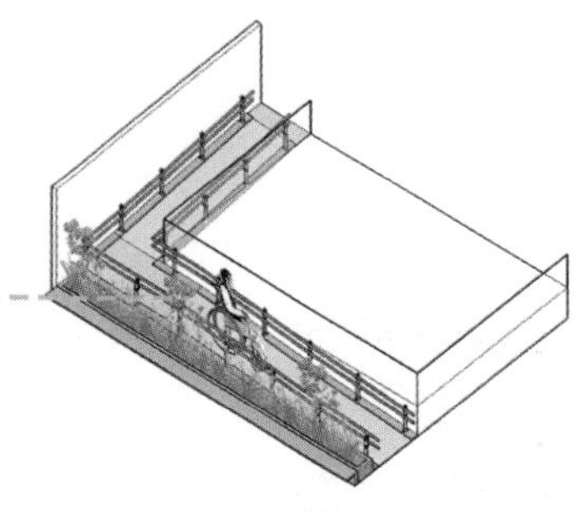

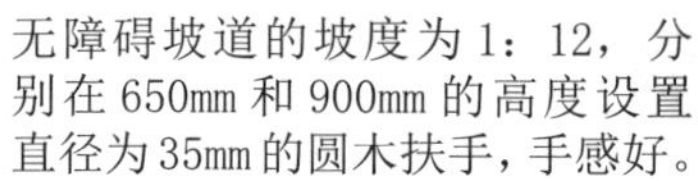

设置无障碍坡道，提高老年人的通达性，同时也缓解室内其他地方地面高差大带来的坡道过长。

无障碍坡道的坡度为1：12，分别在650mm和900mm的高度设置直径为35mm的圆木扶手，手感好。

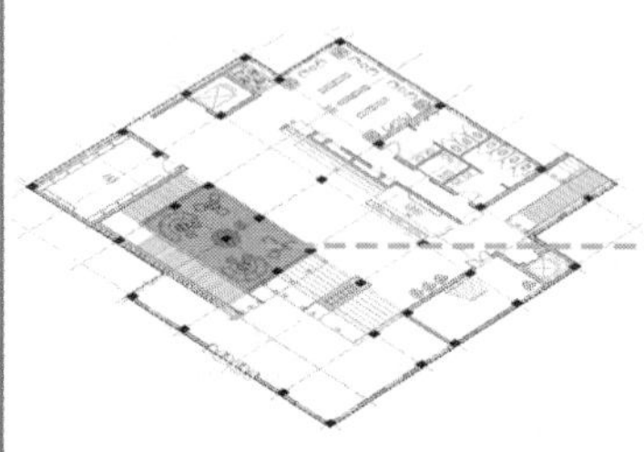

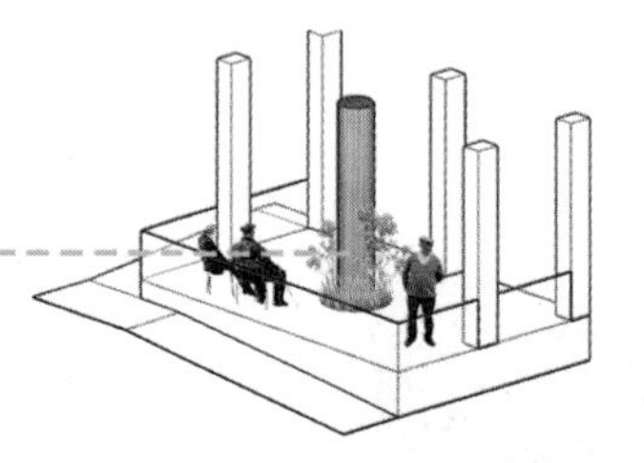

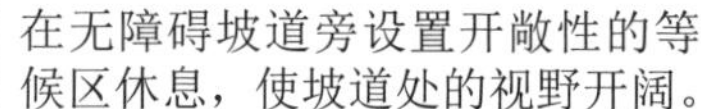

在无障碍坡道旁设置开敞性的等候区休息，使坡道处的视野开阔。

由于建筑结构的限制，将等候区中间的方柱做成景观圆柱，结合绿化提高空间品质。

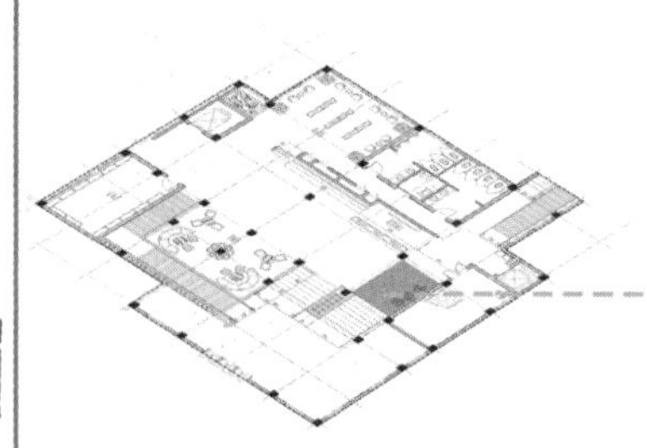

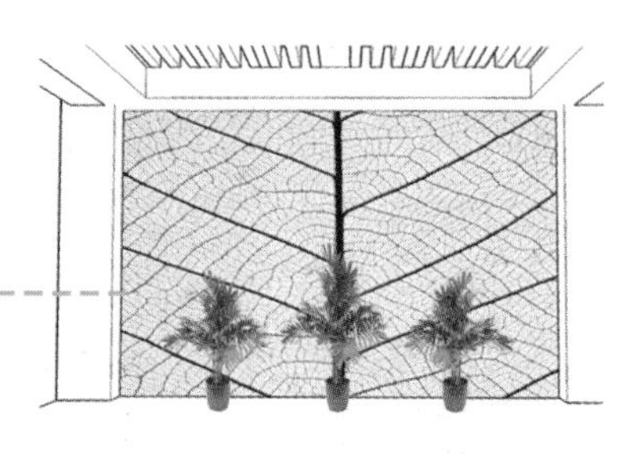

这个位置是室内流线的交汇处，是导向空间的视觉焦点，因此设计成室内景观中心。

以透光性材质作为背景墙，墙面纹理为叶脉，统领以“树”为室内主要设计元素。

总平面图
Site plan

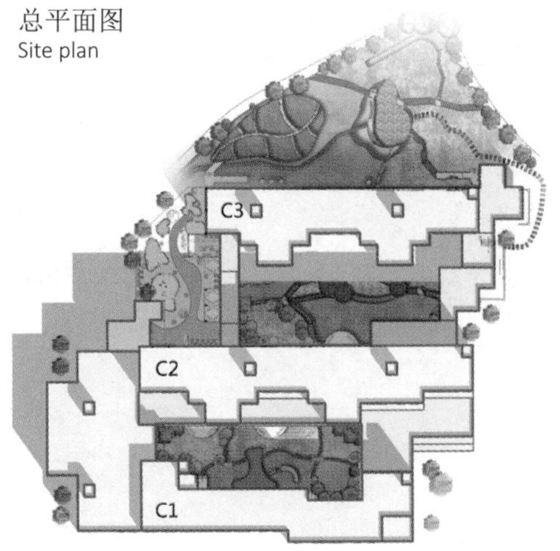

原建筑功能单一，三栋康复楼均为病房，对此，我们呼应最新的娱乐养老理念，增加休闲娱乐空间，丰富老年人的生活。这里不再是养老院，而是老人与朋友们聚会与儿孙们共享天伦之乐的场所。

对于老年人来说，公共娱乐空间是十分重要的。本次设计采用“娱乐养老”理念，专家指出，只有精神上愉悦了，拥有年轻健康的心态丰富的活动，才能颐养天年。

病房入户门设计

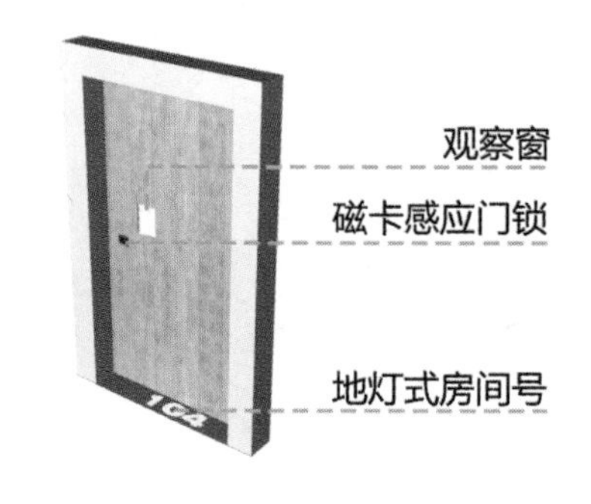

散：漫无目的

聚：聚在一起
共同养老

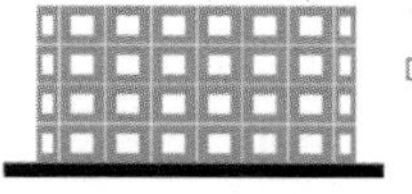

导师点评：

设计作品“团•聚”是对原有老年公寓建筑进行的较大幅度的空间改造的设计方案，设计者针对原有单调乏味的建筑空间进行优化与重组，将串联式的空间调整为组团式空间，设计中充分利用阳光、植物、通风等自然因素对空间进行新的塑造，满足老年人在心理和生理上对空间的感受与需求，为老年人创造了富有生命活力的空间环境，体现了设计者较好的空间把控能力和创新设计能力。在设计细节上还能够关注老年人的需求，无论在室内空间还是室外空间都能够结合老年人的特点以需求为导向深化空间环境设计。全套设计成果图式语言表达的清晰完整，体现设计者较扎实的专业设计基础能力和水平。

吕勤智

Advisor's Comments:

"Communal Living" redesigns the spaces of the buildings to a large extent. The original monotonous architectural spaces are optimized and re-divided. Specifically, spaces originally arranged in a line are divided into groups, and reshaped using natural elements like sunlight, plants and ventilation, for the purpose of meeting the seniors' psychological and physiological needs on an energetic and vibrant environment. The designers' ability to use spaces and innovation can be seen from spatial redesign. Besides, details, no matter in interior design or exterior design, are designed according to the needs of the seniors. The drawings and text clearly and completely deliver the ideas of the designers, indicating the solid basic professional knowledge of the designers.

Lyu Qinzhi

专家点评：

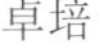

以部分老年人渴望热闹的意愿作为公共空间的设计主线，休闲街空间感强、布局合理、一步一景、层层递进，步行其间让人觉得有很强的叙事性；以另一部分老年人喜爱安静的意愿为起居空间的设计主线，创造了共用起居室的微社区新模式。通过绘制丰富的室内效果图和分析图，在图面上展现出了自己的设计理念和设计工作量。对于原建筑的结构改造也很合理，说明本组师生在设计过程中非常深入地思考和完善了很多的设计细节，最终作品的可实施性较强。

卓培

Expert's Comments:

In the scheme, public areas are designed to satisfy some seniors' preference for a vibrant and busy environment. The pedestrian street features a palpable sense of space and a rational layout, with constantly changing landscape. Walking in the street is like reading a story. Living spaces are designed to meet some seniors' preference for a quiet environment. The new model of shared living room is adopted. From a large number of interior design renderings and analysis graphics, we can see the design concepts of the designers, and the great efforts that they made. The structural redesign of the original buildings is also very rational. Generally speaking, the designers have carefully analyzed many details and developed corresponding solutions, and their final scheme is of high feasibility.

Zhuo Pei

To Live, To Learn

CIID“室内设计 6+1”2016（第四届）校企联合毕业设计

CIID“Interior Design 6+1”2016(Fourth Session)University and Enterprise Joint in Graduation Design

高　　校：北京建筑大学

College: Beijing University of Civil Engineering and Architecture

学　　生：陆昊 魏卿 曾宁

Students: Lu Hao Wei Qing Zeng Ning

指导教师：杨琳 朱宁克

Instructors: Yang Lin Zhu Ningke

参赛成绩：医养园区环境与景观设计组一等奖

Achievement: Frist Prize for Aged Care District Environmental and Landscape Design

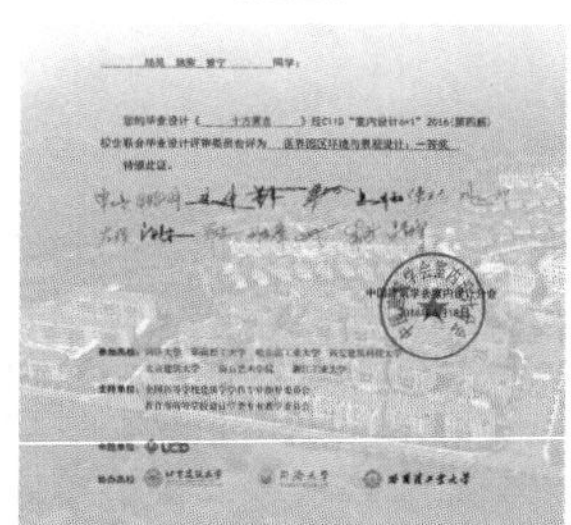

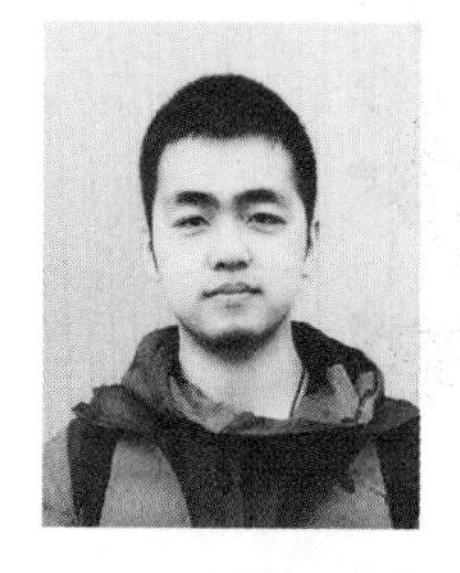

陆　昊
Lu Hao

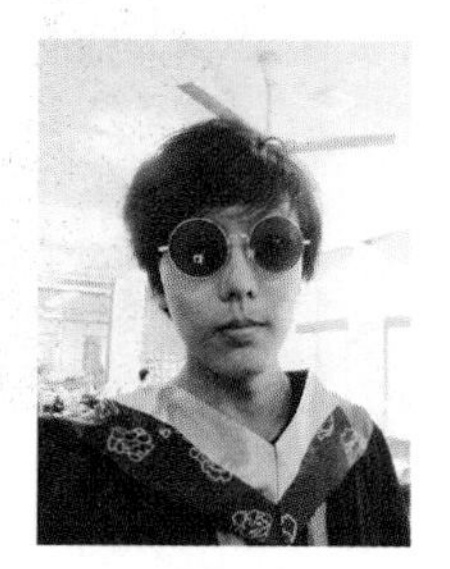

魏　卿
Wei Qing

曾　宁
Zeng Ning

学生感悟：

本次CIID“室内设计 6+1”2016（第四届）校企联合毕业设计让我收获颇丰。首先是知识面的扩展和深化。以前很少涉及老年建筑、室内和景观的调研和设计，这次不但在设计思维的转化上学习到了很多，也在各方面的标准、规范上了解和掌握了很多。其次是设计过程的思路上得到了很大启发，不同于平时课业所涉及的虚拟题目，本次毕业设计更实际、理性，对于前期的工作量，比如实地调研、访谈、总结、文献综述等，有很大的需求。这也让我在面对一个新的设计题目时，对于适合此题目的设计方法逐渐有了更加明晰的掌控。

Students' Thoughts:

I had a lot of gains from the 2016 fourth "Interior Design 6+1" joint graduation design of colleges and enterprises, sponsored by China Institute of Interior Design (CIID). First of all, my scope of knowledge was extended and deepened. I was seldom involved in research and design of architecture, interior and landscape for the elderly. This time I not only learned more about transformation in design thinking, also understood and mastered many standards and specifications in all aspects for this. Furthermore, I had much inspiration on the ideas of design process. Different from virtual topics involved in school, the graduation design is more practical and rational, and also has a lot of demand for the early work such as field survey, interviews, summary, and literature review. This also allowed me to gradually have clearer control over appropriate design methods in the face of a new design topic.

景观分析

建筑围合形成四个院落，一个半开放式院落（西北），C 区北侧为公共绿地。
C1 南侧两个院落对称分布设置为叠水景观。C1 东北院落设置水系小品，西北院落以及北部景观设置为公共绿地。

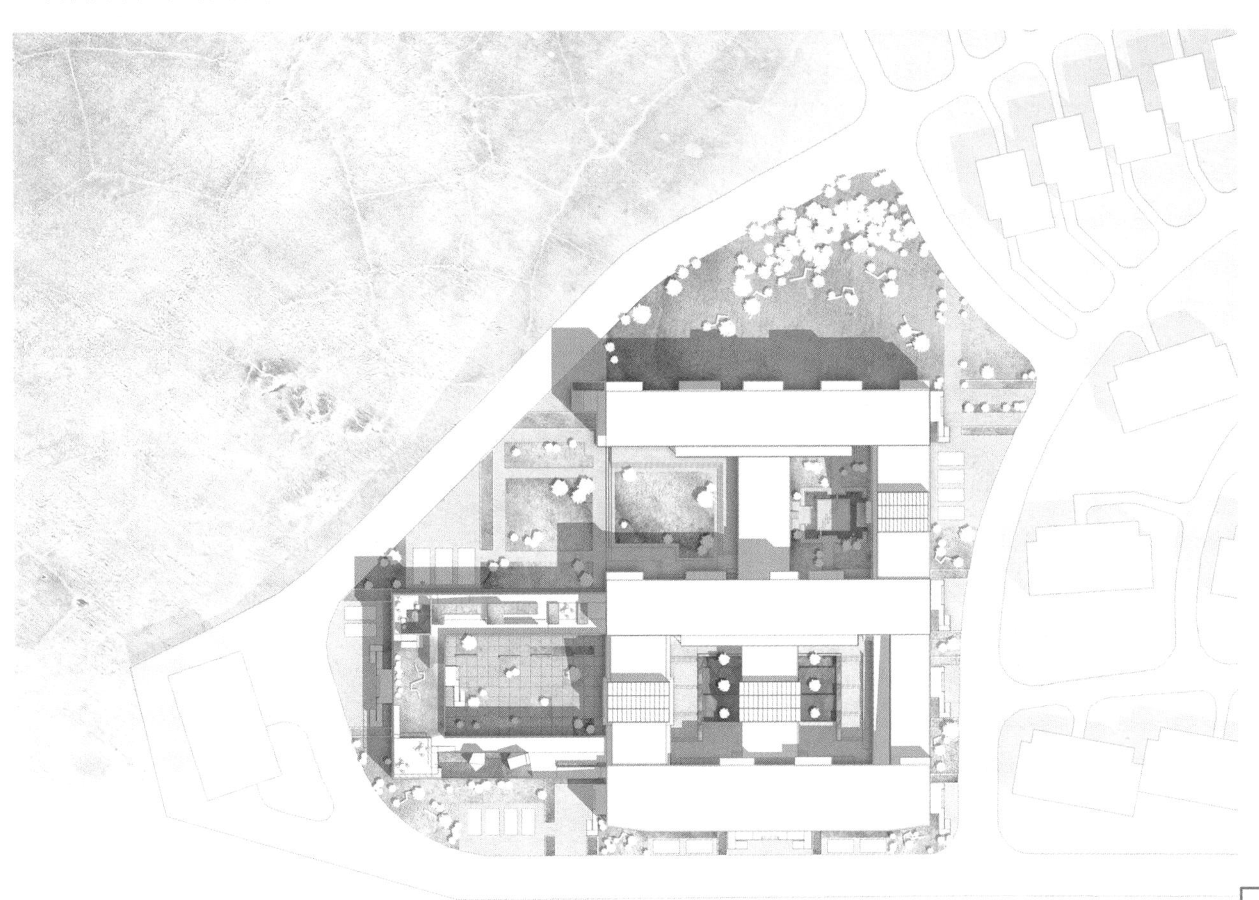

总平面图

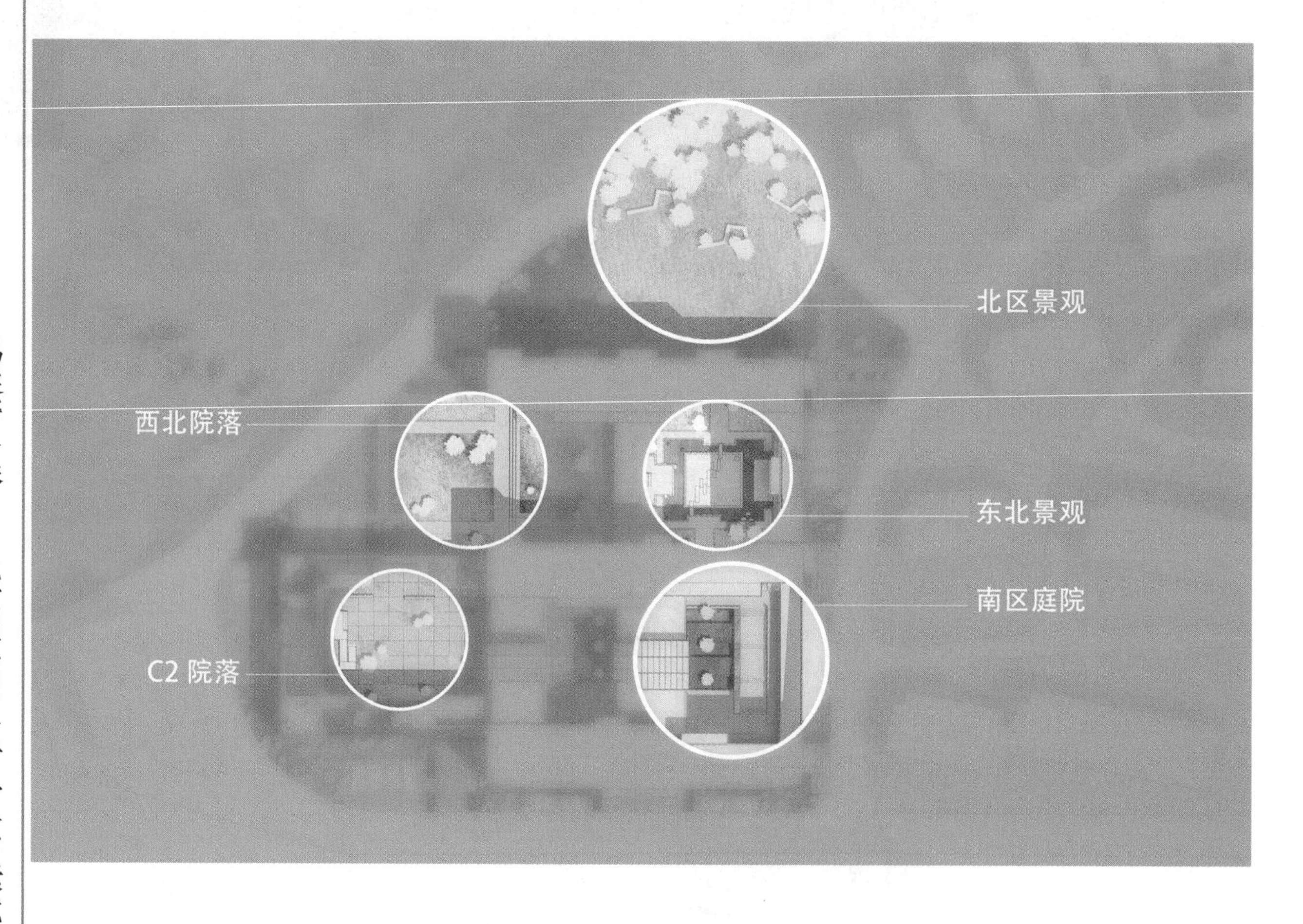
北区景观
西北院落
东北景观
南区庭院
C2 院落

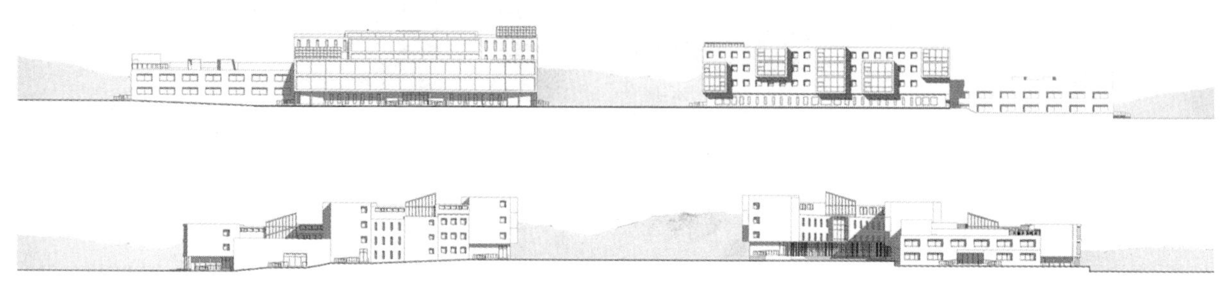

导师点评：

“十方黉舍”，题目别致，立意较好。黉舍有学校、学堂之意，点明设计场所目标为老年大学，宗旨即是设计一处能满足老年人精神需求的场所。设计前期调研扎实，对老年人的生理、心理需求进行了深入解析以指导设计。在建筑一层部分压缩原有居住面积，为老人提供了多个公共活动空间，如活动健身、理疗体验、多功能厅以及图书馆等。

值得一提的是，图书馆与花房的设计。设计者对图书馆的空间设计层次丰富，安逸宁静的氛围符合功能要求。最温暖关怀的设计部分在于 3 栋建筑的每层随处可见公共花房空间，让自理与失智老人都能感受到大自然的气息，调节身心健康，提出几点建议：第一，3 栋建筑一层整体作为公共活动功能使用，过多的活动面积，是否存在利用率低的问题？可以考虑资源有效整合，一个活动室可分时段供多种活动使用。节省出一些一层阳面的空间给失智老人作为居住使用。第二，图书馆的下沉空间设计虽有趣别致，却忽略考虑腿脚不力的自理、半自理老人使用问题。除此之外，一楼的阶梯教室等多个公共区域设计，除了一般性的功能家具、设备布置外。同样，重要的应是更多地考虑老年人的特殊关怀设计，如轮椅坡道。第三，前期调研分析中提到的禅学空间让人意犹未尽，但在最终的设计中没有体现。同时，对居住部分室内空间的设计考虑较少，这是比较遗憾的地方。

总的来说，设计作品对基地三栋楼的整体设计，功能布局明确，交通流线清晰，做到了题目的内涵追求——精神至上的老年家园。

何方瑶

Advisor’s Comments:

"To Live, To Learn" is a very special and meaningful theme. It indicates that the objective of the scheme is turning the retirement community into a school for the elderly where the mental needs of the seniors can be met. In early survey stage, the students carefully studied the physiological and psychological needs of seniors as the basis for later design. On the 1st floors of the buildings, some living spaces are turned to public spaces such as gym, physiotherapy treatment room, multifunctional hall and library.

The laboratory and indoor gardens are the highlights of the design. Diverse functional spaces of the laboratory are rationally designed, creating a comfortable and quiet environment. Indoor gardens on every floor of the three buildings representatively show the designers' care for the seniors. Such gardens create a natural environment for both the able-bodied seniors and seniors with dementia, promoting their health.

Lastly, I'd like to give some suggestions. Firstly, turning the 1st floors of all the three buildings into public areas will actually lower the utilization rate of the spaces. In my view, spatial resources should be effectively integrated. For example, an activity room can be used to hold different activities at different times, saving some sunny spaces for seniors with dementia to live in. Secondly, the idea of designing a sinking space in the laboratory is interesting, but it ignores the access of seniors with disability. Likewise, in several public areas, such as the lecture hall on the 1st floor, in addition to functional furniture and facilities, more importantly, wheelchair friendly facilities like wheelchair ramps should also be designed. Thirdly, the idea of creating a Zen space is mentioned in the early-stage survey. It's a very interesting idea, but it doesn't appear in the final report. Another unsatisfactory thing is that interior design of the living areas is not described in detail.

Generally speaking, the functional layouts and traffic flows of the three buildings are clearly and thoughtfully designed for the purpose of building a spiritual home for the senio.

He Fangyao

专家点评：

同学们解决了中期评审时对佛教中关于禅修的理解太空泛和思路太分散的不足，把握住了对禅修文化的诠释方法，使设计方案具备了一定的思想深度，但细微之处没有能够全面展现出来。在建筑规划和功能布局上，采用拆改结合、部分增建的改造手法，改小了院落的尺度、增加了院落的层次，使得建筑的内部空间感与佛教建筑中庙宇的布局有共通之处，让居住人群在精神层面有所感悟和升华。

卓培

Expert’s Comments:

Compared with the mid-term report which lacks a substantial understanding of Zen meditation and a focus of related design, the final scheme profoundly interprets the Zen meditation culture which makes the scheme more insightful. However, detailed interpretation is still not given. In terms of architectural planning and functional layout, some spaces are removed, some are transformed, and some are newly added. The yards are smaller, but are given more functions and design elements. Interior design borrows some elements of Buddhist temples, promoting the self-cultivation of the occupants.

Zhuo Pei

Awakening the Senses

CIID“室内设计 6+1”2016（第四届）校企联合毕业设计
CIID“Interior Design 6+1”2016(Fourth Session)University and Enterprise Joint in Graduation Design

高　　校：西安建筑科技大学
College: Xi`an University of Architecture and Technology
学　　生：李坷欣　刘雨鑫　曾嘉
Students: Li Kexin　Liu Yuxin　Zeng Jia
指导教师：刘晓军　何方瑶
Instructors: Liu Xiaojun　He Fangyao
参赛成绩：医养园区环境与景观设计组二等奖
Achievement: Second Prize for Aged Care District Environmental and Landscape Design

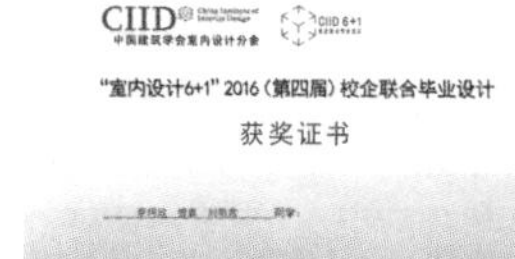

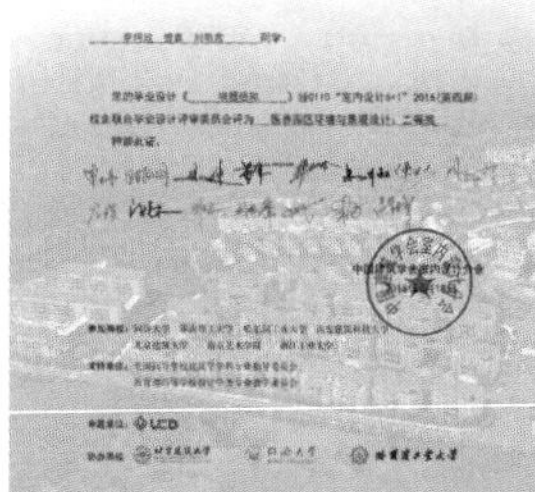

李坷欣
Li Kexin

刘雨鑫
Liu Yuxin

曾嘉
Zeng Jia

学生感悟：

很荣幸能够参加这次的“6+1”室内设计比赛作为我最终的毕业设计，在此次的毕业设计中我们是初次接触老年医养空间的设计，我们从第五维空间即感知出发，探讨了适合老年人医养的生活方式，从中学习到了许多、感受到了许多。首先感谢刘晓军老师和何方瑶老师的指导，让我们明白了很多专业性的知识，同时对我们的感知主题指出了很好的研究方向。通过这次比赛，我们更深刻理解了国内室内设计研究和发展的动向。

Students' Thoughts:

It's an honor to participate in the CIID “Interior Design 6+1” University-Business Graduation Projects Competition. Retirement home spatial design is a brand new topic for us. Themed on the fifth dimension, or the senses of human beings, we discussed lifestyles good for the health of seniors, and have learned a lot during the process. I'd like to give my thanks to Mr. Liu Xiaojun and Ms. He Fangyao for their guidance. They imparted a lot of professional knowledge to us and gave very usefully directions to our theme presentation. Through the competition, I had a deeper understanding of the trends of China's interior design.

楼间中庭景观 || Landscape of the glass atrium

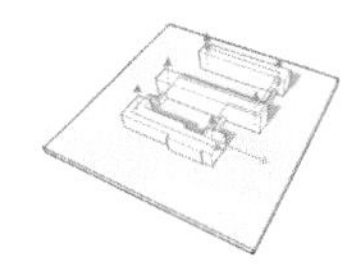

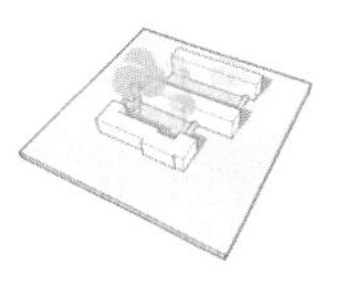

基地位于郊区，在楼间的中庭小院上，为了让自然回归，让老年人更好的感受周边环境，我们在植物配上用野花野草的芬芳，勾起老年人的记忆，让老年人感受不一样的味道。

The retirement community is located at the outskirts. To connect the buildings with surrounding environment, wild plants are grown in the glass atrium, bringing the smell of nature to the seniors.

老年人种植植物，植物被老年人观赏，带来心理感知。

In the process of growing and appreciating plants, a senior will build an emotional connection with plants, and a cycle balancing the energy of the senior and nature will form.

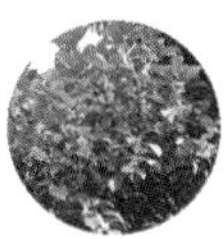

种植金银花、百里香、糙叶黄芪、桂花、萱草等植物，具备药用作用，可以治疗多疾病。种植植物丰富老年人生活，也可以增加生活的趣味性。

Herbs like honeysuckle, thyme, astragalus scaberrimus bunge, sweet-scented osmanthus and day lily are grown in the glass atrium, enriching the life the seniors and providing medicine for them at the same time.

利用高差实现梯田农业

Using the elevation difference realize the terrace agriculture

自然景观结合回应场地

The natural landscape in combination with response to the site

利用水资源条件，开发康复训练项目

Use of water resources condition to develop rehabilitation training program

植物参与疗养和恢复自然环境

Plants in spa respond to the natural environment

这些野生花卉不仅具有良好的景观性价值，同时在调节小气候、水土保持、防风降尘、改善城市生态环境和增强群落稳定性等方面也具有明显的作用。

Aside from pleasing the eyes, such wild plants are also effective in adjusting microclimates, conserving water and soil, controlling wind and dust, improving urban environment, enhancing community stability, etc.

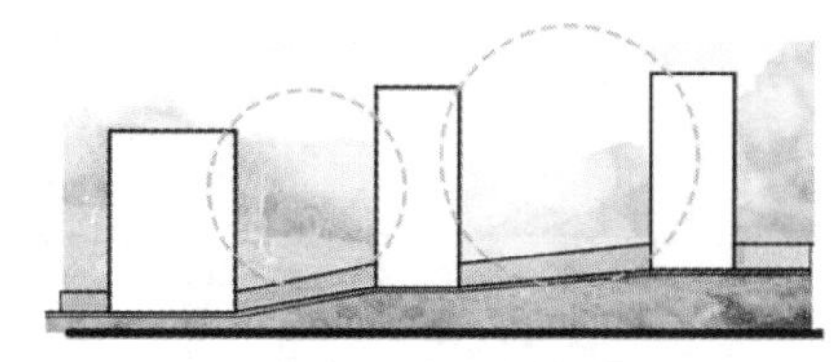

楼间缓坡地势
Gentle slope in the glass atrium

设置雾气喷口，形成雾气后，可以遮挡玻璃连廊，更好地形成一个围合的庭院。

Spray comes out from nozzles to reduce the transparency of the glass corridors, creating a more enclosed yard.

庭院的铺装设置为石子路，踩上去给人松软的感觉，在触觉上让老年人更好地去感知。

Gravel is used to pave the yard, creating a sense of flexibility which stimulates the sense of touch of the seniors.

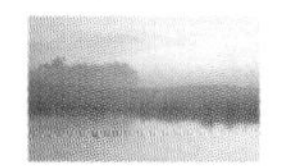

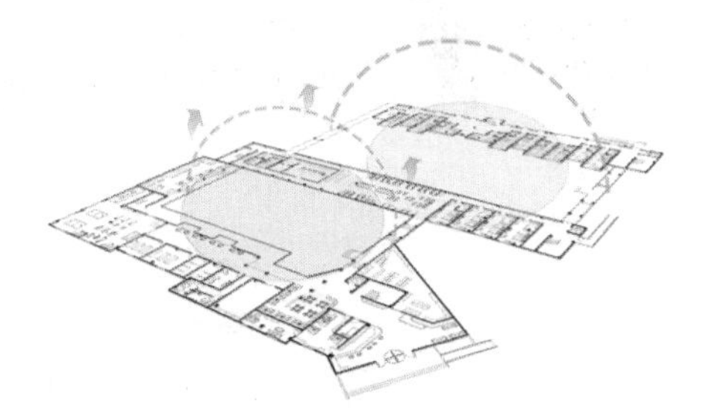

花香辐射范围
Fragrance affected area

顺应地势形成阶梯凳，可供老年人休息。预留的平地，面积较大，老年人可以聚集在此活动、跳舞。

Benches are arranged on the slope for the seniors to have a rest. A large flat area is prepared for the seniors to dance or do other activities.

组团中庭景观 || Landscape of the apartment atrium

公寓楼里每个组团有公共空间。公共空间设置种植池，老年人可以根据自己喜好种植植物，这些植物所散发的气味又是不同的……

Each apartment group has a public area with a planting bed in which the seniors can grow their favorite plants which emit different fragrances.

组团中的公共空间，使用率高，老年人活动的小空间，设置种植池，让他们种植喜欢的植物。

This pubic area, though small, is frequently used by the seniors. Planting beds create an opportunity for the seniors to grow and take care of plants that they like.

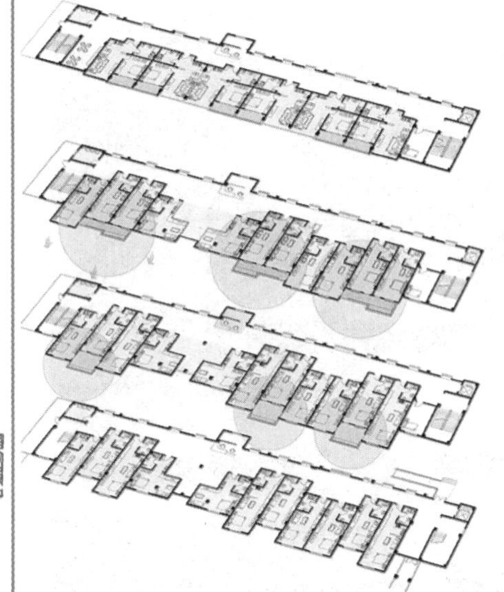

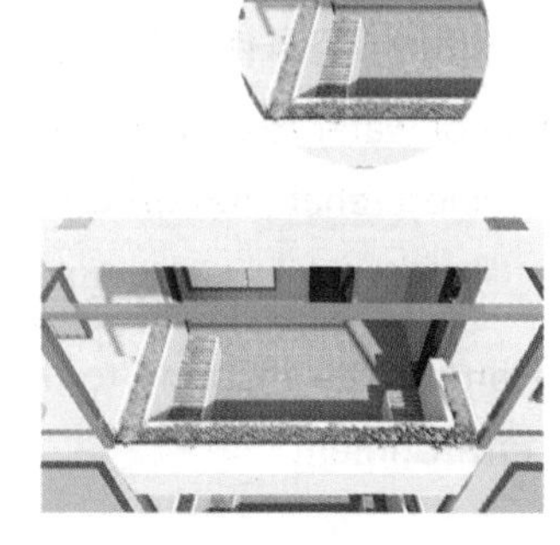

导师点评：

通过“唤醒感知”的立意，以老年人五感渐退为出发点，并通过空间与设施的设计使其延长感知的消退过程，从而实现健康养老的目的。

空间组织合理，在现有的平面布局中，提取部分空间整合出相对独立的公共空间，为原有封闭式单元住宅增加了共融性与交互性的空间，通过对空间使用功能的改变使老年人的行为方式有所转变，以此增进彼此的关系，从而实现唤醒老年人渐渐消退的感知。

渲染图表现细腻、精致，通过使用唐代绘画中的古装人物为表现图内的场景人物，体现了同学们在艺术方面的情趣与追求，同时也取得了一定的视觉表现力。

在空间局部规划与设计方面还缺少较严谨的表现，缺少将老年人的需求要素、行为要素始终如一地融入到整个空间设计的每一处。

马辉

Advisor’s Comments:

Given the fact that seniors are troubled by declining five senses, “Awakening the Senses” tries to, through spatial and facility design, slow down the speed of the decline, thus promoting the health of the seniors.

The scheme features rational spatial organization. Based on the exiting plan layouts, the scheme integrates some spaces into independent public spaces, promoting the communication and interaction among seniors originally living in independent units. The change of spatial functions will lead to the change of the senior’s behavior. In the process of more and more frequent communication, the declining senses of the seniors will be awakened.

The renderings are detailed and exquisite. Painting skills and style of the Tang Dynasty are used to create figures, which indicates the designers’ artistic ambition and creates impressive visual effects as well.

Regional spatial planning and design should be more rigorous. The scheme fails to incorporate the needs and behavioral characteristics of seniors into the entire design process.

Ma Hui

专家点评：

根据本有建筑及空间环境分析利弊，结合老北京人的生活习惯，设计空间内部结构，空间合理整合，这种群落式的生活空间，方便老年人相互沟通和交流。丰富的公共活动空间，拉近了邻里之间的相互关系，形成了老年人的朋友圈，解决了老年人孤单寂寞的实际问题。

明确的交通路线，清晰的人流走向使功能空间联系更加紧密。其间的公共交流空间方便了老年人的活动和休息。

通过各感官的具体分析，结合物理环境进行细致的设计，满足老年人日常生活的实际需求。

方案的整体设计从人本角度出发，诠释了人与人、人与建筑、人与自然的相互情感。

张红松

Expert’s Comments:

Based on an analysis of the original buildings, including their spaces and environments, as well as the lifestyle of the old timers in Beijing, the scheme redesigns the interior structures and rationally integrates the spaces, creating group living spaces which promote the communication among the seniors. Besides, more public spaces are created to build closer neighborly relations, and circles of friends, pulling the seniors away from loneliness.

Clear transport routes and traffic flows remove the boundaries separating various functional spaces. Public spaces are arranged along such routes and flows, for the rest and communication of the seniors.

The five senses are incorporated into the design in the given physical environment of the project, for the purpose of meeting the actual demands of the seniors in daily life.

From a human-oriented perspective, the scheme tries to interpret the relations and emotions between people, between man and architecture, and between man and nature.

Zhang Hongsong

老人與花海 北京曜陽國際老年公寓景觀改造

Seniors & Flowers – Beijing Yaoyang International Retirement Community Landscape Renovation (landscape design)

CIID“室内设计 6+1”2016（ 第四届 ）校企联合毕业设计

CIID“Interior Design 6+1”2016(Fourth Session)University and Enterprise Joint in Graduation Design

高　　校：华南理工大学
College：South ChinaUniversity Of Technology
学　　生：郑潇童 陈谢炜
Students：Zheng Xiaotong　Chen Xiewei
指导教师：谢冠一
Instructors：Xie Guanyi
参赛成绩：医养园区环境与景观设计组二等奖
Achievement：Second Prize for Aged Care District Environmental and Landscape Design

郑潇童
Zheng Xiaotong

陈谢炜
Chen Xiewei

学生感悟：

在这次竞赛中，我们由前期针对目标人群的心理生理特征的调研，发现现存场所的不足之处和老年人的心理诉求。针对调研成果，我们提出老有所养和老有所用的设计定位。试图在现有场地内建设出一个较完整的种植产业链以满足老年人的心理需求，让他们实现晚年的自我价值，并在原有建筑内插入增加新功能，以满足老年人的生理需求，达到老有所养的目的。在这次竞赛的过程中，我们学会了从贴近目标人群的生活，观察目标人群的需求，并通过设计改善目标人群的生活现状。这段竞赛的经历将成为我们未来职业生涯的财富。

Students' Thoughts:

Our early stage survey focused on finding out the psychological and physiological characteristics and needs of seniors, and the problems faced by the project site. According to the survey results, we determined the project theme of “senior care & value realization”. Specifically, on the one hand, a complete system covering all aspects of plant growing is built to satisfy the seniors’ psychological needs of value realization; and on the other hand, more architectural functions are added to meet the seniors’ physiological needs of aged care. The competition gave us an opportunity to closely observe the life and needs of the target group, and improve their living standard using the power of design. The experience of participating in the competition has become our wealth that benefits our future careers.

前期调研分析 || Early-stage survey and analysis

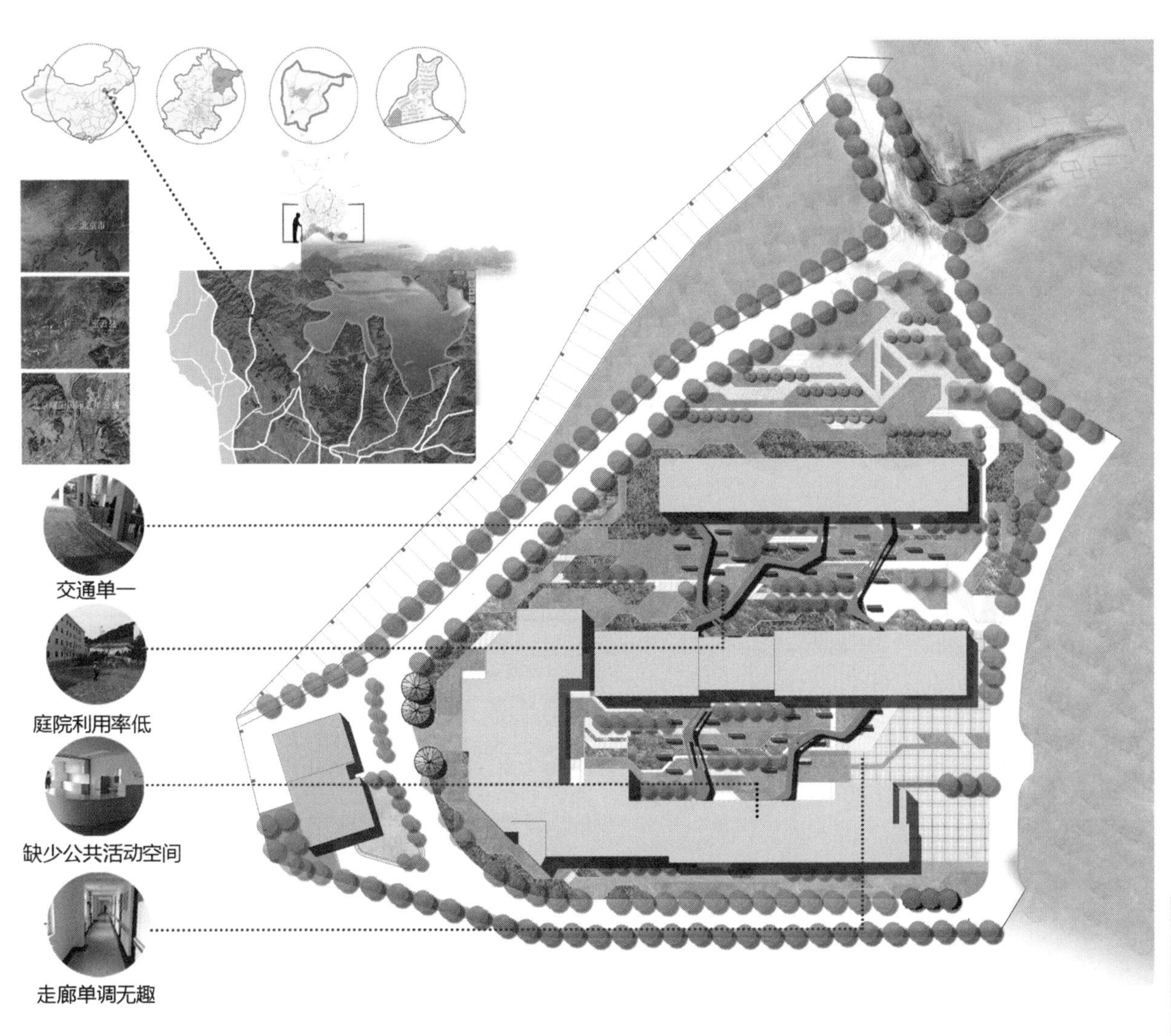

改造对策 || Renovation measures

两个庭院之间缺少联系	改变建筑形式：削弱中间建筑体积
庭院利用率低	改变庭院功能：大量增加种植面积
无公共活动空间	改变功能配置：进行功能置换，增加活动空间
交通形式单一	改变路径：增加桥连接建筑与建筑

努力建设成社区农场，更专注打造一种田园氛围。在农场的养老产业中，老年人可亲自参与或监督整个农场的生产、生活环境的建设工作，从吃到住，完全可以满足农场老年社会中所需的一切，老年人组成的团队，甚至可以走向外面的社会，产生更大的价值和影响力。

使用人群分析 || Target population analysis

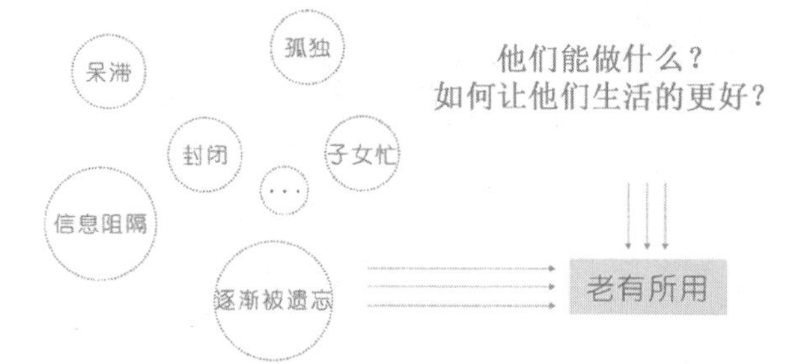

分析得到种植是比较适合老年人这个群体的，一方面让老年人从心理上有个寄托，另一方面通过种植，收获，提高老年人自身价值。所以我们采用了种植的方法来实现“老有所用”的目标。

Site analysis results show that growing vegetables is suitable for seniors living in the retirement community. On the one hand, it gives a kind of emotional sustenance to the seniors, and on the other hand, the seniors can prove their value during the process of growing and harvesting.

(1) 自给自足，丰富精神世界。
(2) 植物有精神方面治愈作用，可以疏导负面情绪。
(3) 可食用到有机蔬菜。
(4) 工作强度可根据自己能力来。
(5) 部分植物可以当药材且具有观赏价值。

1. Growing vegetables can satisfy both the physical and spiritual needs of the seniors.
2. Plants have the function of helping release negative emotions.
3. Organic vegetables are available.
4. The time spent on growing vegetables can be decided by the seniors themselves.
5. Herbs and ornamental plants can be grown.

垂直绿化墙旋转分析

垂直绿化墙

	优点	缺点
	植物生长速度快 便于预防自然灾害	温度较高 通风较差 维护成本较高
	方便老人种植	阳台使用率降低 互动性差
	光照雨水充足 面积充沛 屋面平坦	老人参与度低 互动性低 存在安全隐患
	维护成本低 互动性高 参与度高 观赏价值得以发挥	光照不足 二层以上老人参与度低
	参与度高 互动性强 光照较充足 观赏性强	维护成本较高

增加垂直绿化，垂直绿化墙可以跟随太阳旋转，植物可以充分吸取阳光。一方面增加了绿化面积，解决了园区利用率低的问题，另一方面老人们可以对垂直绿化墙进行照顾实现老有所用的目的。

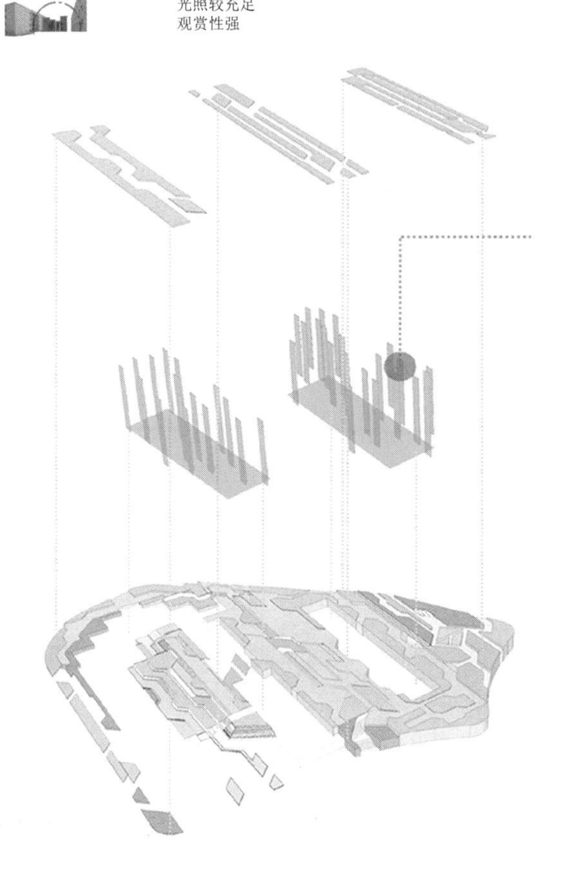

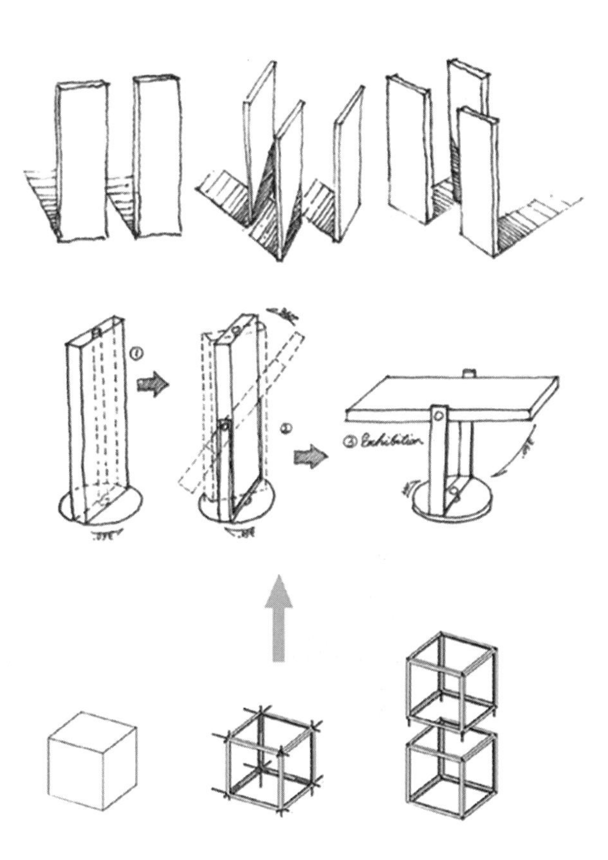

垂直绿化墙技术分析

将活的植物像贴墙纸一样直接铺贴在垂直面上，无论室内还是室外，无论是否有阳光，其上的植物都可以像在大自然中一样长期生长。

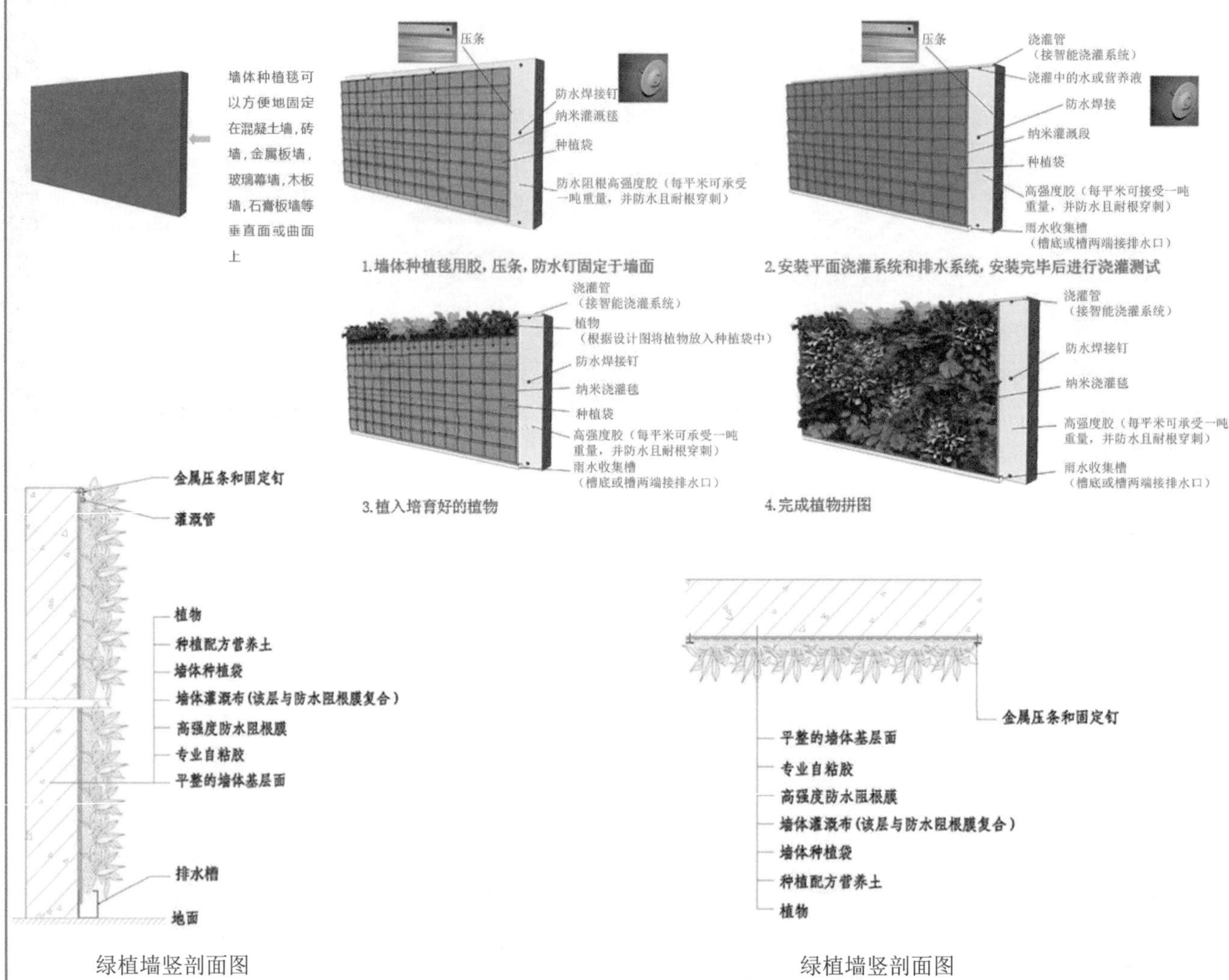

绿植墙竖剖面图　　　　绿植墙竖剖面图

过程分析图 || Process analysis diagram

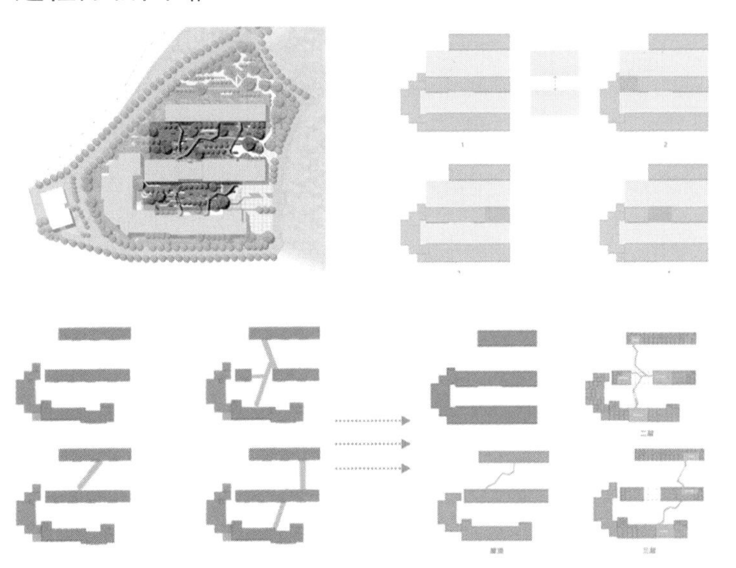

为了加强庭院与庭院之间联系，我们把中间建筑进行挖洞，把两个庭院加强联系，解决了庭院缺少联系，单调无趣的问题。

The building in the middle is holed to connect the separated yards, enhancing the communication between the two yards, and enriching the life of the seniors.

为了照顾园区内更多的垂直绿化墙，由直线桥变为曲桥。一方面解决了交通单一，建筑与建筑之间缺少联系的问题，同时老人可以站在不同标高的桥上进行浇水，达到老有所用目标。

Straight bridges are turned to curve bridges linking different buildings, so that the communication among the seniors is promoted. Besides, the seniors can stand on bridges of different heights to water the vertical gardens, realizing their value more conveniently.

功能置换 || Function replacement

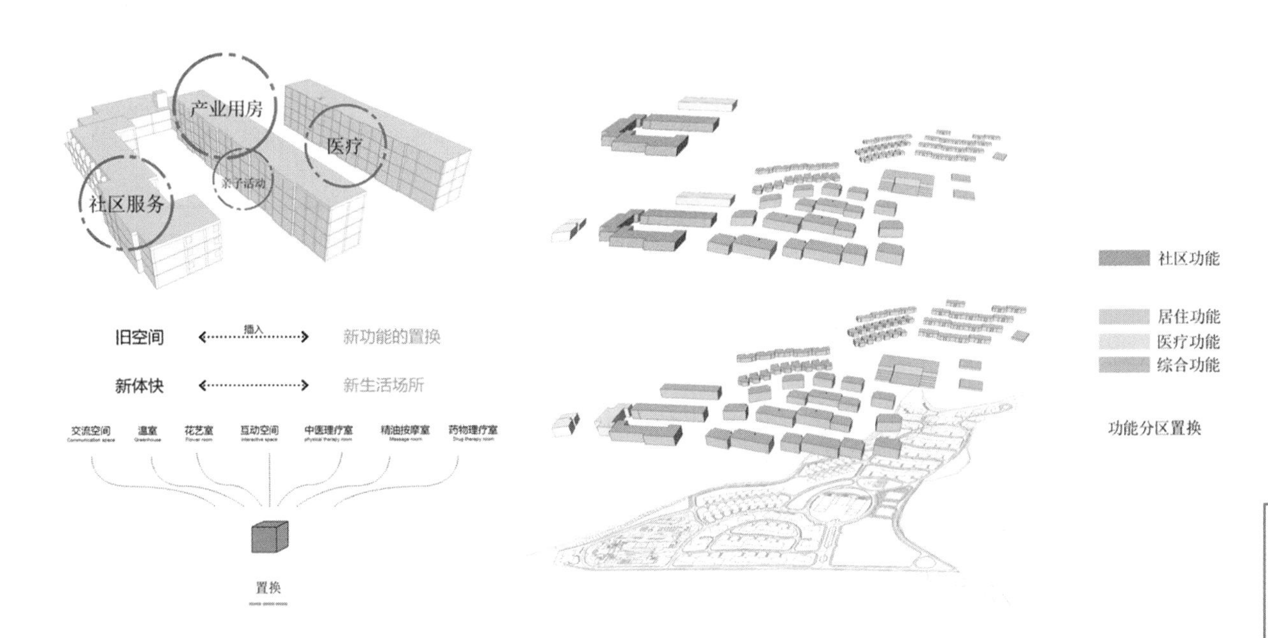

立面图 || Elevation

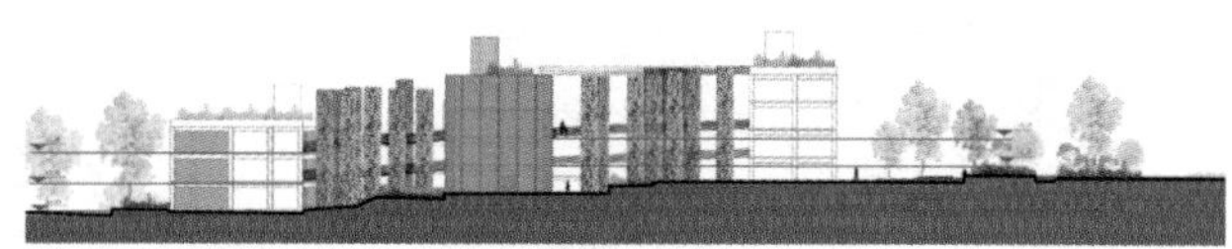

功能分区图 || Functional zoning

楼层平面图 || Floor plan

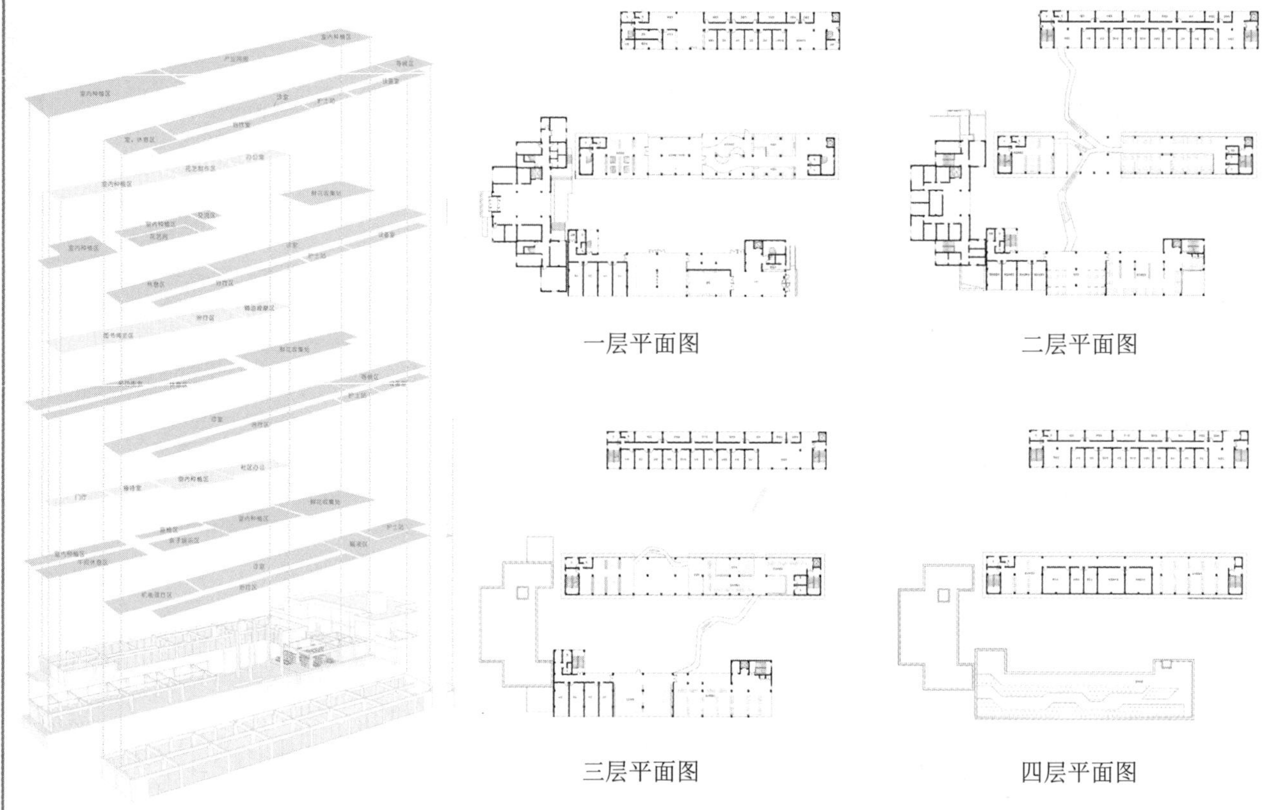

剖面图 || Section

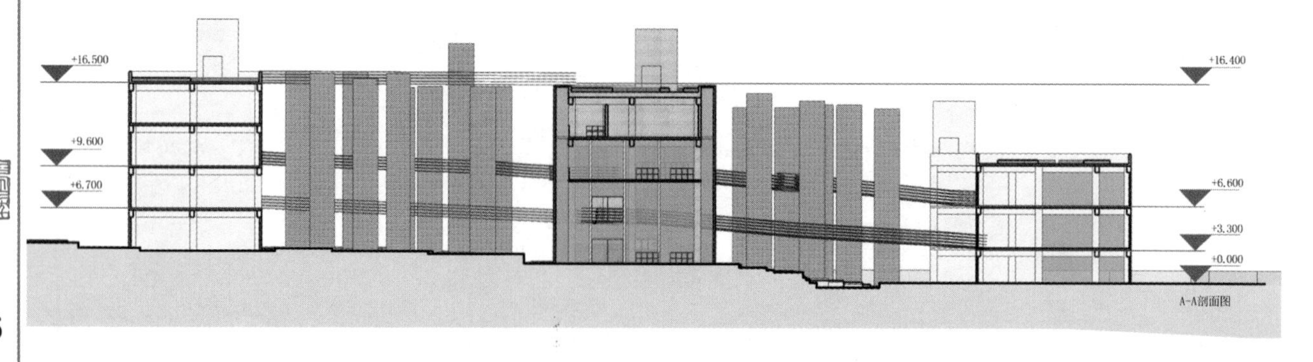

导师点评：

该组方案根据老人行为和心理特征，利用垂直绿化种植技术和联系前后大楼的趣味廊桥，将此老年公寓的环境景观改造成社区农场，打造出一种田园氛围，设计构思有一定的创意。新功能定位下的老人农场变身为鲜花种植园，老人身处其间可以天天种花、赏花，这个理想的图景看似美好，但不太接地气，方案缺乏对北京这样一个典型北方城市气候特征的深入研究，同时垂直绿化技术和廊桥景观建筑设计都未做进一步展开，致使整个概念缺乏较强的说服力。

左琰

Advisor's Comments:

According to the behavioral and psychological characteristics of seniors, the scheme designs vertical green walls, and innovative gallery bridges that link buildings, trying to transform the outdoor spaces of the project into a community farm, creating a kind of idyllic atmosphere. I think the concept is creative. According to the scheme, the farm is for flower planting, so that growing and appreciating flowers can be a part of the seniors' daily life. The idea sounds great, but is not quite practical, because the climatic characteristics of Beijing as a typical northern China city are not carefully analyzed. Besides, the scheme fails to introduce the ideas of vertical green walls and innovative gallery bridges in detail, making the two ideas less convictive.

Zuo Yan

专家点评：

（1）解决问题的意识较为突出。方案是针对“北京曜阳国际老年公寓”景观改造进行的创意设计。在大量有关对“地点”“人群”“现存问题”“改造对策”等设计所对应的问题上寻求回答如何解决养老机构的“老有所用”和“老有所养”，并给出了具体的解决方法。

（2）宏观规划意识地明确。“田园、公园化农场”，作为养老公寓景观策略的长远营建目标，主旨愿景是让老人“参与生产”“体验建设”来发挥老人的价值与作用，实现“老有所乐”，通过“置换”功能空间，优化空间布局，增加绿地与植物种植面积，使其景观在休闲和审美的基础上又突出了生产及医疗的价值，并落实在形成景观设计的产业经营格局上，进而成为这一景观建设方案中最具创新意识的核心所在。

（3）强化老人参与和体验的生活态度。方案对种植植物的位置、光照、通风、装置、维护及老人参与的可能性方面提出了自己的看法，并在老人“互动”“交流”的具体需求上，有针对性的加以解决，值得肯定。

（4）不足之处：一是学生对养老机构的设计规范、政策、理论、技术、实际案例等方面接触不多，对原建筑功能的改变，使相关设计务虚在给定的程序逻辑上。二是对现有空间或预设的老人生活所构成的相关条件要素的多元性及制约性了解也不够，基础材料准备支撑不足，影响了深度的设计思考。三是在造物与环境层面上过多关注形态与美学本身，忽略了特定养老机构不同健康类型的老年人安全、行走（无障碍）、休息、劳作（操作）的可实施性，以及种植和维护过程中技术设施设备应用的可能性。

董赤

Expert's Comments:

1. Strong awareness of solving problems. The task is developing a creative solution to the landscape renovation of the Beijing Yaoyang International Retirement Community. According to the answers to a number of questions like "what's the project location?", "what's the target group?", "what are the current problems?", and "what are the renovation measures?", the scheme gives solutions to how to realize the theme of "senior care & value realization".

2. Clear macroscopic planning. The scheme sets a clear long-term goal of transforming the retirement community into an "idyllic garden-like farm", with the view of encouraging the seniors to "participate in production", and "experience farm construction", thus realizing their value and "enjoying their life" in the process. Through the "replacement" of functional spaces, spatial layout is optimized to grow more plants, including herbs, so that the recreational, aesthetic, participation and medical needs of the seniors are all satisfied. Industrial management of landscape is the most creative core concept of the scheme.

3. Encourage the seniors to participate in and experience landscape construction. The scheme gives opinions on the locations, lighting, ventilation, facilities and maintenance of landscape, and the possibility of the seniors' involvement. It also gives solutions to specific issues like the "interaction" and "communication" of the seniors. This is worthy of some praise.

4. Deficiencies: Firstly, since the designers are not familiar with the design specifications, policies, theories, techniques and cases of retirement homes, when it comes to the functional redesign of the buildings, creative and flexible application of such knowledge is not realized. Secondly, they fail to have an in-depth understanding of the diversity and conditionality of the conditional elements of existing spaces and presupposed lifestyle of the seniors, and the basic materials required are not enough, which affect the depth of their design. Thirdly, they pay too much attention to the forms and aesthetics of hardware and environmental design, ignoring the practical needs of seniors in different physical conditions on safety, movement (wheelchair facilities), rest, work (operation), and the application possibility of technologies, facilities and equipment for planting and maintenance.

Dong Chi

Creative Retirement Community

CIID“室内设计 6+1”2016（第四届）校企联合毕业设计

CIID"Interior Design 6+1"2016(Fourth Session)University and Enterprise Joint in Graduation Design

高　　校：哈尔滨工业大学

College：Harbin Institute of Technology

学　　生：李佳楠 巴美慧 李皓 李曼园

Students：Li Jianan　Ba Meihui　Li Hao　li Manyuan

指导教师：周立军 兆翚 马辉

Instructors：Zhou Lijun　Zhao Hui　Ma Hui

参赛成绩：医养园区环境与景观设计组三等奖

Achievement：Third Prize for Aged Care District Environmental and Landscape Design

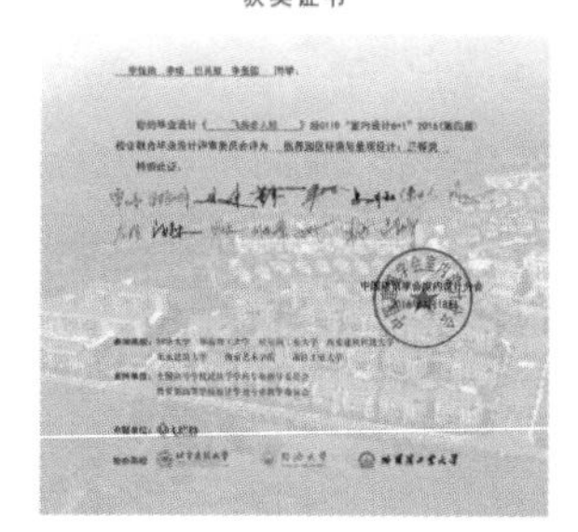

李佳楠
Li Jianan

巴美慧
Ba Meihui

李皓
Li Hao

李曼园
Li Manyuan

学生感悟：

通过本次设计，我们明确了中国养老服务机构的现状，了解了老年人在养老机构的生活方式及各方面需求，并深入研究了老年人的生理特点及生活习惯，对养老院室内外空间设计提出一些新思路，也为目前存在的问题提出一些解决办法，意图为老年人提供更加贴心的照顾与呵护。

社会对老年人与养老院存在诸多偏见，认为老年人应该静养不宜多动，养老院则是像医院病房一样的存在。我们通过设计表明了自己的观点，即养老院应该为老年人营造家一样的归属感和邻里空间感。医养建筑的“养”不是疗养，而是提供“养”分使其身心更加健康。

Students' Thoughts:

During the survey stage, we probed into the current situations of China's retirement homes, and the various needs, physiological characteristics and lifestyles of seniors. Based on the survey results, we rolled out some new ideas about the interior and exterior spatial design of the project, and provided solutions to some current problems, for the purpose of providing more thoughtful care to the seniors.

Our society has some erroneous views of seniors and retirement homes, thinking that seniors should exercise less, and retirement homes are like hospitals. We expressed our opinions through our design, that is, retirement homes should be home-like spaces that give seniors the sense of belonging and community. Aged care means not just physical care – metal care is more important for the health of seniors.

飞跃：偏见 \ 世俗 \ 现状

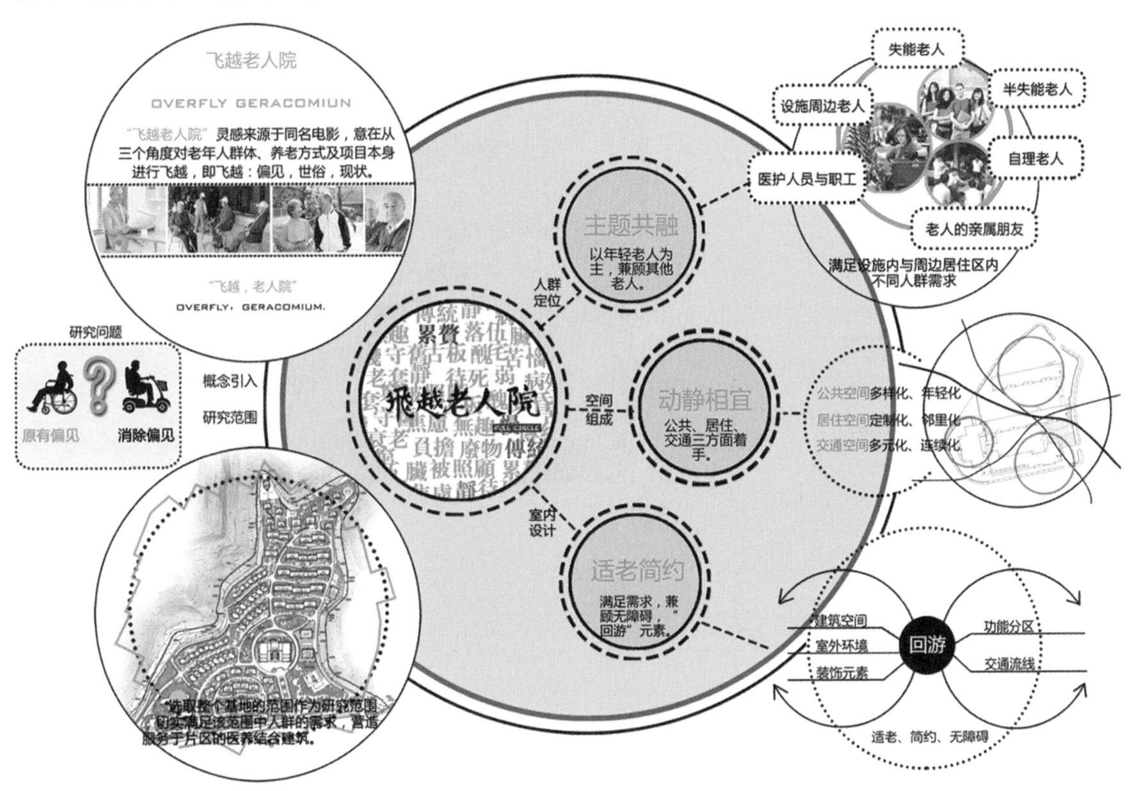

体块生成

Step1：局部推拉邻里空间
Step 1: Neighborhood spaces created through "local structural push and pull"

Step2：增设轨道新型交通
Step 2: Newly added creative rail transport system

Step3：景观桁架
Step 3: Landscape trusses

Step4：根据轨道交通设置节点空间
Step 4: Arrangement of nodes based on the rail transit system

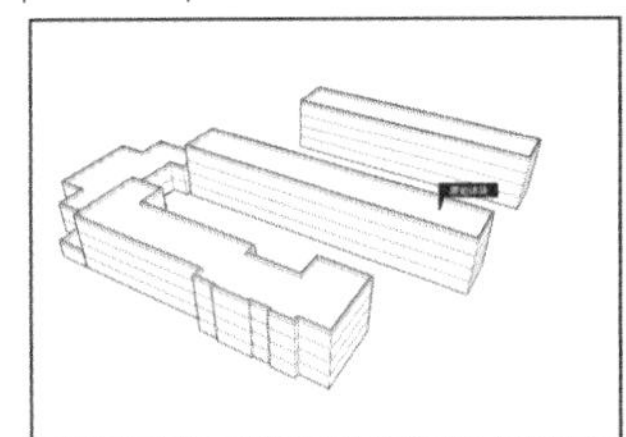

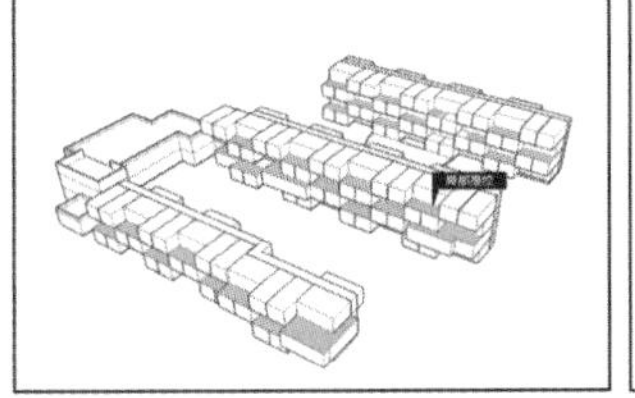

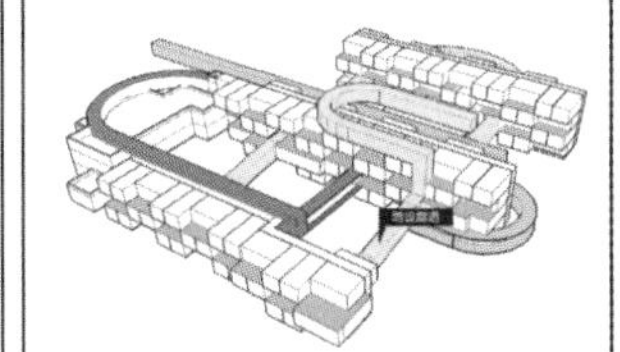

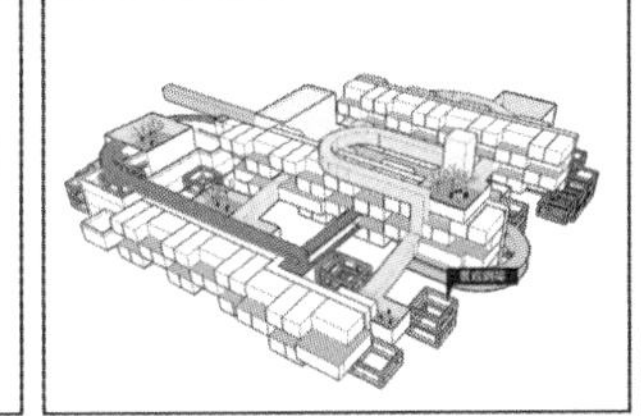

方案的体块生成，我们在旧有的建筑体块上进行局部推拉，形成邻里空间。
The original buildings are reshaped through "local structural push and pull" to create neighborhood spaces.

在此基础上增设轨道，建立新型交通体系，提高整个场地的通达性。
A creative rail transport system is built based on the reshaped buildings to improve the accessibility in the target area.

空间结构

在B楼二层中设置了一个通高的大空间为本次室内设计中公共空间的重点，尺寸为1500mm×1600mm×9900mm。

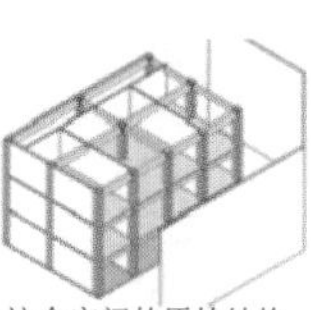

这个空间的原始结构为普通框架混凝土结构，由梁板柱组成其结构体系，结构高度一层为3300mm。

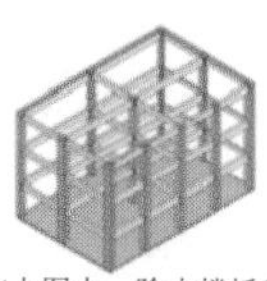

在上图中，除去楼板和山墙，可以清晰的看到这个体量的结构。由柱子和梁搭接而成的框架结构。

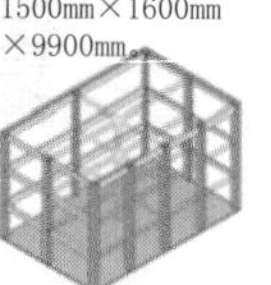

在设计中，为了保证大空间的完整性，我们去掉了大空间中间的一颗柱子，以及附着在它上的梁。

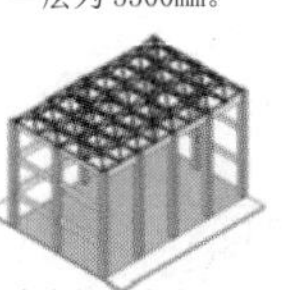

大空间的产生，使得屋顶没有职称结构，因此设计网架结构来解决因为失去中间柱所带来的影响。

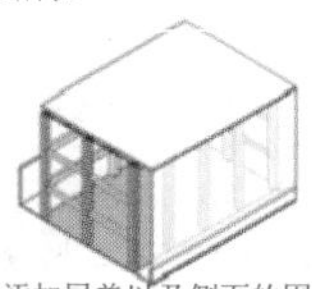

添加屋盖以及侧面的围合结构，在阳面的一侧设置成长玻璃幕墙，给空间增加阳光，并用百叶来调节。

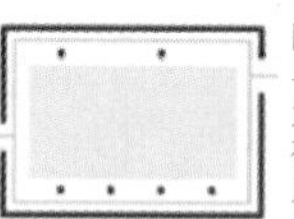

在解决了结构的问题后，我们首先得到一个完整的大空间，它的周围有一圈柱子，以及两侧各有两个出入口。

在没有任何限制条件下，人们在这个空间中的活动轨迹是随机而无序的。这在一定程度上来说交通破坏了空间的完整性。

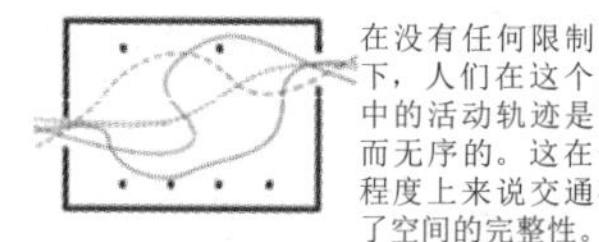

因此，我们在柱子与墙之间的空间作为主要的交通空间，在地面上加以导引系统的标识路线，让人们在这个空间中贴边行走。

当交通让出了中间的空间，可以在中设置一定的活动，比如羽毛球场，层高正好符合要求，且有空间在一侧这是看台。

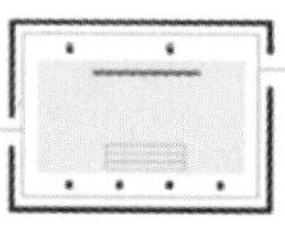

大空间 的存在，必然会造成利用率的降低，如何解决这一问题。预示我们设置了活动展板来满足不同的功能，例如演出功能。

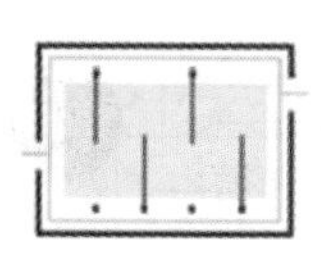

将活动展板变化位置，可设置成展览空间，在大空间中定期布展供老年人学习欣赏，多种功能的结合令空间增加利用率。

平面布局

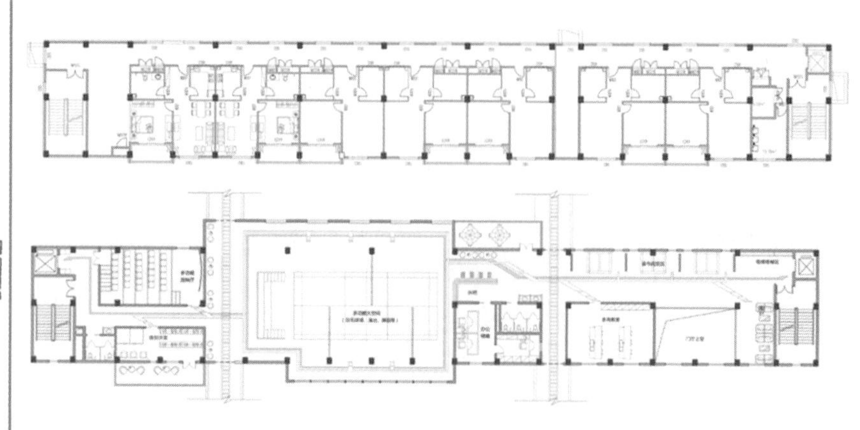

居住空间

——介助老人居住空间公共空间的连续与分离

——介助老人居住空间交通流线分类设计

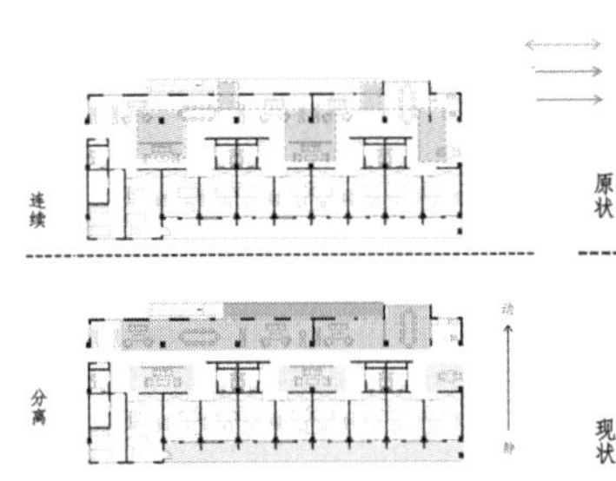

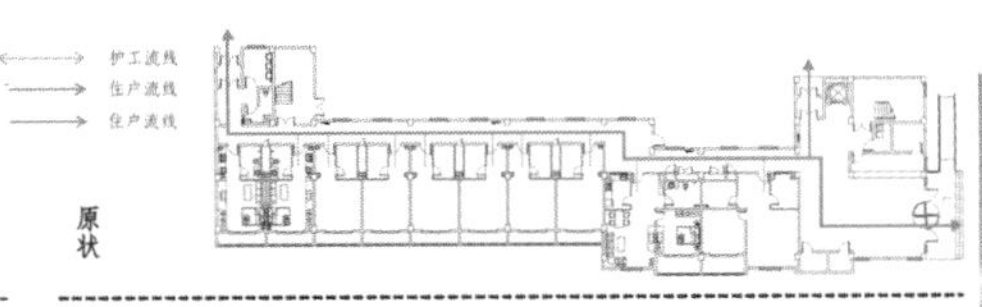

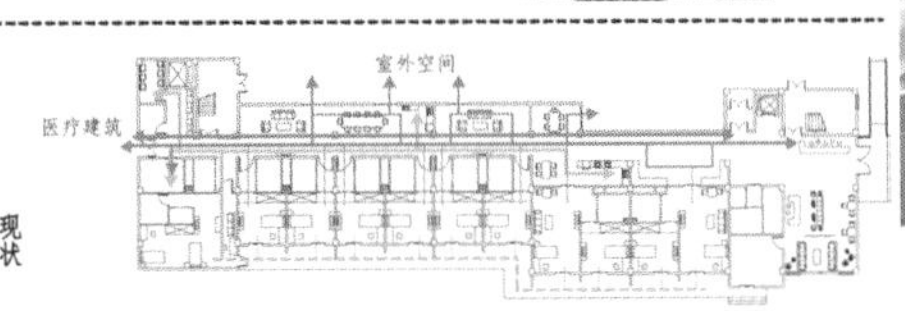

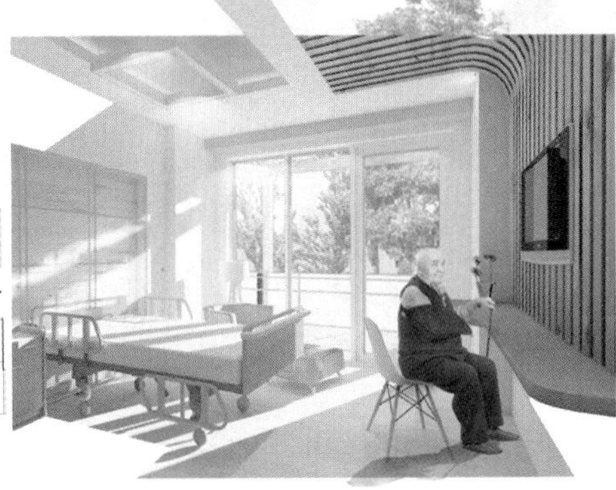

医疗空间

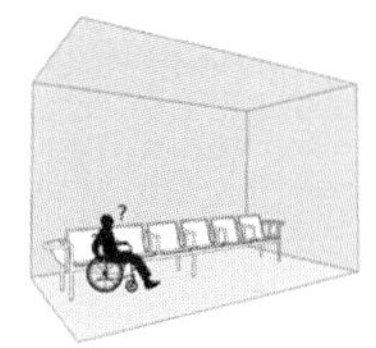

传统医院候诊室座椅功能单一，缺乏扶手，没有留出轮椅停留的空间。

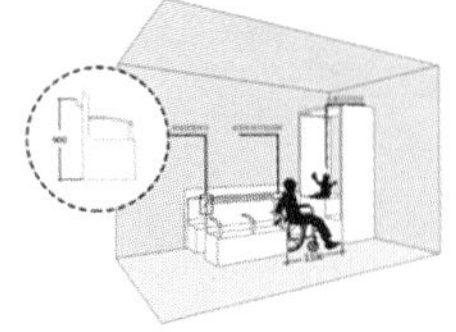

设计后门厅内椅背设置扶手，供后排老年人站立时依靠，作为设置轮椅停留空间，并且加入儿童活动空间。

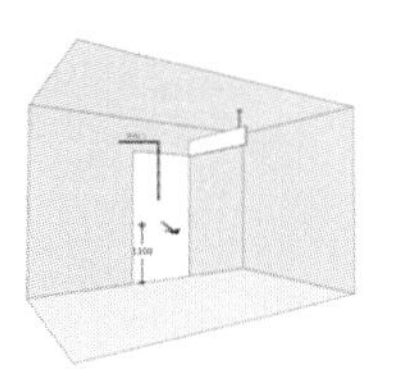

传统医院内缺乏无障碍设计，家具陈设不具有适老性，门多为平开门，导引系统仅体现在墙面。

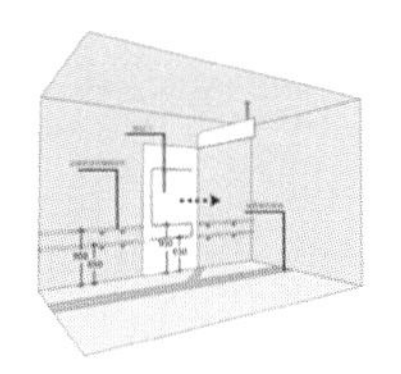

设计后采用双层无障碍扶手，门等家具设计针对于老年人的双层扶手，并且采用推拉式，地面增加导引指示线。

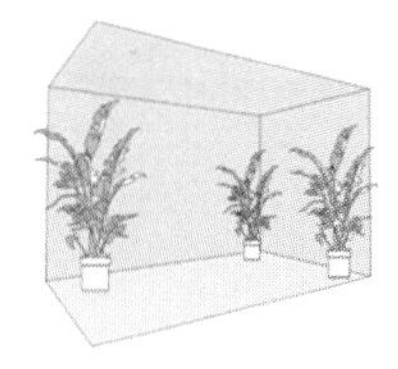

传统景观分散布置，仅仅起到装饰作用，基本无其他功能。

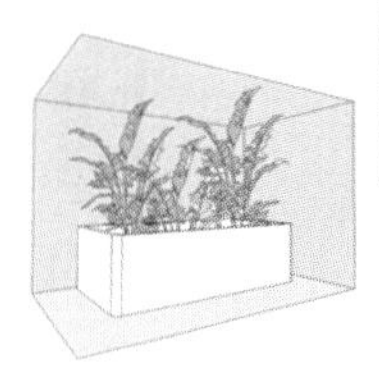

设计后将绿植景观集中化，形成中央景观池，营造绿意盎然之感，调节区域小气候，消除患者紧张的心理感受。

飞跃\室内外公共空间 FLY-Public space

在室内外空间做了多个供老人活动的节点，有大的，小的，有开敞的，有封闭的，有被功能属性的，也有留白给老人自由发挥的。

拓宽走廊并使其与公共空间关系多样化，让其成为充满生活情趣的步道。

共享空间为老人们提供了一个可以种植花草怡情，做饭聊天，下棋的平台。

老人的居住空间将向阳的一面尽量的用玻璃带来更多的阳光，并提供宽阔的视野。

北侧空地建造一个老人集体经营的小型农场来种植四季时蔬；室外庭院采用“食物地景”同时每个居住单元和共享空间也可让老人种植并独立经营。

最终形成从建筑到内庭、室外庭院再到后院层层递进富于变化的空间类型。

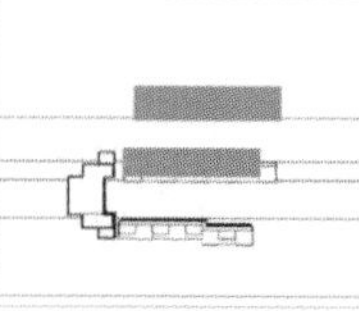

现状：四个建筑+两个院子，通过两个连廊相互联系。户外空间无趣且不实用。冬天和雨天，室外空间基本利用率为0。

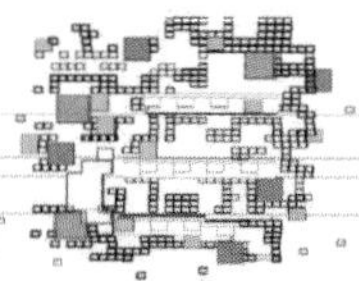

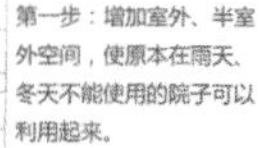

第一步：增加室外、半室外空间，使原本在雨天、冬天不能使用的院子可以利用起来。

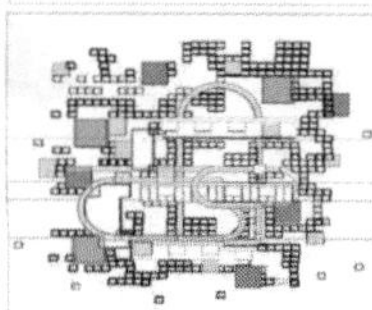

第二步：通过有序而便利的轨道小车，可以到达室内外的各个空间场所，在增加老人活动体育的同时，增加空间的利用率。

导师点评：

设计作品创意独特，灵感来源于电影“飞越老人院”，试图从多个角度入手进行建筑改造飞越。设计前期，针对使用者需求、场地周边环境、投资者经营模式均进行了较为深入全面的理性分析。

设计有效地将原有建筑进行改造，增加居住部分的采光面积，公共活动空间达到资源整合，整个设计功能布局合理，立面设计深入，对建筑室内大空间设计的多种活动功能模式进行演绎，如体育、舞蹈、展览等多功能，教室也可多用于烘焙、休息、交流等。

针对老年人生活起居所需，通过人体“五感”进行标识设计，并注重老年人的心理健康、室内各处尺寸亦尊重老年人而特别设计，体现设计者的关怀。设计者的细心之处还体现在阅读空间的飞越，老人们不但有相对独立的阅读空间，更多提供了互相交流的平台，设计出闲情雅致的场景。

最后应提到的是，环绕建筑顶部的绿色长坡道提升了整个建筑空间的环境景观效果，为老年人提供健步走的空间以及广阔的视野平台，但从北京的地域性气候考虑，长坡在漫长冬季的实用性还有待深入考虑。

刘晓军

Advisor's Comments:

The concept of the scheme is creative. Inspired by film Full Circle, it tries to, from different perspectives, transform the buildings in an creative way. In the early survey stage, the designers profoundly and rationally analyzed the requirements of the target group, the surroundings of the project site, and the operating mode adopted by the investor.

The transformation of the buildings is successful. Daylight area is increased, public spaces are integrated, spatial functions are rationally redesigned, façade design is specific, and large interior spaces are designed to conduct various activities such as sports, dancing, exhibition, baking, rest, communication and so on.

Signs that can be recognized by the "five senses" are designed to help the seniors in daily life. Besides, the dimensions involved in interior design are determined according to the psychological and physiological needs of the seniors. Care for seniors can be seen in many details.

Creative design of reading spaces also reflects the designers' special attention to details. In addition to independent reading spaces, shared spaces for exchanging opinions are also designed, giving the reading area a comfortable and relaxed atmosphere.

Lastly, the green long slope running around the roofs of the buildings improves the overall landscaping effects. It offers a space for the seniors to walk, and also a rest platform with a broad view. One problem is that given the long and cold winters of Beijing, the feasibility of the slope still needs to be further discussed.

Liu Xiaojun

专家点评：

设计组对于项目地及现有建筑进行分析的同时对国内养老现状也进行详尽的调研，并且以调研成果作为本次设计的基础，体现了严谨的设计流程及对项目的整体控制能力。公共空间设计理念有一定的创新，对空间布局收放有度，公共功能空间安排合理。在区域空间的贯穿链接上大胆的采用了立体的轨道交通的理念，虽然在实际实施过程中会遇到一些困难，但动线规划更加灵活，扩大了老人们的活动空间，提升了入住者的生活品质，最大限度提升了空间的利用率，有效的增加了空间的使用面积。

设计组的发表形式新颖，表现出整体的专业水平与合作精神，充分发挥想象力达到了思想的飞越。设计作品的工作量大，模型和图纸表现完整，是一件高品质的优秀毕业设计作品。

王兆明

Expert's Comments:

The scheme was developed based on a detailed survey and in-depth analysis of project site, existing buildings, and the current aged care situations worldwide. The design procedure is rigorous, and the designers' overall control of the project is successful. Ideas concerning public space design are creative, spatial layout is appropriate, and functional design of public spaces is rational. The idea of connecting regional areas with a sail system is bold, but implementation will not be easy. Traffic flows are designed flexibly that the activity areas of the seniors are expanded, the quality of their life is improved, and space utilization rate is maximized.

The scheme was presented in an innovative manner, which showed the professional level and spirit of cooperation of the design team. I see active imagination throughout the entire project process. A lot of design work has been done, and the models and drawings fully reflect their ideas. I think the scheme is an excellent high-quality design work.

Wang Zhaoming

Six Harmonies

CIID“室内设计 6+1”2016（第四届）校企联合毕业设计

CIID“Interior Design 6+1”2016(Fourth Session)University and Enterprise Joint in Graduation Design

高　　校：南京艺术学院
College: Nanjing University of the Arts
学　　生：张梦迪　龚月茜　戚晓文
Students: Zhang Mengdi　Gong Yuexi　Qi Xiaowen
指导教师：朱飞
Instructors: Zhu Fei
参赛成绩：医养园区环境与景观设计组三等奖
Achievement: Third Prize for Aged Care District Environmental and Landscape Design

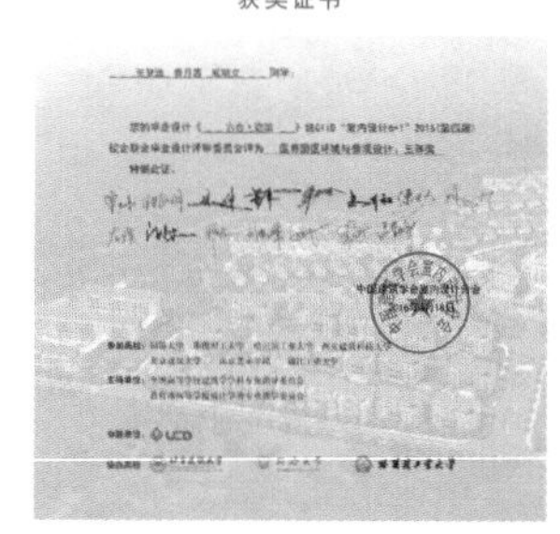

张梦迪
Zhang Mengdi

龚月茜
Gong Yuexi

戚晓文
Qi Xiaowen

学生感悟：

经过大半年的奋战，从北京到上海再到哈尔滨，我们经历了很多，也收获了很多。在没有参加“6＋1”之前，我们觉得毕业设计只是对这几年来所学知识的单纯总结，但是通过这次锻炼，我们发现我们的认识过于片面。毕业设计不仅仅对前面所学知识的一种检验，更是对自己能力的一种提高。通过这次互动使我们明白了自身的知识和经验还是缺乏的。虽然在设计的过程中，我们遇到了很多的困难，方案改了一次又一次，也被批评了很多次，但是在方案完成的那一刻，我们是快乐的。只有坚持，才会有真正的收获。

Students' Thoughts:

From Beijing to Shanghai to Harbin, during the competition which lasted for more than half a year, we experienced a lot and also learned a lot. Before participating in the competition, we thought that a gradation project is just a test of the knowledge that we'd learned in the university years. Later we realized that our understanding is too partial – a graduation project is not only a test, but also a process of improvement which helps us realize our lack of experience and knowledge. It's true that the process of design was hard. We met a lot of difficulties, changed our scheme again and again, and were criticized for many times. However, at the moment when the scheme was finally completed, we were very happy and excited. Success cannot be made without insistence.

设计前入口 || Entrance before design

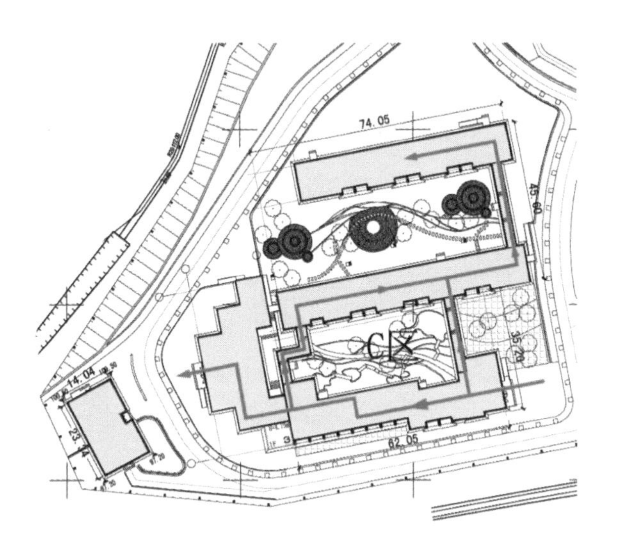

设计后入口 || Entrance after design

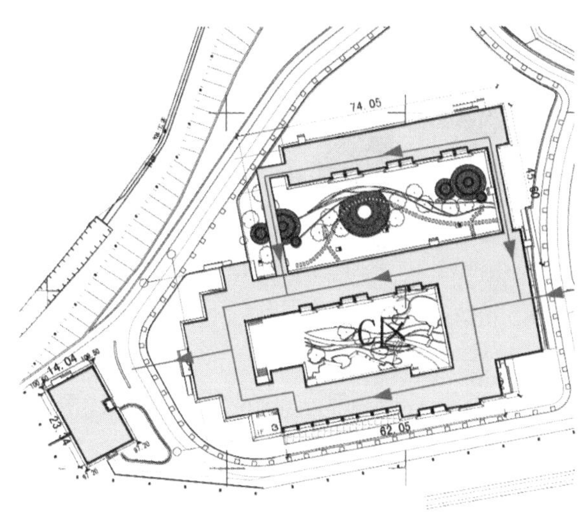

形式创意 || Formative creation

我们将空间尽可能地开放，就是尽可能地让身在其中的老人能够看到对方，为人与人之间的沟通创造机会。此外，多人参与的活动，如乒乓球、健身操等，也能更多地鼓励老人与老人之间的交流。

We try to create open spaces as many as possible, creating opportunities for the seniors to see and communicate with each other. For example, places for group activities such as table tennis and setting-up exercise are designed, with the view of encouraging communication among the seniors.

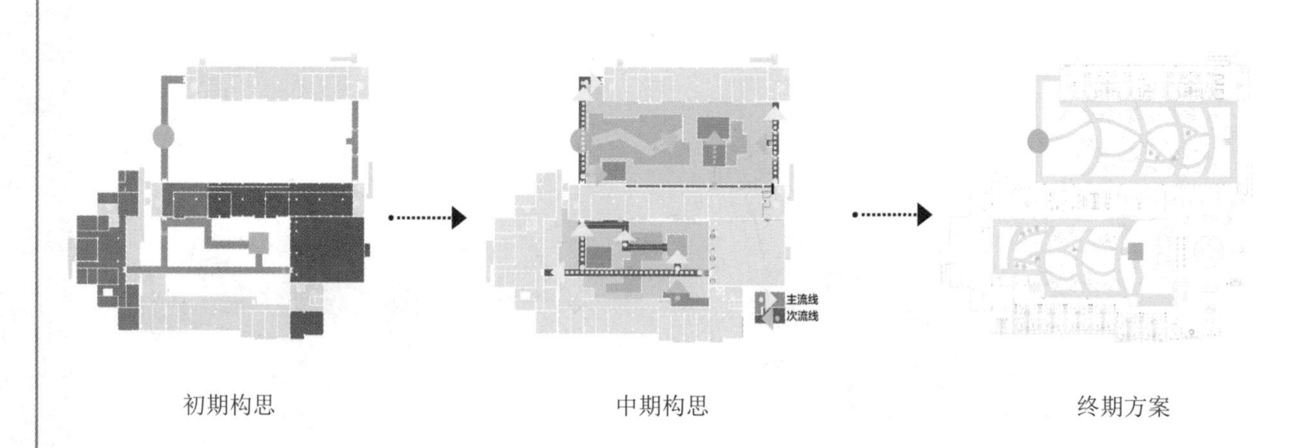

初期构思　　中期构思　　终期方案

区块分析

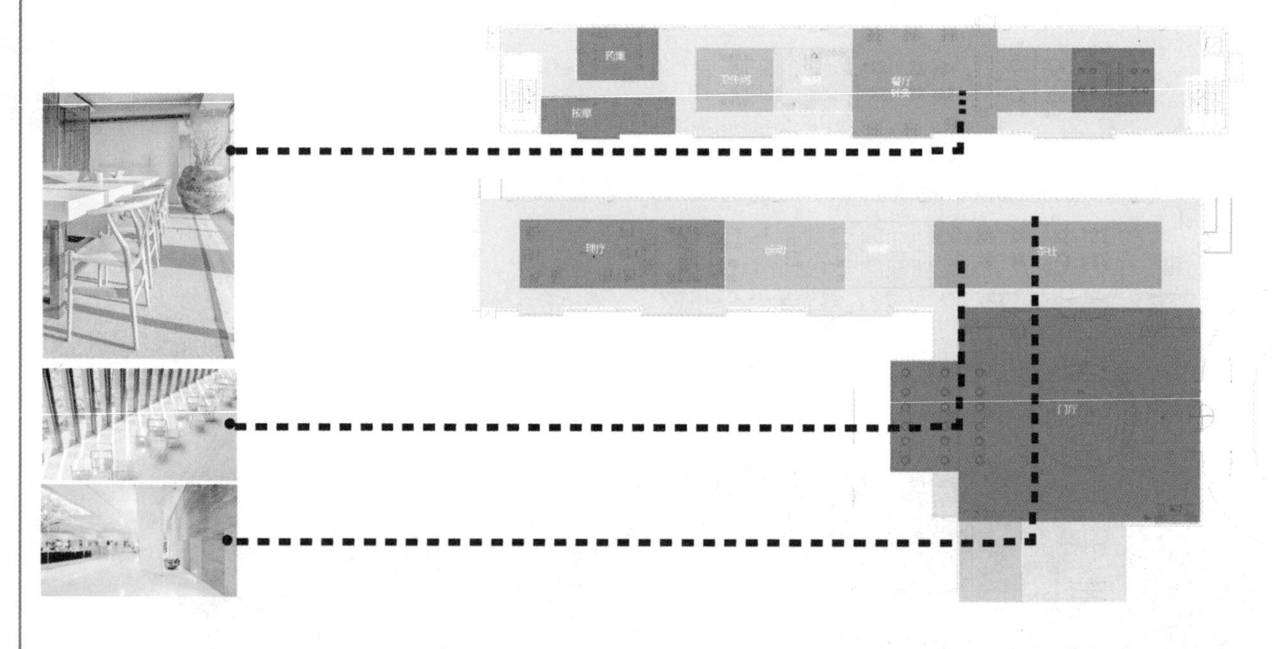

在通过对老年的详细调研和系统的分析后，我们决定将原本比较单一且各自独立的医院模式病房区进行改变，变成温暖如家、方便老人相互交流的空间。且将三栋住宅楼之间的两处庭院加设廊道、亭、榭，划分出更多层次的空间从而改善相对孤立的状态。这些改变都将是建立在对老年人的活动方式和建筑原本特质的思考产生的结果，对我们的设计来说，老年人的养老模式将不再仅仅受限于孤零零的病房中，也不会固步自封于老去的恐慌中，而是与其他老人一起成为伙伴，完成人生最灿烂的绽放。

导师点评：

“六合·宓馆”命题新颖，充满东方意味，设计方案从中国传统中医养生进行切入，“顺气一日分为四时，中医养生、养气、气和而神形、阴阳平衡”，并提出宅物、宅己、宅生、合形、合气、合心的设计思想，方案带入感极强。

方案充分考虑了普通健全老人、残障老人、旅游人群、家属子女、医疗服务人员、物业管理人员六类受众人群，并创新的将老人分为展示型老人和观赏型老人，注重搭建展示公共平台、方便老年人交流互动，采用竹、木、石、土、水的设计元素，使得空间充满淡雅、传统的中式味道。平面方案规划合理，注重阴阳结合，病房区域分类设置，合理布局，入口门厅重点设计，效果尤为突出。庭院设计有机结合太极八卦的意向，设置2处凉亭，体现中式建筑的特征。

方案设计新颖，紧扣中式建筑、园林、室内的特征，有一定创新性。但方案图纸部分有所欠缺，制图规范性不足，需进一步提升。

朱宁克

Advisor's Comments:

"Six Harmonies" is a typical oriental theme. Adopting TCM as the highlight, the scheme incorporates such TCM concepts as "variation of qi in a day", "health maintenance", "cultivation of qi", "qi-spirit harmony", and "yin-yang balance" into the design. It also discusses the relationship between homes and objects, man, and life, and the coexistence of different forms, types of qi, and ideas in the context of design. The scheme is very attractive.

The needs of six groups including able-bodied seniors, seniors with disability, tourists, occupants' families, medical personnel, and property management are discussed in the scheme. The scheme innovatively divides seniors into seniors preferring to show and seniors preferring to watch, and corresponding spaces for performance and communication are designed. Besides, elements like bamboo, wood, stone, earth and water are employed to create an elegant environment filled with traditional Chinese culture. The plan design features yin-yang integration, ward classification, and rational spatial division. The entrance hall is a highlight of plan design. The design of the yards applies the concepts of Taiji and Eight Diagrams, with two pavilions presenting yin and yang in each yard.

The scheme is creative, incorporating traditional Chinese culture elements into architectural, landscape and interior design. The drawings should be more normative, and relevant improvements are needed.

Zhu Ningke

专家点评：

该方案通过对原有建筑的门厅、公共活动区域，各层病房和室外庭院进行改造，试图改善目前的居住环境。门厅的改造，引入视觉冲击力较强的体块，其实质的功能、空间感受尚待商榷；公共活动区域的引入能增加使用人群的交流，但应做到动静分区，设置更符合老年人生理及心理需求的内容；病房的分区规划有一定的合理性，建议根据各种设定人群的需求推导病房护理单元可提供的服务，从而进行更为细致的空间调整和划分；室外庭院的形式及材料的选择等，可紧扣提出的“太极”“中医”“道家”这几个主题进行深入设计。方案思路明确，各图纸在规范表达及突出表现方面欠佳，版面略松散。

李莉

Expert's Comments:

In the scheme, the halls and wards of the buildings, public areas, and yards are redesigned, for the purpose of creating a better living environment. The hall design, including functional and spatial design, is impressed, but its actual effects still need further discussion. More public areas are created to promote the communication among the seniors, but areas for active and quiet activities should be separated to meeting the physiological and psychological needs of different occupants. The zoning of wards is kind of rational. My suggestion is that the ward area can be divided in a more specific way that seniors needing different medical services are arranged in different zones correspondingly. As for the design of the yards, forms and the selection of materials should be themed on Taiji, TCM and Taoism. The scheme expresses design ideas clearly and logically, but the drawings are less normative, impressive, and compact.

Li Li

Opera and Clouds

CIID"室内设计 6+1"2016(第四届)校企联合毕业设计

CIID"Interior Design 6+1"2016(Fourth Session)University and Enterprise Joint in Graduation Design

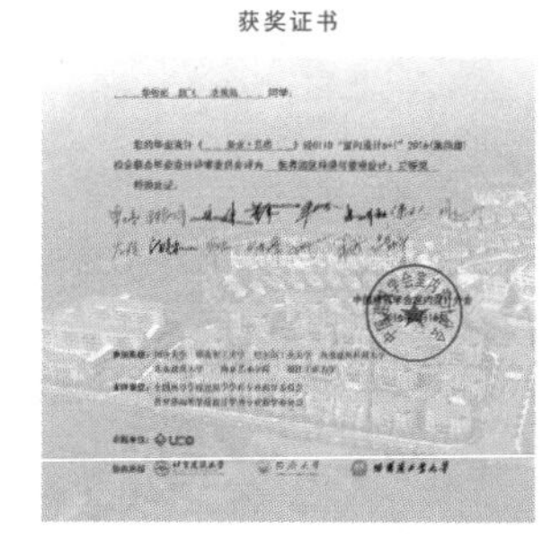

高　　校：南京艺术学院

College: Nanjing University of the Arts

学　　生：华彬彬 魏飞 冼佩茹

Students: Hua Binbin Wei Fei Xian Peiru

指导教师：朱飞

Instructors: Zhu Fei

参赛成绩：医养园区环境与景观设计组三等奖

Achievement: Third Prize for Aged Care District Environmental and Landscape Design

华彬彬
Hua Binbin

魏飞
Wei Fei

冼佩茹
Xian Peiru

学生感悟：

我们经过前期对本次课题的调研和资料收集，将对象定位在有自理能力的久居老人和失智老人身上。整个老年公寓的定位为以疗养为主的医养结合的老年公寓。通过参加这个联合毕业设计，尝试了一次自己从未涉及过的领域，使我接触到了建筑设计、室内设计、景观设计，并且从中学到了很多知识，学会用建筑的思维去思考室内设计，并从室内设计角度考虑到了信息的传达，在设计过程中了解了许多京剧文化、养老机构的规范等多方面的知识。

Students' Thoughts:

Based on early stage survey results and other information that we collected, we chose able-bodied seniors and seniors with dementia as the target group, and gave the functions of medical care and recuperation to the buildings. The competition gave me an opportunity to touch a new design field, and in this process, I learned a lot of new knowledge about architectural design, interior design, landscape design, as well as Peking opera and regulations concerning retirement homes. I also learned how to understand interior design from the perspective of an architect, and how to deliver information from the perspective of an interior designer.

区位分析 ‖ Location analysis

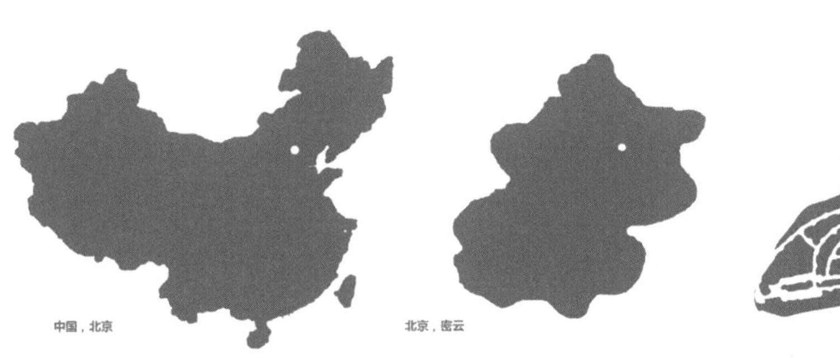

中国，北京　　北京，密云

密云曜阳国际老年公寓

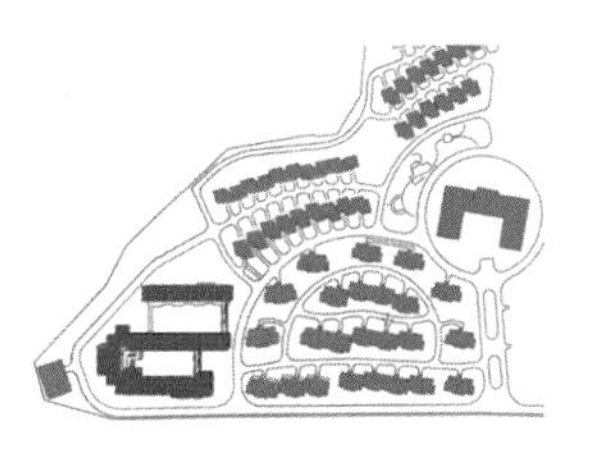

设计用地，C区护理中心

1．主体建筑　　2．花园平台
3．厨房后场　　4．水池
5．水上廊道　　6．广场
7．草坪　　8．汀步
9．古牌坊　　10．阶梯

景观设计中，我们考虑到原有建筑的地势，所以改造成阶梯式的可供休息娱乐的阶梯草坪。

Lawn terraces as a part of landscape are designed according to the original terrains of the buildings, creating a space for the seniors to rest or entertain.

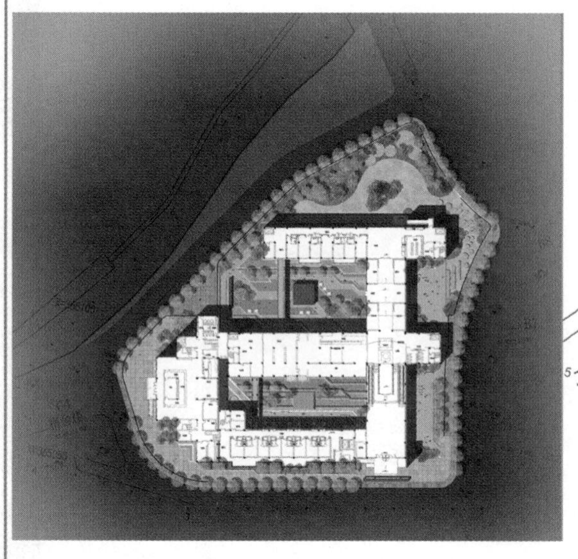

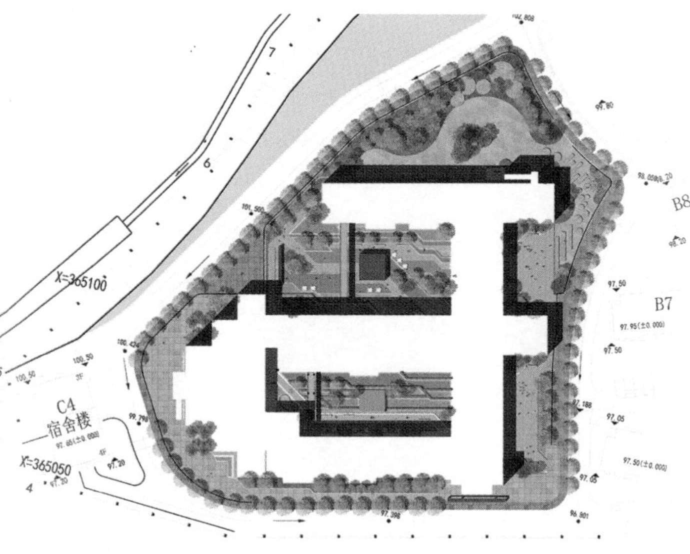

周边分析 || Surrounding analysis

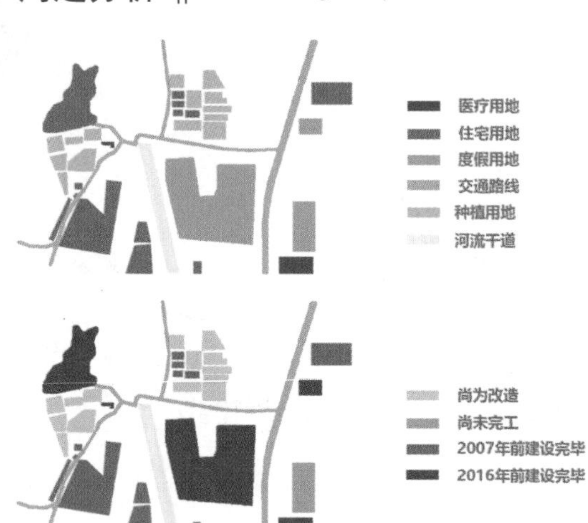

在景观中间设置了平台，可供平时表演娱乐等活动，并且在两栋建筑间添加了连廊，具有一定的通达性。

A platform is arranged in the center of the landscape area for performance and entertainment. Besides, an exterior corridor is added to connect the two buildings, promoting the transport and communication between the buildings.

人群定位 || Target group

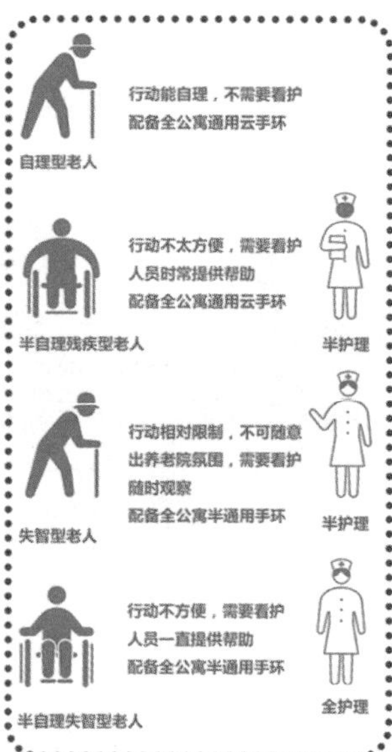

立面分析 || Façade analysis

导师点评：

“梨舍．云居”作为以中国传统京剧元素为展示设计主题的老年公寓改造方案，设计者从改变空间水平及垂直交通的角度去改善空间流线的通达性，试图改变建筑室内空间形态的联结关系。设计方案的特点在于将京剧元素符号化运用到空间标识系统及色彩方案中，以点、线、面的构成语言呈现在室内空间。在老年居室空间的设施设计中引用了蒙特里安的色彩方案，并加以变化和延伸，旨在突出居室室内与公共空间中西合璧的视觉感受。整体方案在概念主题的逻辑关系、整体格调把握，受众群体——老年的生理、心理适宜性分析、空间设施的舒适性方面待进一步调整和深入。

任彝

Advisor's Comments:

"Opera and Clouds" renovates the retirement community through exhibition design themed on Peking opera elements. The scheme tries to create smoother traffic flows horizontally and vertically, thus changing the connection relations between interior spaces. The highlight of the scheme is incorporating Peking opera elements into the design of the sign system, and into interior color design in the form of points, lines and surfaces. Mondrian's color scheme is applied in interior facility design. Through the change of color and spatial extension, Chinese and western cultural elements are harmoniously mixed, creating visual impacts. Further adjustments and deeper analysis are required in terms of the logical relationship between the theme and the content, overall style control, the physiological and psychological suitability of the target group (the seniors), and comfort of facilities.

Ren Yi

专家点评：

“梨舍•云居”的设计者以中国的国粹，京剧元素为主线展开设计概念，将京剧的内涵文化思想意识同老年公寓环境进行了巧妙融合，主题明确清晰，方案结构完整，并且在多方面考虑到老年人生理、心理的需求，进而延续到空间布局及装饰，突出了医养结合的特色，同时也考虑到经营方的投入与产出的关系，是有一定可实际实施的设计作品。

设计者“精心”选定的主题，创造了设计的亮点也局限了设计本身，在局部空间设计及色彩使用上，还是较多的“想当然”占据了主导，养老设施的特殊性促使设计者，应时刻将自己作为老年人去思考的同时进行设计。方案最终的呈现还具有一定的可操作空间，还需要对老年人生活规律及习惯进行进一步的调研。

陈天力

Expert's Comments:

Adopting Peking opera, the quintessence of Chinese culture, as primary design element, "Opera and Clouds" skillfully incorporates Peking opera culture into the environmental redesign of the retirement community. The theme is clear, and the scheme structure is complete. Besides, the physiological and psychological needs of the seniors are considered in the design of various aspects, including spatial layout and decoration. The scheme realizes the integration of medical and aged care resources, and analyzes the input-output relationship of the project. It's feasible to some extent.

The carefully selected theme is the highlight of the design, but also a restriction. Regional spatial design and color selection are kind of divorced from reality. After all, the design of retirement homes requires designers to think and design from the perspective of seniors. Generally speaking, the final scheme is feasible to some extent, but a further survey on the lifestyles of seniors is still required.

Chen Tianli

Memories in Courtyard Houses

CIID“室内设计 6+1”2016(第四届)校企联合毕业设计
CIID"Interior Design 6+1"2016(Fourth Session)University and Enterprise Joint in Graduation Design

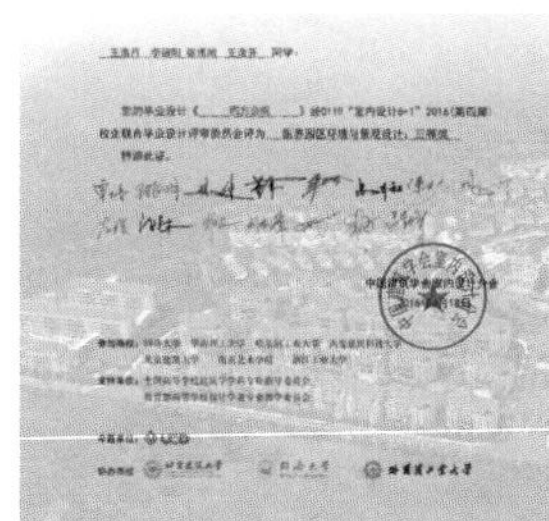

高　　校：北京建筑大学
College: Beijing University of Civil Engineering and Architecture
学　　生：王海月 李丽阳 王逸开 张博闻
Students: Wang Haiyue　Li Liyang　Wang Yikai　Zhang Bowen
指导教师：杨琳 朱宁克
Instructors: Yang Lin　Zhu Ningke
参赛成绩：医养园区环境与景观设计组三等奖
Achievement: Third Prize for Aged Care District Environmental and Landscape Design

王海月
Wang Haiyue

李丽阳
Li Liyang

王逸开
Wang Yikai

张博闻
Zhang Bowen

学生感悟：

这次联合毕业设计中，在几次交流中看到其他学校对于相同题目的不同想法，大家的想法都很棒，使我们扩展了很多眼界。同时，这次毕业设计从单纯一个点的想法开始引出，在不断丰富方案的过程中，我们的设计经历了本心——偏离主体的发散（中期）——回归本心完善思考（末期）的纠结过程，指导老师与评委老师给予的意见与建议提醒我们回归到了本心，帮助我们改善了整个设计，对我们来讲是一个非常宝贵的经历。

Students' Thoughts:

During this joint graduation design, I met different ideas from other schools for the same topic in the exchange activities. These great ideas expanded our outlook. At the same time, this graduation design started with ideas for a single point, and our design experienced an entangling process, from original center (early stage), to divergence deviating from the subject (mid stage) and to improving thinking returning to the original center (late stage). With their sincere opinions and suggestions, our coaching counselor and judge experts reminded us of returning to the original center and helped us improve our design. This is absolutely valuable experience for everyone of us.

空间营造

■ 创造大量的公共空间

■ 引进北京四合院文化

鼓励老人从自己独立的空间里走出来，进入到公共的开放空间。通过空间的引导，促进老人之间的交流，缓解老人因缺乏与子女之间的交流而产生的孤独感

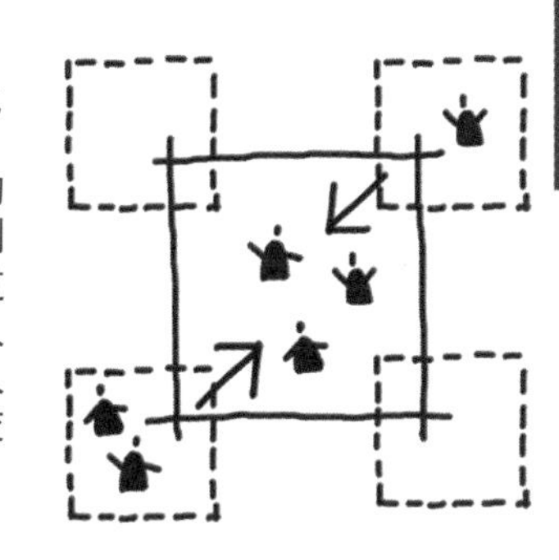

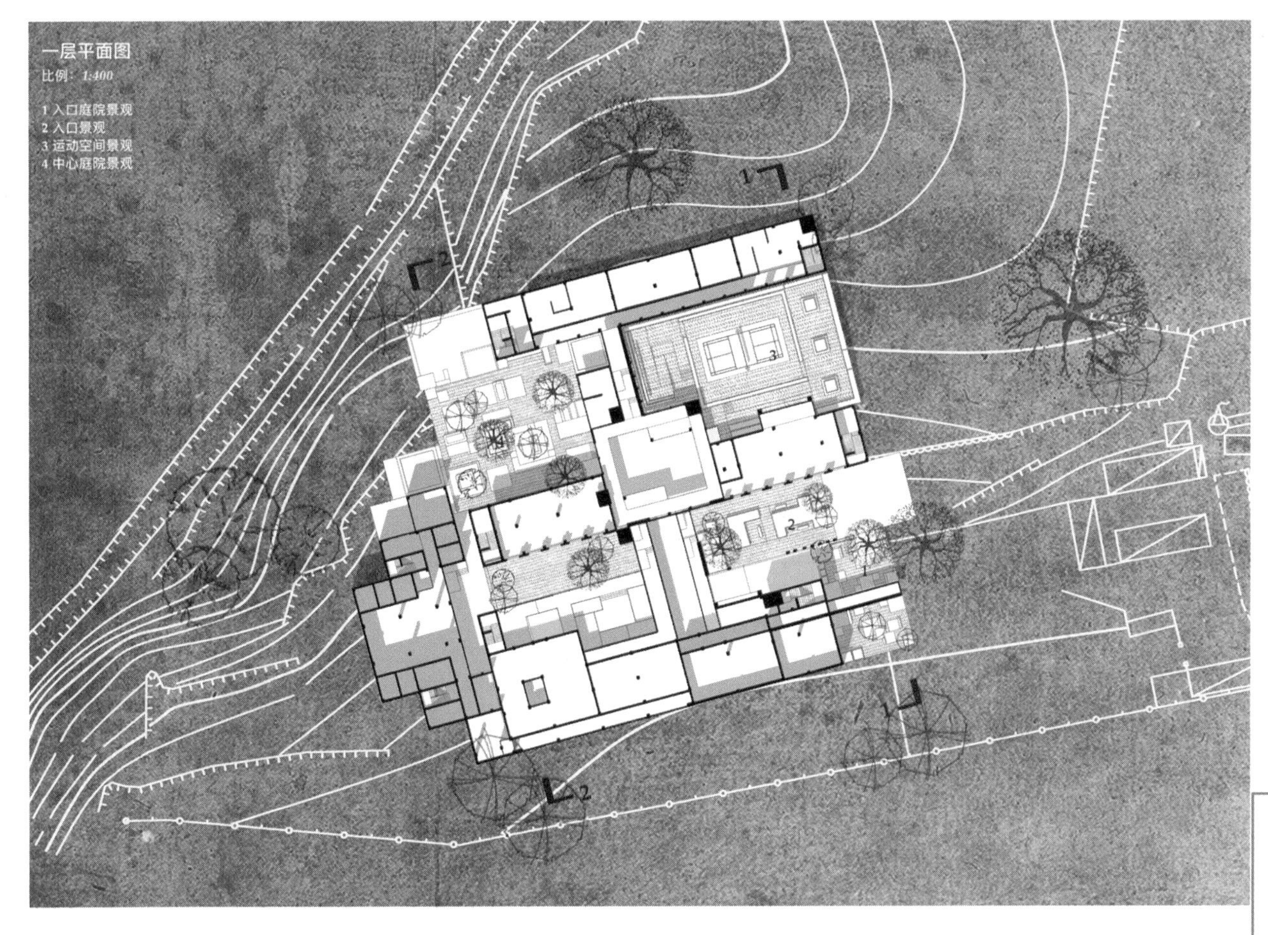

流线分析图

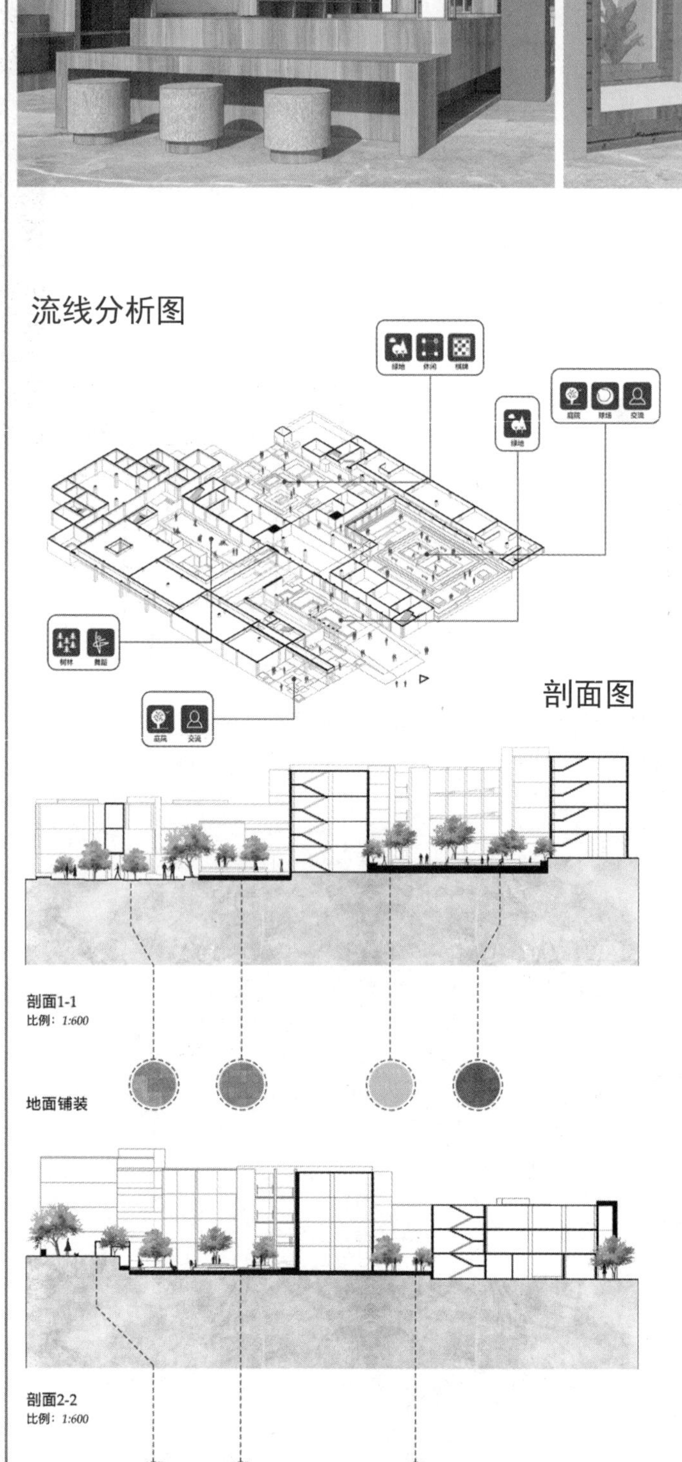

色彩选择

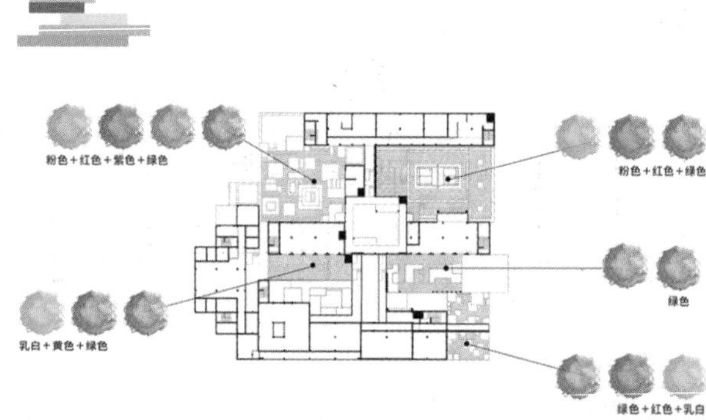

植物选择

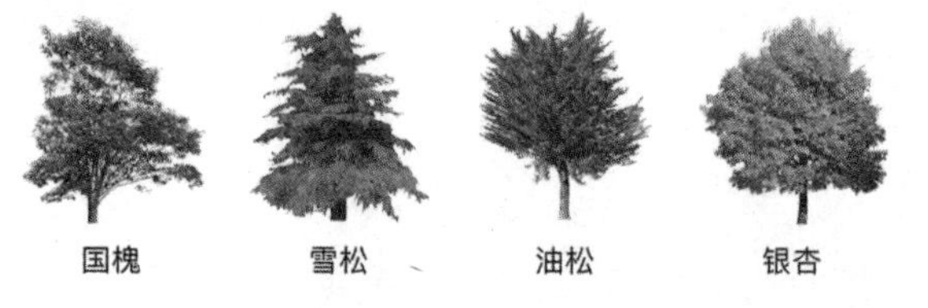

迎春花

高0.3～5m，花单生在去年生的枝条上，先于叶开放，有清香，金黄色，外染红晕，花期2~4月。

榆叶梅

高2～3m，花单瓣至重瓣，紫红色，1～2朵生于叶腋，花期4月。

西府海棠

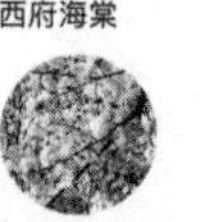

高2.5～5m，花瓣近圆形或长椭圆形，长约1.5厘米，基部有短爪，粉红色，花期4～5月。

山里红

高6～8m，伞状花序有小花10~12朵，白色或淡红色，花期5月。

太平花

高1～2m，花5～9朵成总状花序；乳白色，微芳香，花瓣4，卵圆形，花期4～6月。

紫丁香

高1.5～4m，圆锥花序，近球形或长圆形，花淡紫色、紫红色或蓝色，花期5～6月。

玫瑰花

最高可达2m，花瓣倒卵形，重瓣至半重瓣，花有紫红色、白色，花期5～6月。

木槿花

高3～5m，花大，单生叶腋，直径5～8厘米，单瓣或重瓣，有白、粉红、紫红等色，花瓣基部有时红或紫红，花期6～9月。

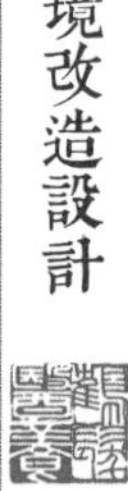

导师点评：

“四方合院”设计者以老北京文化为场所记忆，在对老年人心理行为分析的基础上对场地空间功能梳理、汇集，对传统老年疗养院进行了重新定义，在既有空间格局中植入合院、胡同式载体，形成了既融合又独立，富有趣味性内部空间的集合式疗养院。整个方案概念表达逻辑清晰，形式与功能相对统一，空间组织聚散富于变化，成果呈现系统完整。特别是在适老性设计及人性空间设计上做了较为精彩的解答，如动线组织上减少竖向阻隔，以水平流线为主；在大空间中穿插如茶馆、大舞台这样一些充满京味的公共空间等。如在概念深化，细节处理等方面进一步思考优化，则整体方案将更具说服力及合理性。

黄炎

Advisor's Comments:

Based on an analysis of seniors' psychological and behavioral characteristics, "Memories in Courtyard Houses" tries to integrate and re-divide the spaces of the project, and redefine the functions of retirement homes. Courtyard houses and hutongs are incorporated into the existing spatial layouts, creating connected but also independent interesting group living spaces. The scheme features clear and logic expression, consistency between forms and functions, flexible spatial organization, and systematic and complete results presentation. Senior-oriented spatial design is the highlight of the scheme. For example, the locations of barriers on traffic flows are changed so that they are parallel to the flows, and public spaces with the cultural characteristics of Beijing like teahouses and performance stages are arranged in large spaces. With more specific concept elaboration and further detail design, the scheme will be more convictive and rational.

Huang Yan

专家点评：

(1) 梳理研究并审视养老模式的既有问题，建立以（老人）“行为类别”为基础的设计逻辑；
(2) 归纳诸行为的类型，依此界定能分别承载这若干类行为的“公共子空间”群；
(3) 以这样的经过“群化”的子空间为节点，重组并达成总体的养老系统；
(4) 同样，以这样的子空间为节点，叠加并达成总体的营运与服务系统；
(5) 以上的设计逻辑，使空间不再抽象，改建不再“想象”，界面不再“风格”，功能不再“飘浮”。
(6) 从此方案中人们能感受到：“为他人”而非“我认为”的设计；“为具体人群”而非“实现我的理想”的设计；
(7) 此方案较为缜密、周到；较为别致、简捷；另外，具有“健全、平衡”的时代气息。

赵健

Expert's Comments:

1. The scheme designs based on a detailed analysis of the problems existing in current aged care models, and the behavioral characteristics of "different types of seniors".
2. "Public sub-spaces" are created according to the behavior of different types of seniors.
3. An aged care system is built by taking various public sub-spaces as spatial nodes.
4. Likewise, an operating and service system is built based on such public sub-spaces as spatial nodes.
5. The above design concepts ensure formative spaces, targeted redesign, useful boundaries and practical functions.
6. The designers think and design from the perspective of and also for the users.
7. The scheme is meticulous, thoughtful, unique and compact. It's also "healthy and balanced", reflecting the characteristics of our times.

Zhao Jian

Regional Self-Sufficient Life in Groups

CIID“室内设计 6+1”2016（第四届）校企联合毕业设计

CIID"Interior Design 6+1"2016(Fourth Session)University and Enterprise Joint in Graduation Design

高　　校：　西安建筑科技大学

College:　Xi`an University of Architecture and Technology

学　　生：　张峰　颜强　赵凯文

Students:　Zhang Feng　Yan Qiang　Zhao Kaiwen

指导教师：　刘晓军　何方瑶

Instructors:　Liu Xiaojun　He Fangyao

参赛成绩：　医养园区环境与景观设计组三等奖

Achievement:　Third Prize for Aged Care District Environmental and Landscape Design

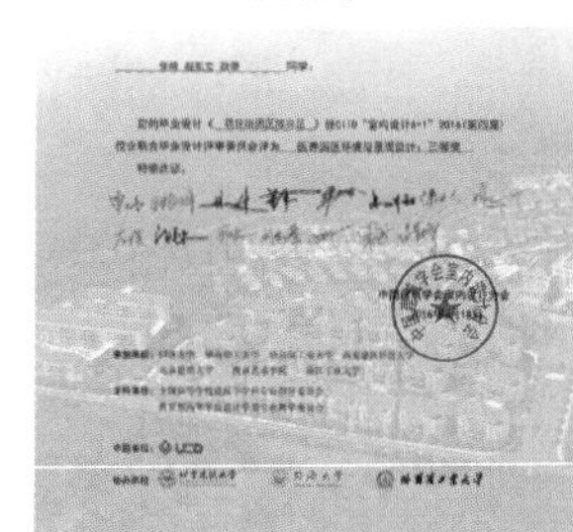

张　峰
Zhang Feng

颜　强
Yan Qiang

赵凯文
Zhao Kaiwen

学生感悟：

如果说毕业季充满了即将步入社会的紧张和师生分别的感伤，那么这次的“6+1”联合毕设之旅便让我们有一种新的情绪和感受：从 3 月的开题相聚，再到 4 月的汇报熟知，最后 6 月的答辩。我们跟 6 所学校的同学和老师共同走过了一段难忘的旅程。我们相信必定是怀揣着共同的目标和理想才让彼此有机会相聚。这次关于养老居住环境改造的命题，也让我们对设计有了新的认识——社会问题和社会需求才是设计者的设计动机和动力。此次毕业设计有遗憾但更多的是成长，是结束同样也是开始。路漫漫其修远兮，吾辈将上下而求索！

Students' Thoughts:

Usually, graduation means feeling nervous about starting a new journey and feeling sad about leaving teachers and classmates. This year's CIID "Interior Design 6+1" University-Business Graduation Projects Competition gave us new feelings. From thesis proposal presentation in March, to mid-term reporting in April, and to defense in June, we experienced an unforgettable journey with students and teachers from six other universities. We believe that it was our common goals and ideals that gathered us together. Through the retirement community renovation project, we realized something new: solving social problems and meeting social needs are the real motives and driving forces for a designer. The project recorded our regrets, but more our improvements. Its conclusion means a new start for us. The road ahead is long, and we will never stop walking forward.

在景观规划上围绕组团和自足的概念展开，满足观赏性和功能性的同时，加强了景观的协同参与性和实用性。从平面上将建筑所处的环境划分了 3 个景观节点，并且用一条环形的园路进行连接；之前建筑规划出来的屋顶空间同样作为景观进行处理。

Plants are grown by seniors in groups according to their needs, so that the plants are appreciated and used as a result of their own efforts. Three landscape nodes are selected on the site plan and linked by a circular garden path. Besides, roofs of the buildings are used to build roof gardens.

总平面图 || Site plan

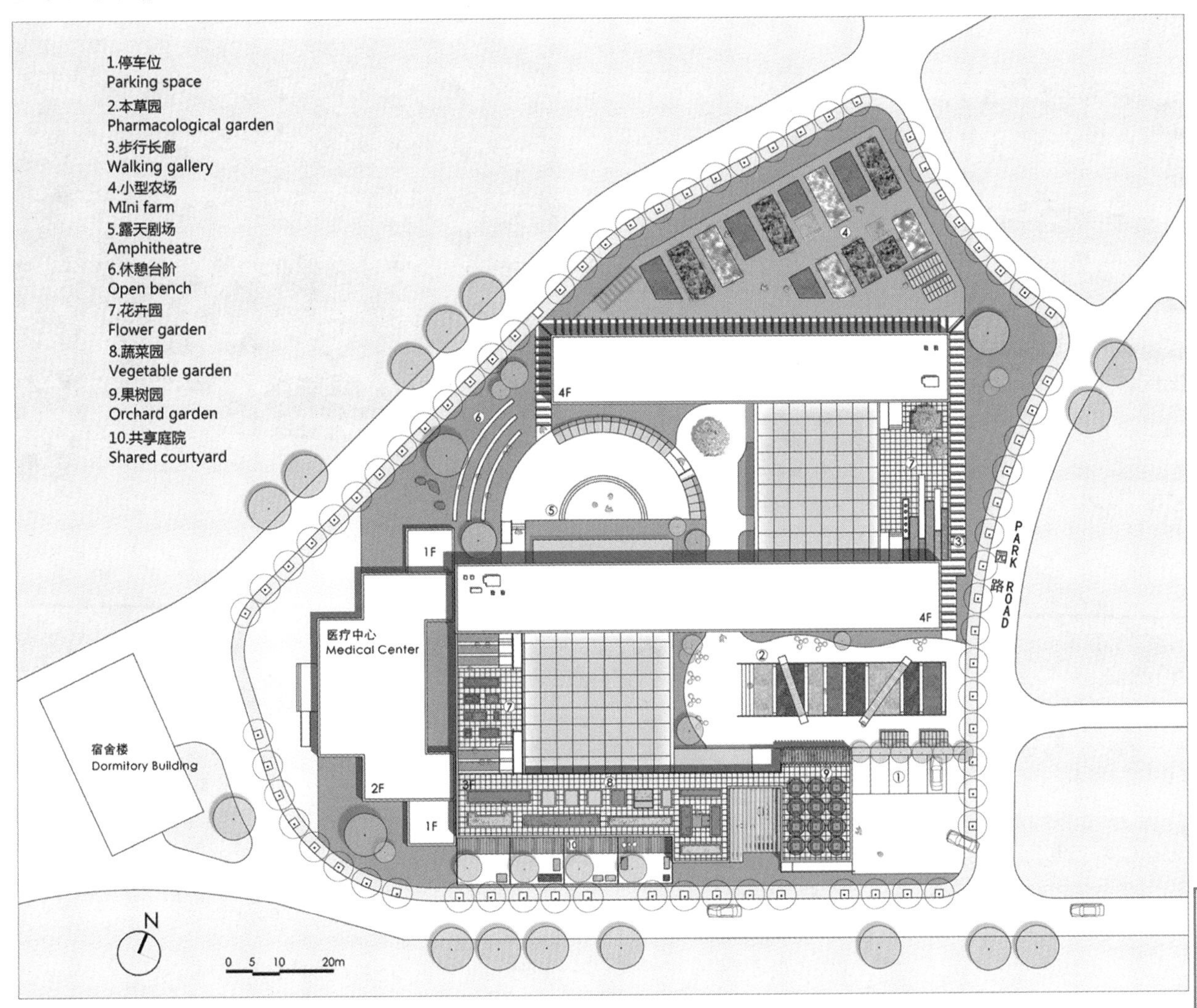

小型农场 ‖ Small farm

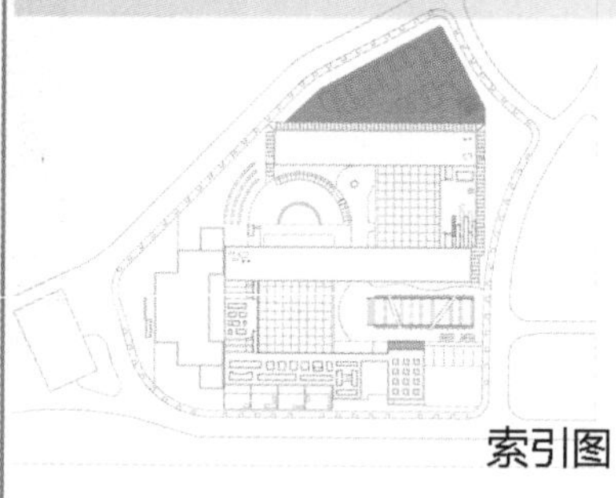

① 廊架
② 枣树
③ 菜地
④ 廊架
⑤ 家禽养殖
⑥ 花池
⑦ 树莓
⑧ 玻璃房

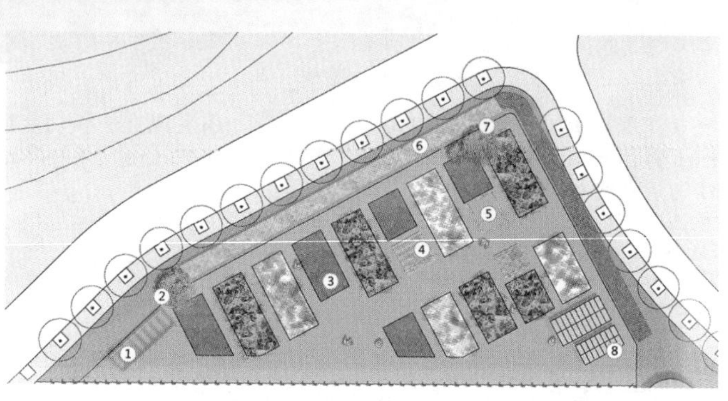

种植分析
Planting analysis

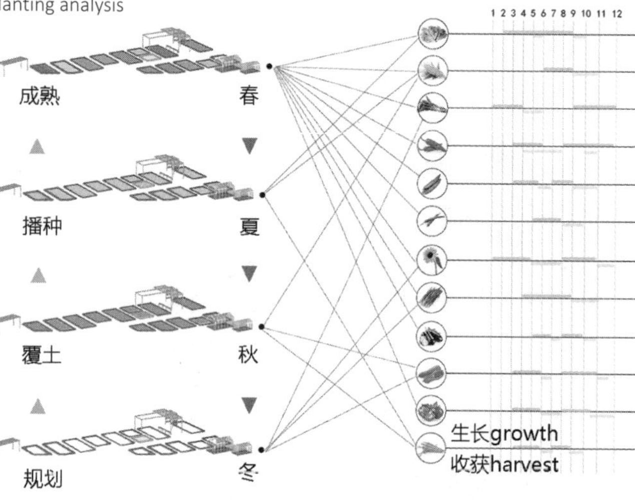

• 春季：第一阶段，2月初播种。首次种植采用玻璃房育苗然后移栽的方法。

• Spring: Seeds are sowed at the beginning of February in a glass nursery, and then transplanted.

• 夏季，蔬菜疯长，密度较大观赏效果佳。雨水增多，病虫害加剧，应做好园区保养工作。

• Summer: Flourishing vegetables are attractive and eye-pleasing. Since ample rain will lead to plant diseases and insect pests, control measures should be taken.

• 秋天是趣味性最强的季节可组织采摘类科普活动，吸引老人的家人和社会大众。

• Autumn: Harvesting and science communication activities can be organized to attract the families of the occupants, and the public.

• 冬季，可替换为可露天越冬的蔬菜品种，只需定期除草施肥，灌溉。

• Winter: Winter vegetables can be grown with periodical weeding, fertilization and irrigation.

本草园 || Herb garden

索引图

一个长 33m、宽 8m 的花园的中央，8 种不同药用植物整齐的排列在一个下沉的种植池中。比如说端午节用来洗浴熏蒸的艾草等等。平时老人种植和管理该花园，在开展活动中也可供孩子和访客来参与种植和学习认知。两条轻质栈道横跨了花园，提供和植物亲密接触的机会。

Eight types of herbs (e.g. wormwood which is used to prepare bath water in the Dragon Boat Festival) are grown in a sinking planting bed in the center of a 33 m x 8 m garden. The garden is managed by the seniors, and visitors like children can also come to grow plants and learn relevant knowledge. Two light plank roads stretch across the garden, shortening the distance between people and plants.

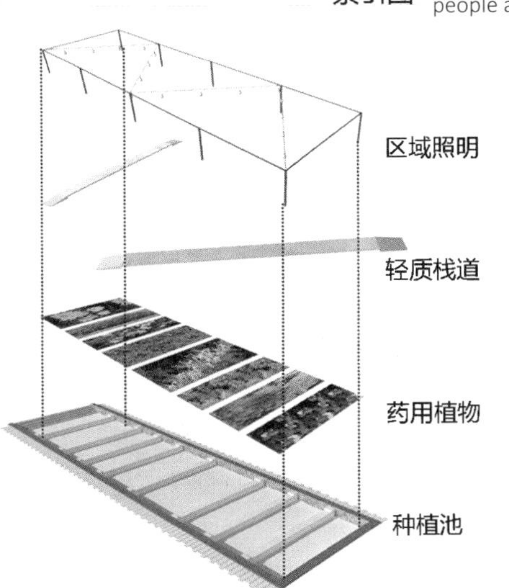

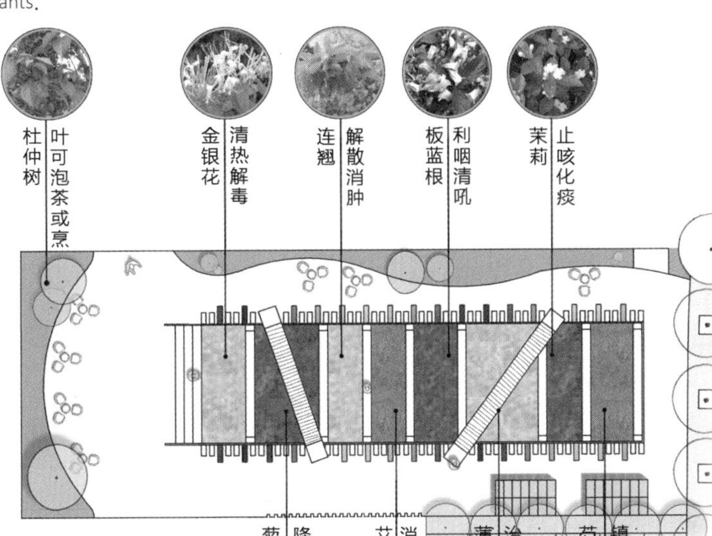

花园一周安置了钢架和吊灯可以在使用过程中设置护栏来提高安全性。

Steel frames and droplights are arranged around the garden. Fences can be erected to ensure the safety.

铜牌记录着植物的具体信息。让花园如同一本记载着中药的古老书籍。

Copper plates record the information of various plants, making the garden a book of herbs.

屋顶花园 || Roof garden

屋顶花园一共划分了 3 个活动和休憩空间：花卉园、蔬菜园和果树园。供于老人不同的选择。弥补了公寓室外活动空间面积的局限性。

Each roof garden is divided into three areas for activity and rest: floral area, vegetable area and fruit tree area, for the seniors to choose. The roof gardens create more spaces for outdoor activity.

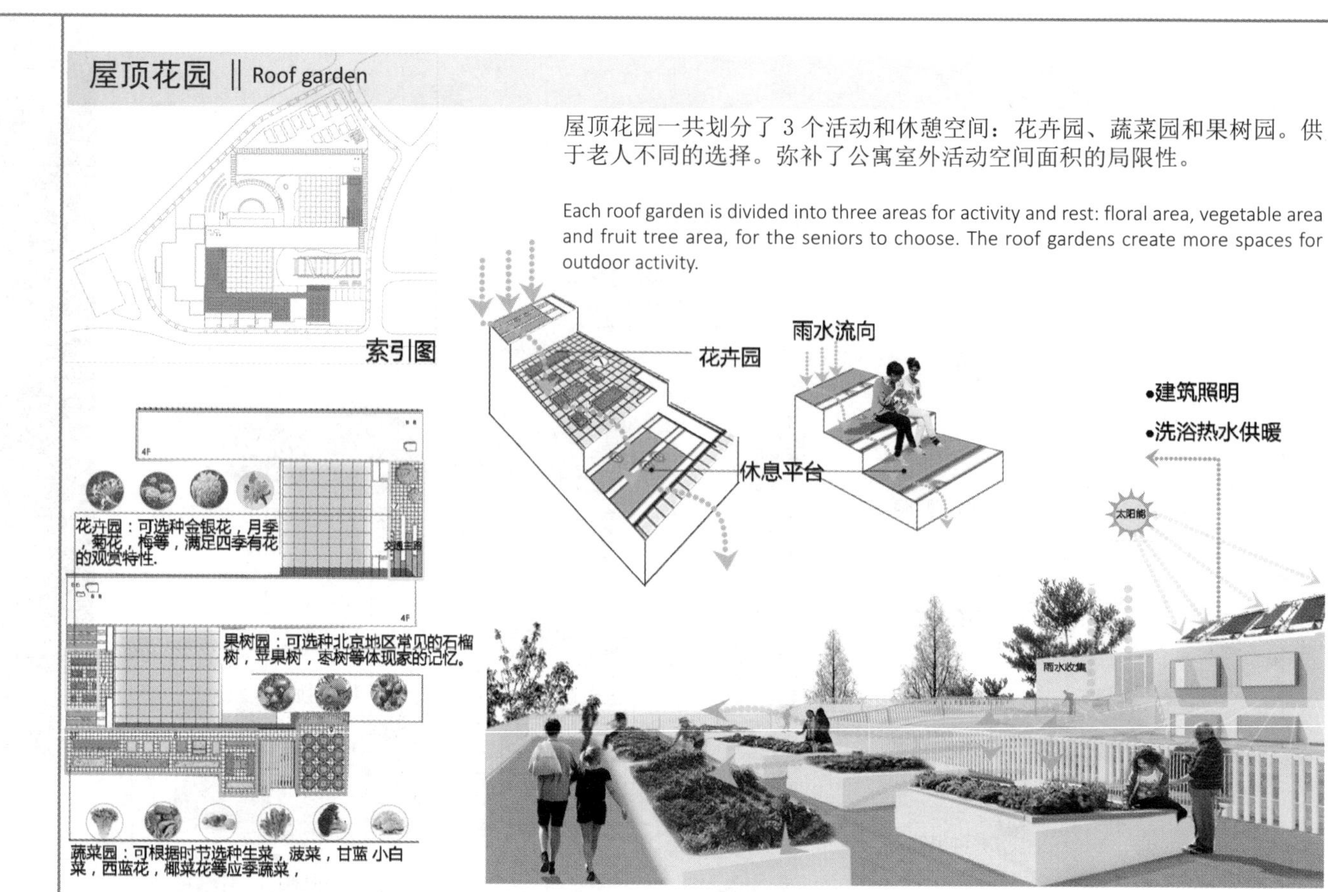

露天剧场 + 行走长廊 || Open-air theater + outdoor corridor

一个半圆的小广场为老人提供了室外活动的场地。该处拥有很强的关注度和聚会氛围。
A small semicircular square is designed to meet the need of the seniors on outdoor activity. The square is very attractive to the seniors, and is an ideal place for gathering.

索引图

150m 的行走长廊围绕着建筑和绿地将本草园和露天剧场连接起来，为老人在饭后或清晨提供一个散步的长廊。
A 150 m long outdoor corridor runs around the buildings, linking the Herb Garden with the open-air theater. The seniors can take a walk along the corridor in the morning or after dinner.

地面的铺装用舒缓视觉的色彩点缀。
Materials with comfortable visual effects are used to pave the square and corridor.

导师点评：

这个作品在思考老年人应该居住在怎样的地方？他们之间的关系怎么样？他们群体行为的独立性和相互心理的交融性的关系。学生们通过对老年人的行为和心理探索过程来转译这个特殊体验空间的逻辑关系，强调老年人优雅、独立生存的重要性，表现了超出一般的“世外桃源”。

整个作品系统完整，表达成熟。从建筑改造、庭院景观设计、室内设计、平面导向设计到药物园植物配置、阳光利用、雨水回收等各方面的探索，无不体现学生们对老年人的关注和爱护，以及源于对自然的观察和感悟，也体现了学生们扎实的专业基础和融会贯通的学习能力。

朱飞

Advisor's Comments:

The scheme tries to answer some questions: What's the living environment that the seniors really want? What's the relationship among the seniors? What's the relationship between behavior independence and the need for communication of the seniors? Through probing into the behavioral and psychological characteristics of seniors, the designers translated the logic relationship of a special experience space. Stressing on the importance of an elegant and independent lifestyle of seniors, the scheme creates an "Arcadia" for the seniors.

The scheme features high completeness and mature expression. From architectural renovation, yard landscape design, interior design, and traffic flow design, to the plant selection, sunlight exposure and rainwater harvesting of the herb garden, we can see the students' care and love for seniors, their observation and understanding of nature, their solid basic professional knowledge, and their ability to apply different kinds of knowledge in design.

Zhu Fei

专家点评：

从完成作品中可以看到，同学们深入地思考了群居老年之间生活交流的心理和行为特点。从前期调研、概念的建立、解题思路的推导、功能分析的细节表述和最终作品的绘制等各个方面，让整体方案设计的主题和要点能够逐层次地展示出来，并且创造了一种针对老年人群的具有人文关怀特色的小组团居住模式，值得大家认真品味和思考。

卓培

Expert's Comments:

From the scheme we can see that the students have an in-depth study of seniors' desire and need for communication, and their behavioral characteristics. From early-stage survey, conceptual framework formation, filling of the conceptual framework, detailed expression of functional analysis results, and the preparation of the final documents, the theme and highlights of the scheme are gradually specified and become more and more clear. The idea of group living is human-oriented, and is worth further thinking.

Zhuo Pei

遵從老年人居住養老的設計原則，開展將本地建築的地方文化氣息與室內設計相結合的探討。將C區改造成適宜老年人生活娛樂和文化交流的功能空間；基於老年人的生理心理和行爲特點，分別進行更新功能空間區劃、動線設計、室內設計、綠色環保設計等適合老年人生活及康復的功能空間。

零伍

05

醫養建築改造與室內設計

Reconstruction and interior design of medical and support buildings

CIID“室内设计 6+1”2016（第四届）校企联合毕业设计

CIID"Interior Design 6+1"2016(Fourth Session)University and Enterprise Joint in Graduation Design

鹤发医养卷

——北京曜阳国际老年公寓环境改造设计

White hair volume medical support

—Beijing Yao Yang International Apartments for the elderly environmental reconstruction design

CIID“室内设计 6+1”2016（第四届）校企联合毕业设计

CIID“Interior Design 6+1”2016(Fourth Session)University and Enterprise Joint in Graduation Design

高　　校：　哈尔滨工业大学
College：　Harbin Institute of Technology
学　　生：　李佳楠　巴美慧　李皓　李曼园
Students：　Li Jianan　Ba Meihui　Li Hao　Li Manyuan
指导教师：　周立军　兆翚　马辉
Instructors：　Zhou LIjun　Zhao Hui　Ma Hui
参赛成绩：　医养建筑改造与室内设计组一等奖
Achievement：　Frist Prize for Aged care Building Renovation and Interior Redesign

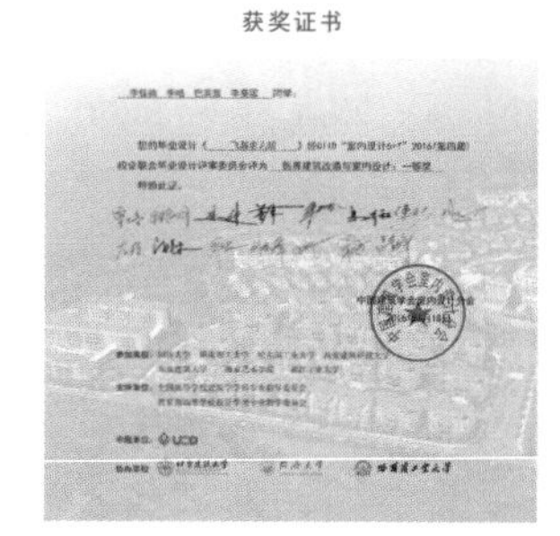
“室内设计6+1” 2016（第四届）校企联合毕业设计

获奖证书

李佳楠
Li Jianan

巴美慧
Ba Meihui

李皓
Li Hao

李曼园
Li Manyuan

学生感悟：

通过本次设计，我们了解了中国养老服务机构的现状，和老年人在养老机构的生活方式及各方面需求，并深入研究了老年人的生理特点及生活习惯，对养老院室内外空间设计提出一些新思路，也为目前存在的问题提出一些解决办法，意图为老年人提供更加贴心的照顾与呵护。

社会对老年人与养老院存在诸多偏见，认为老年人应该静养不宜多动，养老院则是像医院病房一样的存在。我们通过设计表明了自己的观点，即养老院应该为老年人营造家一样的归属感和邻里空间感。医养建筑的“养”不是疗养，而是提供“养”分使其身心更加健康。

Students' Thoughts:

During the survey stage, we probed into the current situations of China's retirement homes, and the various needs, physiological characteristics and lifestyles of seniors. Based on the survey results, we rolled out some new ideas about the interior and exterior spatial design of the project, and provided solutions to some current problems, for the purpose of providing more thoughtful care to the seniors.

Our society has some erroneous views of seniors and retirement homes, thinking that seniors should exercise less, and retirement homes are like hospitals. We expressed our opinions through our design, that is, retirement homes should be home-like spaces that give seniors the sense of belonging and community. Aged care means not just physical care – metal care is more important for the health of seniors.

概念立意 ‖ Conceptual framework

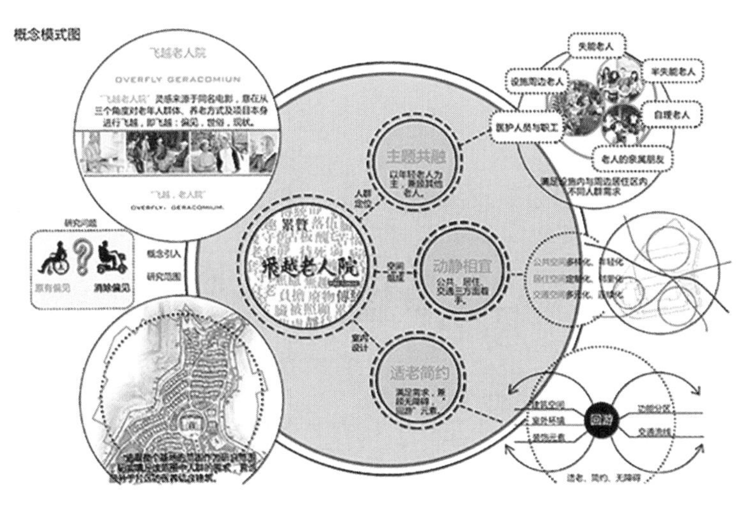

总平面图 ‖ Site plan

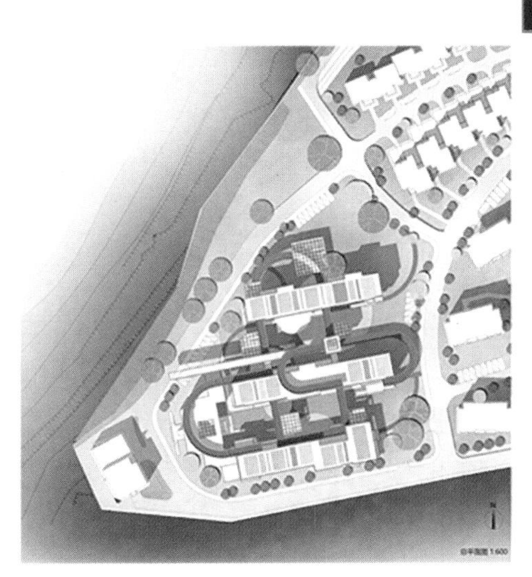

设计主旨

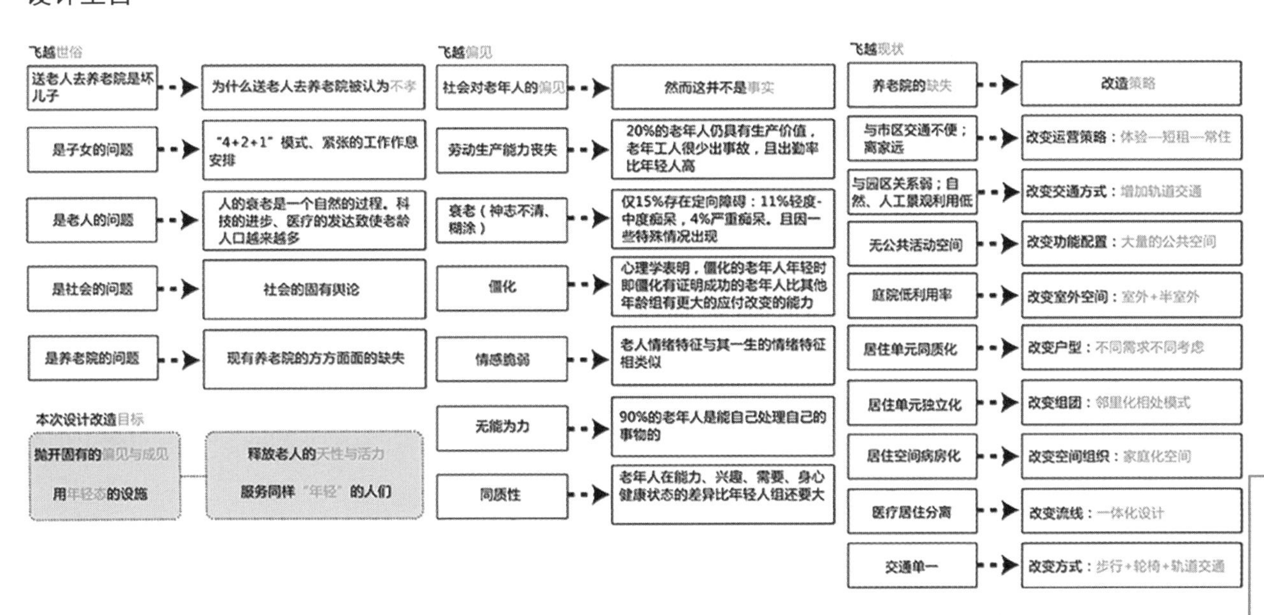

平面图 || Plan

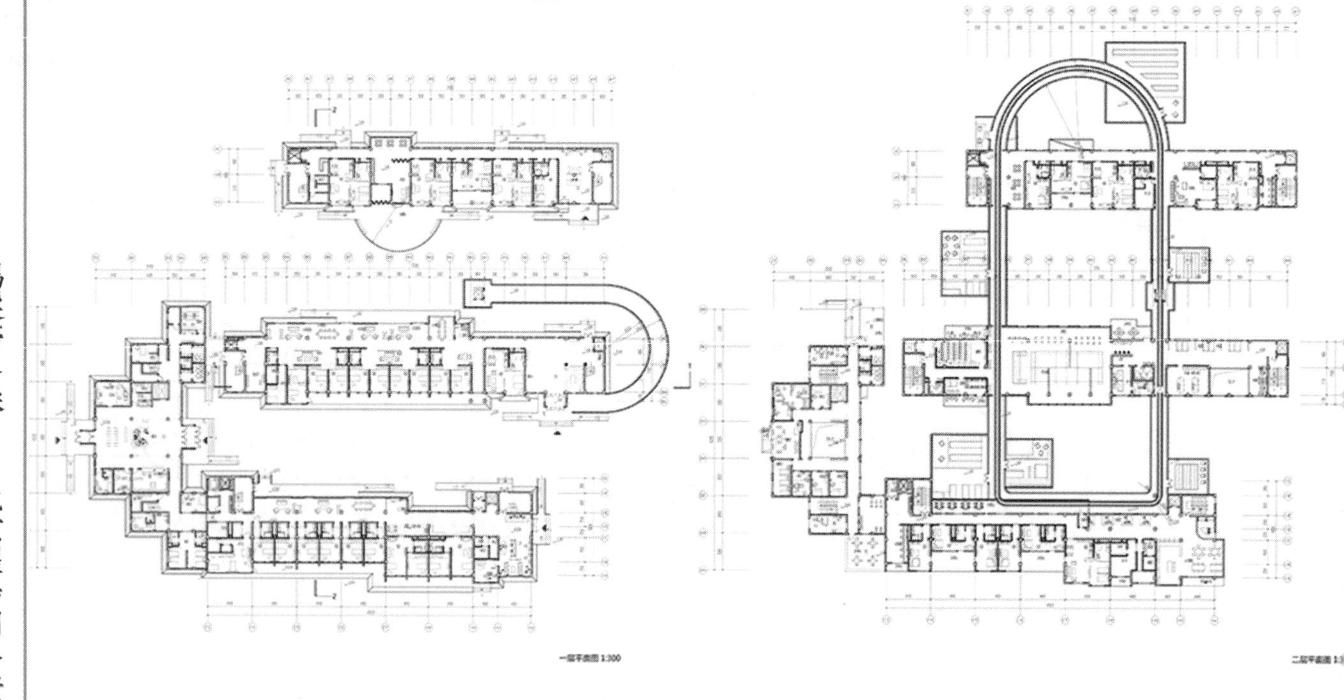

剖面图 || Section

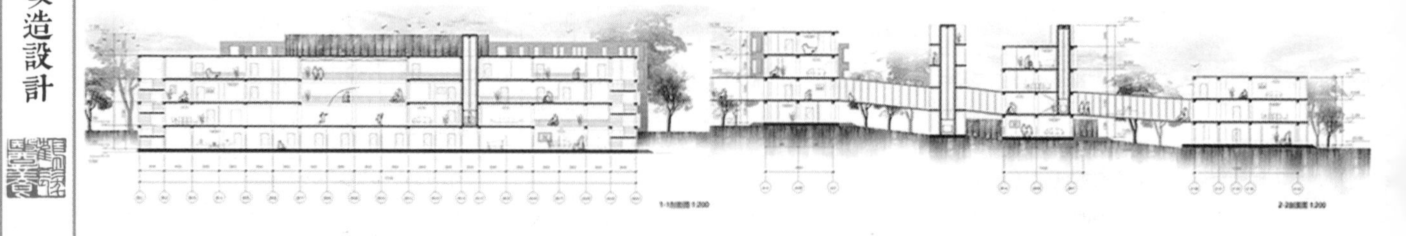

醫養建築改造與室內設計

结构模式的飞越 || Creative structural design

大空间结构的飞越

在B楼二层中设置了一个通高的大空间为本次室内设计中公共空间的重点，尺寸约为1500*1600*9900

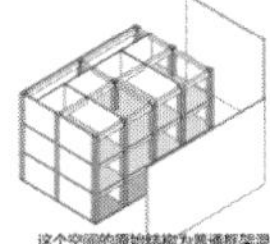

这个空间的原始结构为普通框架混凝土结构，由梁板柱组成其结构体系。结构高度一层为3300mm。

在上图中，除去楼板和山墙，可以清晰的看到这个体量的结构。由柱子和梁搭接而成的框架结构。

在设计中，为了保证大空间的完整性，我们去掉了大空间中间的一颗柱子，以及附着在它上的梁。

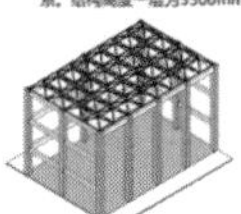

大空间的产生，使得屋顶没有承称结构，因此设计网架结构，来解决因为失去中间柱所带来的影响。

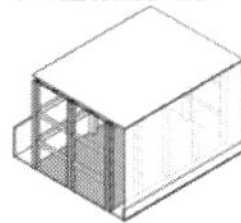

添加屋盖以及侧面的围合结构。在阳面的一侧设置成长玻璃幕墙，给空间增加阳光，并用百叶来调节。

大空间模式的飞越

在解决了结构的问题后，我们首先得到一个完整的大空间，它的周围有一圈柱子，以及两侧各有两个出入口。

在没有任何限制条件下，人们在这个空间中的活动轨迹是随机而无序的。这在一定程度上来说交通破坏了空间的完整性。

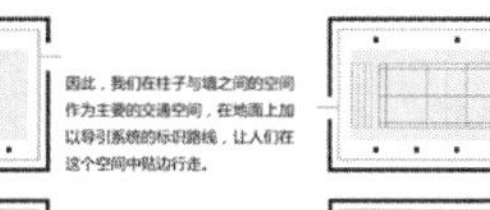

因此，我们在柱子与墙之间的空间作为主要的交通空间，在地面上加以导引系统的标识路线，让人们在这个空间中贴边行走。

当交通让出了中间的空间，可以在其中设置一定的活动，比如羽毛球场，层高正好符合要求，且有空间在一侧设置看台。

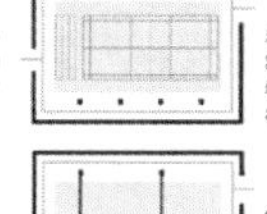

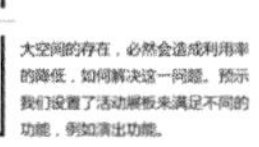

大空间的存在，必然会造成利用率的降低，如何解决这一问题。预示我们设置了活动展板来满足不同的功能，例如演出功能。

将活动展板变化位置，可设置成展览空间，可在大空间中定期布展，供老年人学习欣赏。多种功能的结合让空间增加利用率。

平面布置的飞越 || Creative plan design

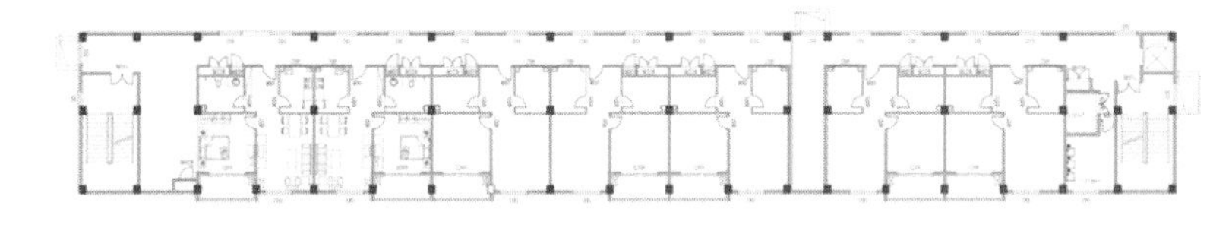

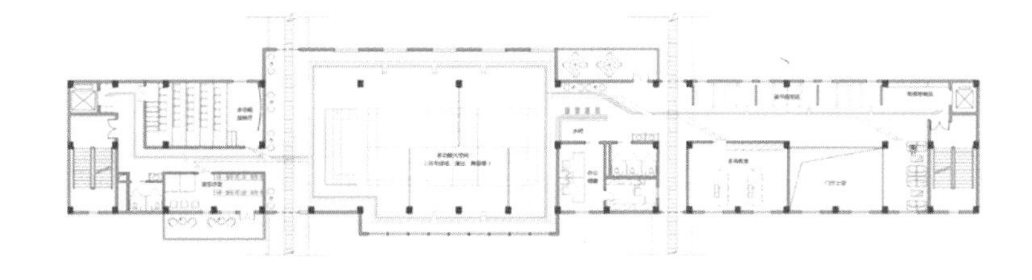

休闲区

阅读空间的飞越 || Creative reading space design

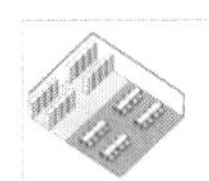

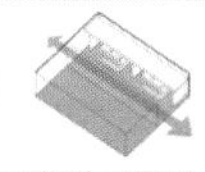

传统：一个屋子两部分，一为书架区域，一为阅读区域。
设计中：将书架和座椅结合，不在另设房屋，与走廊相邻

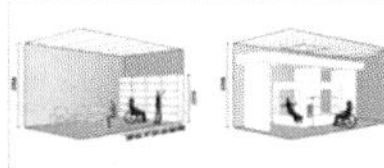

传统：书架高2200mm，需要取书回座位，不安全距离。
设计中：书架高1200mm，便于获取，方便轮椅者使用。

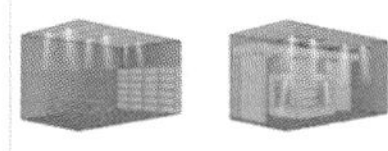

传统：一层灯光照明，灯管离桌面约为2500mm左右。
设计中：三级灯光照明，给予老年人更充足的光照。

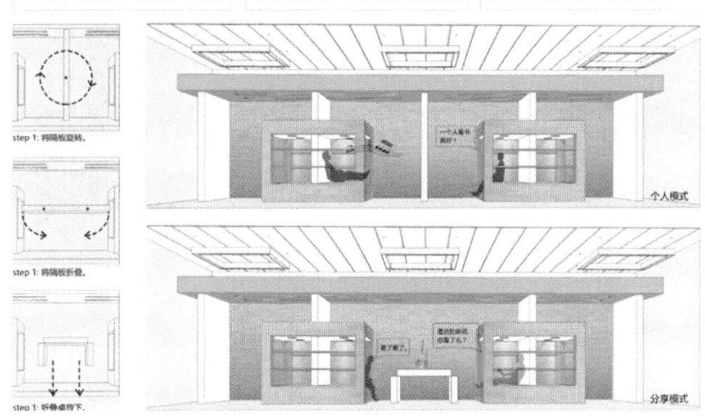

医疗空间的飞越 || Creative medical space design

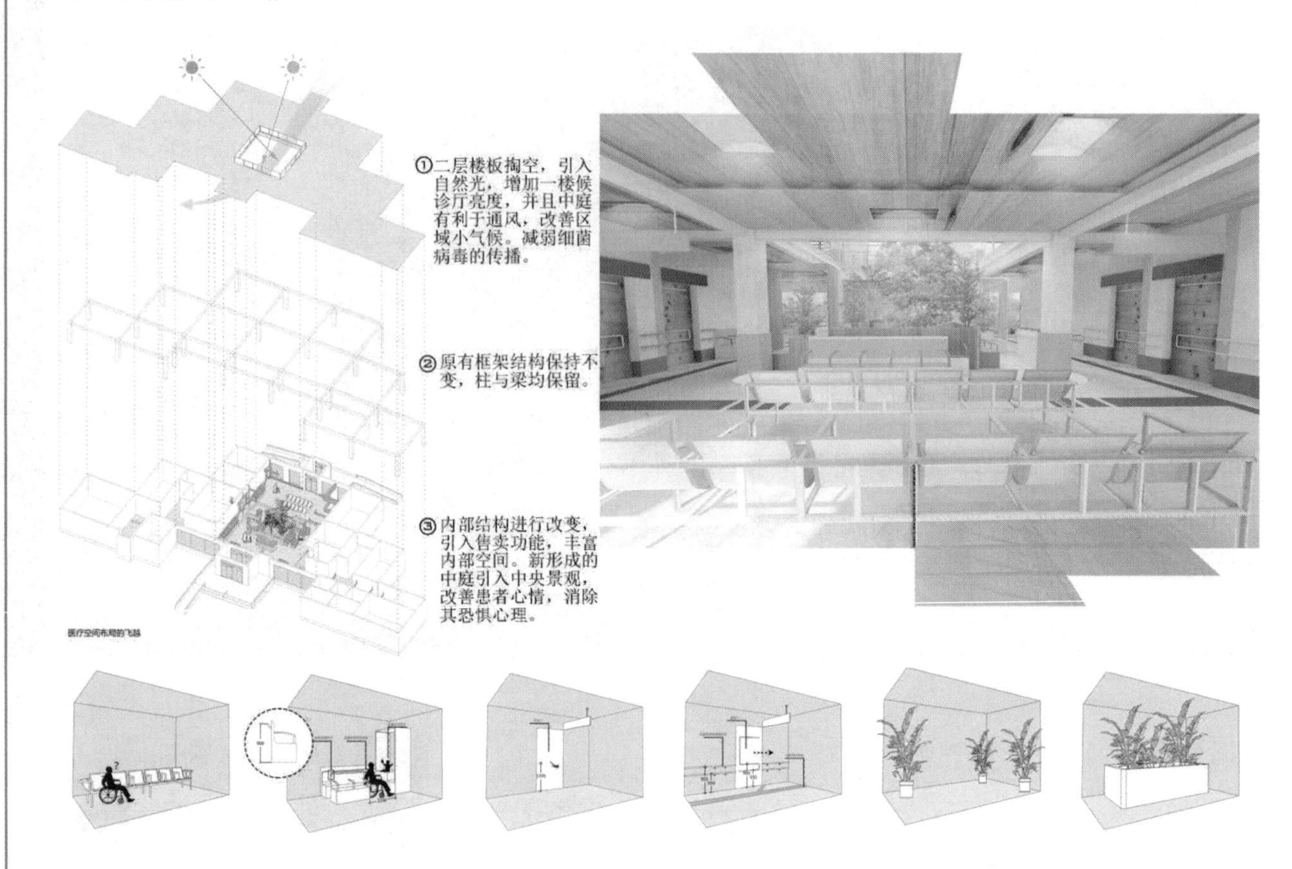

医疗空间布局的飞越

居住空间的飞越 || Creative living space design

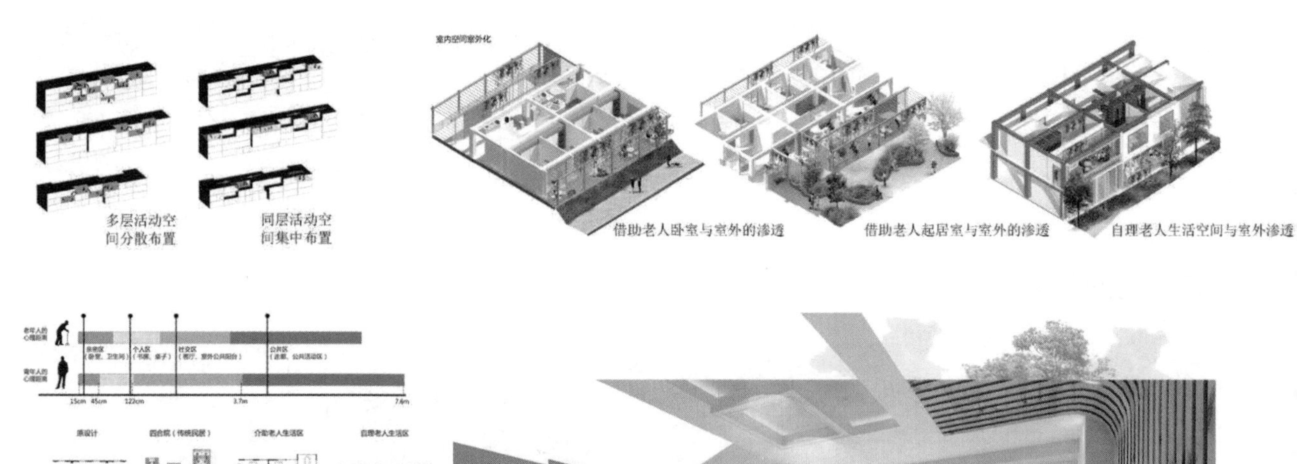

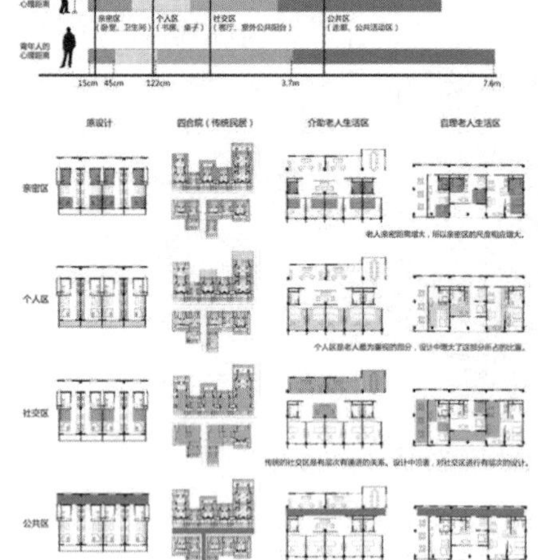

导师点评：

设计作品创意独特，灵感来源于电影“飞越老人院”，试图从多个角度入手进行建筑改造飞越。设计前期，针对使用者需求、场地周边环境、投资者经营模式均进行了较为深入全面的理性分析。

设计有效地将原有建筑进行改造，增加居住部分的采光面积，公共活动空间达到资源整合，整个设计功能布局合理，立面设计深入，对建筑室内大空间设计的多种活动功能模式进行演绎，如体育、舞蹈、展览等多功能，教室也可多用于烘焙、休息、交流等。

针对老年人生活起居所需，通过人体“五感”进行标识设计，并注重老年人的心理健康、室内各处尺寸亦尊重老年人而特别设计，体现设计者的关怀。设计者的细心之处还体现在阅读空间的飞越，老人们不但有相对独立的阅读空间，更多提供了互相交流的平台，设计出闲情雅致的场景。

最后应提到的是，环绕建筑顶部的绿色长坡道提升了整个建筑空间的环境景观效果，为老年人提供健步走的空间以及广阔的视野平台，但从北京的地域性气候考虑，长坡在漫长冬季的实用性还有待深入考虑。

刘晓军

Advisor's Comments:

The concept of the scheme is creative. Inspired by film Full Circle, it tries to, from different perspectives, transform the buildings in an creative way. In the early survey stage, the designers profoundly and rationally analyzed the requirements of the target group, the surroundings of the project site, and the operating mode adopted by the investor.

The transformation of the buildings is successful. Daylight area is increased, public spaces are integrated, spatial functions are rationally redesigned, façade design is specific, and large interior spaces are designed to conduct various activities such as sports, dancing, exhibition, baking, rest, communication and so on.

Signs that can be recognized by the "five senses" are designed to help the seniors in daily life. Besides, the dimensions involved in interior design are determined according to the psychological and physiological needs of the seniors. Care for seniors can be seen in many details.

Creative design of reading spaces also reflects the designers' special attention to details. In addition to independent reading spaces, shared spaces for exchanging opinions are also designed, giving the reading area a comfortable and relaxed atmosphere.

Lastly, the green long slope running around the roofs of the buildings improves the overall landscaping effects. It offers a space for the seniors to walk, and also a rest platform with a broad view. One problem is that given the long and cold winters of Beijing, the feasibility of the slope still needs to be further discussed.

Liu Xiaojun

专家点评：

设计组对于项目地及现有建筑进行分析的同时对国内养老现状也进行详尽的调研，并且以调研成果作为本次设计的基础，体现了严谨的设计流程及对项目的整体控制能力。公共空间设计理念有一定的创新，对空间布局收放有度，公共功能空间安排合理。在区域空间的贯穿链接上大胆的采用了立体的轨道交通的理念，虽然在实际实施过程中会遇到一些困难，但动线规划更加灵活，扩大了老人们的活动空间，提升了入住者的生活品质，最大限度提升了空间的利用率，有效的增加了空间的使用面积。

设计组的发表形式新颖，表现出整体的专业水平与合作精神，充分发挥想象力达到了思想的飞越。设计作品的工作量大，模型和图纸表现完整，是一件高品质的优秀毕业设计作品。

王兆明

Expert's Comments:

The scheme was developed based on a detailed survey and in-depth analysis of project site, existing buildings, and the current aged care situations worldwide. The design procedure is rigorous, and the designers' overall control of the project is successful. Ideas concerning public space design are creative, spatial layout is appropriate, and functional design of public spaces is rational. The idea of connecting regional areas with a sail system is bold, but implementation will not be easy. Traffic flows are designed flexibly that the activity areas of the seniors are expanded, the quality of their life is improved, and space utilization rate is maximized.

The scheme was presented in an innovative manner, which showed the professional level and spirit of cooperation of the design team. I see active imagination throughout the entire project process. A lot of design work has been done, and the models and drawings fully reflect their ideas. I think the scheme is an excellent high-quality design work.

Wang Zhaoming

Memories in Courtyard Houses

CIID“室内设计 6+1”2016（第四届）校企联合毕业设计
CIID“Interior Design 6+1”2016(Fourth Session)University and Enterprise Joint in Graduation Design

高　　校：北京建筑大学
College: Beijing University of Civil Engineering and Architecture
学　　生：王海月 李丽阳 王逸开 张博闻
Students: Wang Haiyue Li Liyang Wang Yikai Zhang Bowen
指导教师：杨琳 朱宁克
Instructors: Yang Lin Zhu Ningke
参赛成绩：医养建筑改造与室内设计组一等奖
Achievement: Frist Prize for Aged care Building Renovation and Interior Redesign

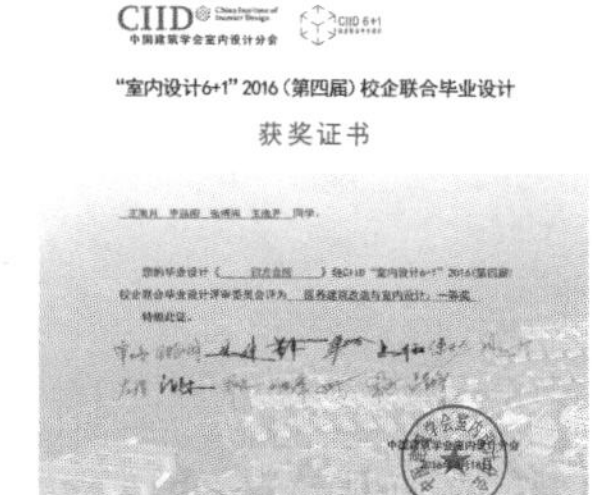

王海月
Wang Haiyue

李丽阳
Li Liyang

王逸开
Wang Yikai

张博闻
Zhang Bowen

学生感悟：

最大的感悟是这次联合毕设的交流带给我们的，在几次交流中看到其他学校对于相同题目的不同想法，大家的想法都很棒，使我们扩展了很多眼界。

同时，这次毕业设计从单纯一个点的想法开始引出，在不断丰富方案的过程中，我们的设计经历了本心——偏离主体的发散（中期）——回归本心完善思考（末期）的纠结过程，指导老师与评委老师给予的意见与建议提醒我们回归到了本心，帮助我们改善了整个设计，对我们来讲是一个非常宝贵的经历。

Students' Thoughts:

During this joint graduation design, I met different ideas from other schools for the same topic in the exchange activities. These great ideas expanded our outlook. At the same time, this graduation design started with ideas for a single point, and our design experienced an entangling process, from original center (early stage), to divergence deviating from the subject (mid stage) and to improving thinking returning to the original center (late stage). With their sincere opinions and suggestions, our coaching counselor and judge experts reminded us of returning to the original center and helped us improve our design. This is absolutely valuable experience for everyone of us.

让我们由老人的一天开始讲起……

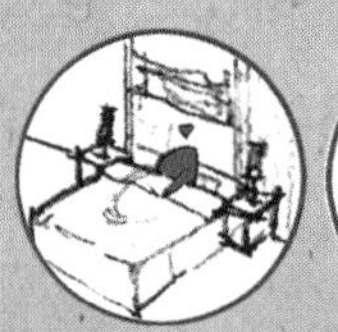

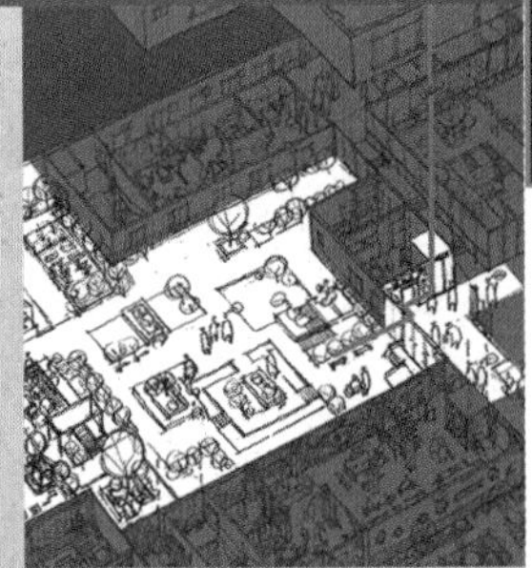

■早上 6 点，张大爷开始了他的一天生活，在床上洗漱完毕后，出门买了早点，回屋，在电视里今日新闻背景下吃完早饭，修整完毕下楼，到小花园，沿着遛弯小道边走边锻炼，欣赏风景，逗逗院子里的鸟。8 点半走时，院子里下象棋的老人们已经开局，比的旗子碰撞，围观的给双方支招。张大爷坐大厅的电梯回屋小憩。

6:00-8:30AM

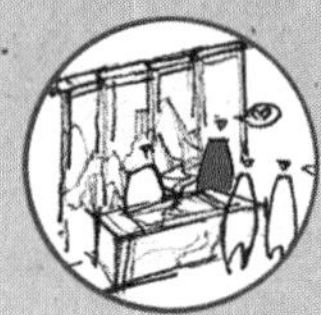

9:00-11:00AM

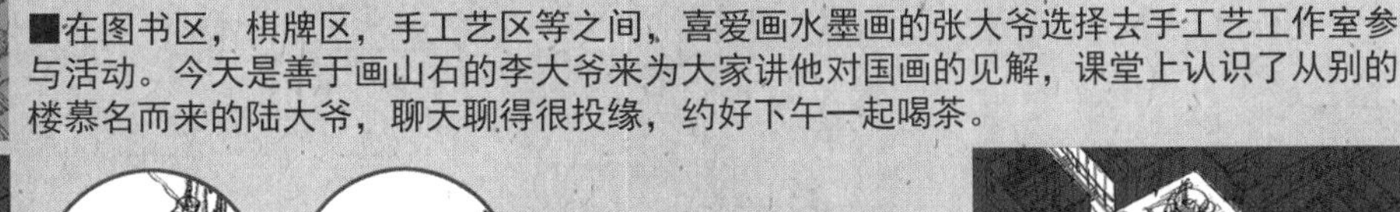

■在图书区，棋牌区，手工艺区等之间，喜爱画水墨画的张大爷选择去手工艺工作室参与活动。今天是善于画山石的李大爷来为大家讲他对国画的见解，课堂上认识了从别的楼慕名而来的陆大爷，聊天聊得很投缘，约好下午一起喝茶。

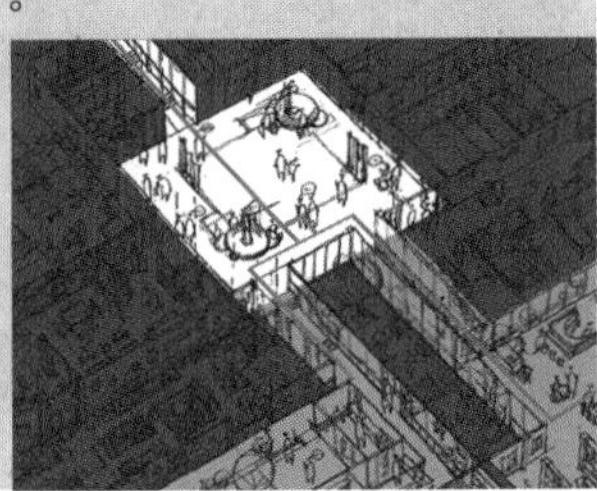

11:00-12:30AM

■从手工艺区出来，在广场碰到从图书馆出来的邻居王大爷，寒暄几句，交流下今天上午开心的事情，一起去餐厅吃午饭。

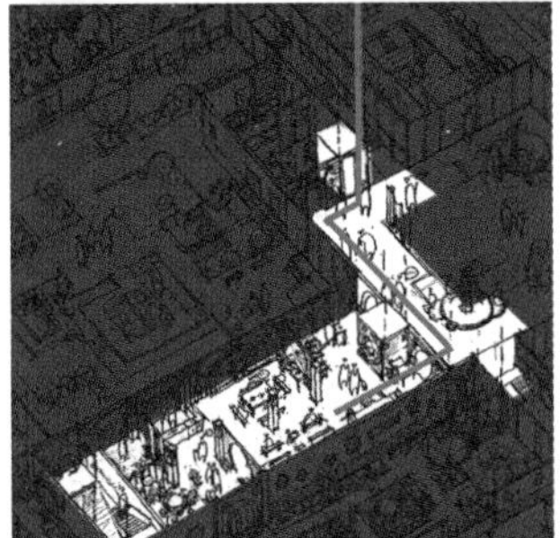

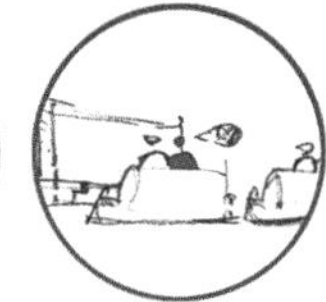

■ 12:30 吃过午饭回屋午休，午休后，下午 3 点张大爷准时来到茶馆与陆大爷见面，聊到关于过去好玩的事，对桌的魏大爷对这个话题非常感兴趣，凑过来聊起了自己的经历。伴着茶馆的京韵大鼓，愉快的下午结束了。

■ 5:00 回去后，张大爷歇了一会，5:30 与同屋的人一起做了一顿丰盛的晚餐，7:00 伴着电视中开讲的新闻联播，几个人围坐在一桌享用晚餐。

■晚饭后，张大爷与同屋同时代的几位老大爷一起看最新的电视剧，聊一些过去的事儿。

■ 10 点左右，大家各自回屋，洗漱睡觉，期待新的一天。

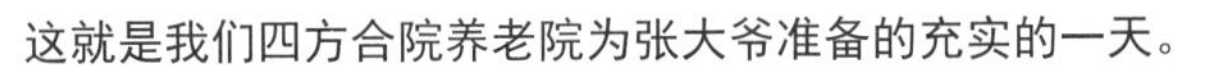

这就是我们四方合院养老院为张大爷准备的充实的一天。

方案推导——由三个方面组成

1 场地分析

■仅住宿功能的楼房建筑，住宿房间与房间单独存在，楼与楼之间也仅有半室外窄廊连接，仅景观在之间缓冲，而景观设计交流成分少，整组建筑带给人很强的孤独感。

■修改方向：
带来一些活力因子，减少孤独感；加入活动空间，增强住户间交流，不仅于住宿功能；住宿之间打破冰冷的竖向隔断，促进互通。

去养老化
+
原建筑
增加活力因子
增强互通

有住宿功能的活力街区
+
老北京情怀

2 地域特质

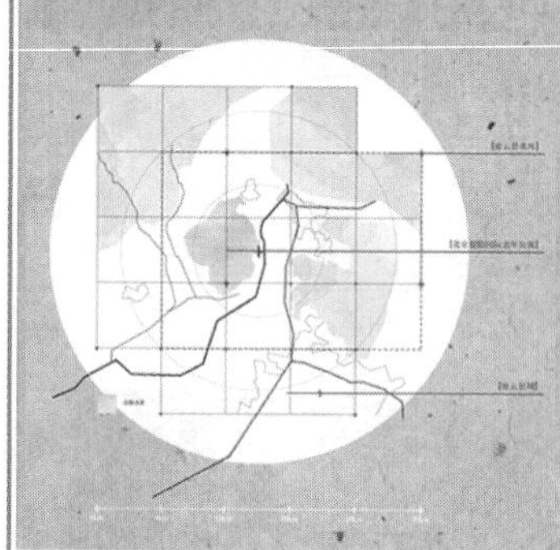

■修改方向：
地处北京，客户群体也为在北京生活的老年人，因此应迎合北京的文化，有北京生活的影子。

修改方向■
去养老化，减少明显的老年公寓制度，缓解老人对于老年公寓的负面心理，减少孤独感，使老人体验新的快乐。

3 现今养老模式的社会分析

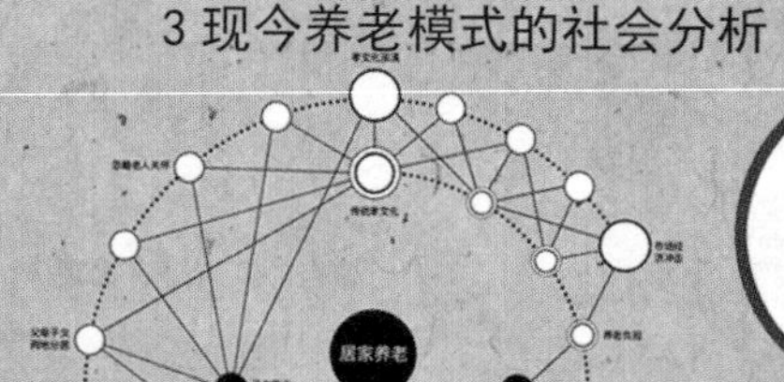

中心理念

合人

合心

合院

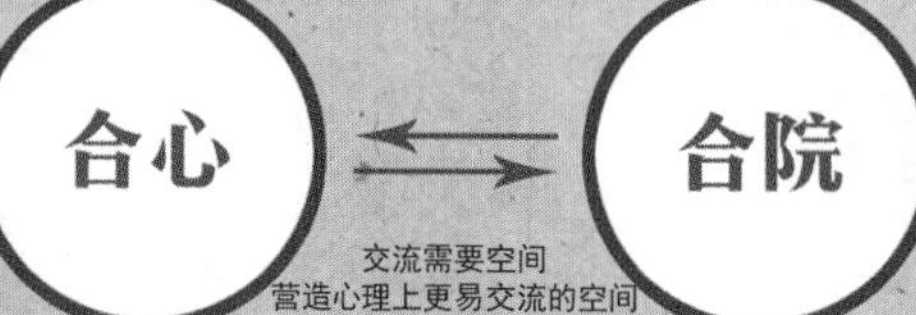

四：文化“四方老人，汇集于此”
方：方位“水平空间，纵向生态”
合：形式“合院大宅”
院：空间“聚宅为院，聚院为集”

彼此相连的合院文化

建筑创造空间，空间聚合人群

■老人不再是孤立的个体，层层集体将其包围，集体中同样爱好、同样经历的人一同找到生活的乐趣。来养老院不等于是一种对自己年岁的审判；应是过来体验家的感觉，找到同样经历的老人一起分享，找到以前想要做的事，找到一起分享故事的人，找到自己曾经的情怀，让自己生活过得更开心的地方。

合院生成

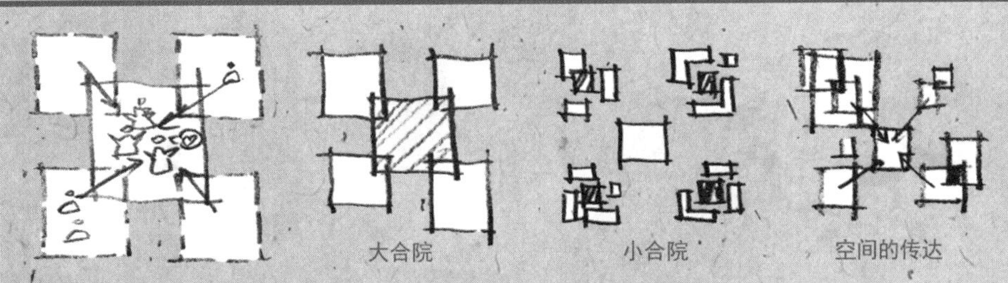

■建筑本身是一个大合院，大合院中又形成各种小合院。
■在空间中建立传递性，促进不同区块中老人的交流，中心区域作为空间传达的交汇点，使空间活动进行新的升华。

■通过空间形制与氛围的引导，鼓励老人从自己的独立空间中走出来，进入到公共的开放空间，创造心理融合的契机。

功能汇集 ——基于老人行为模式调研建立功能空间，汇集老人的爱好范围。

功能整理

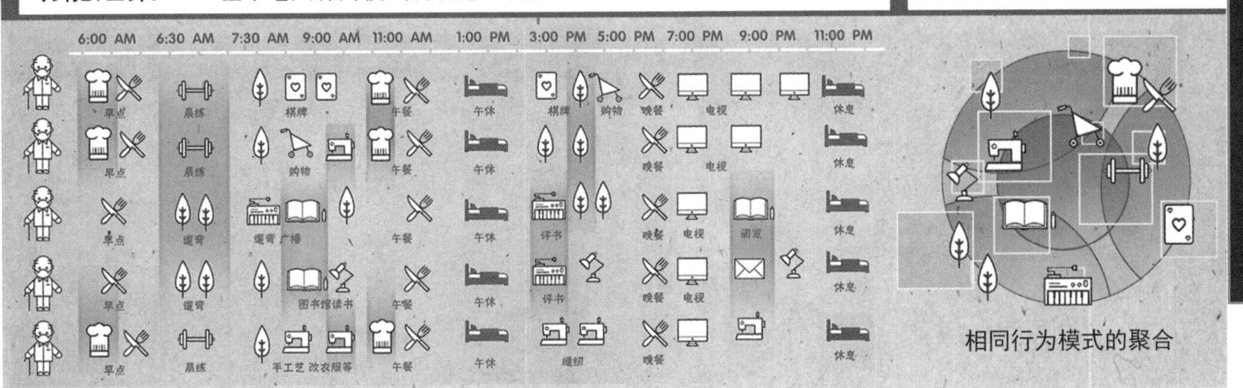

体块划分

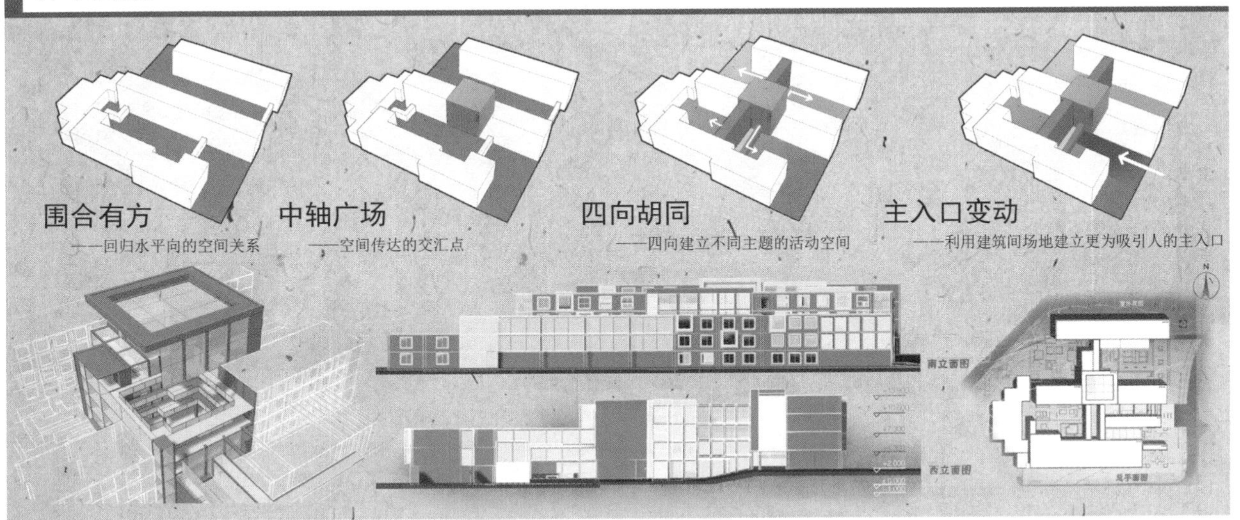

建筑设计

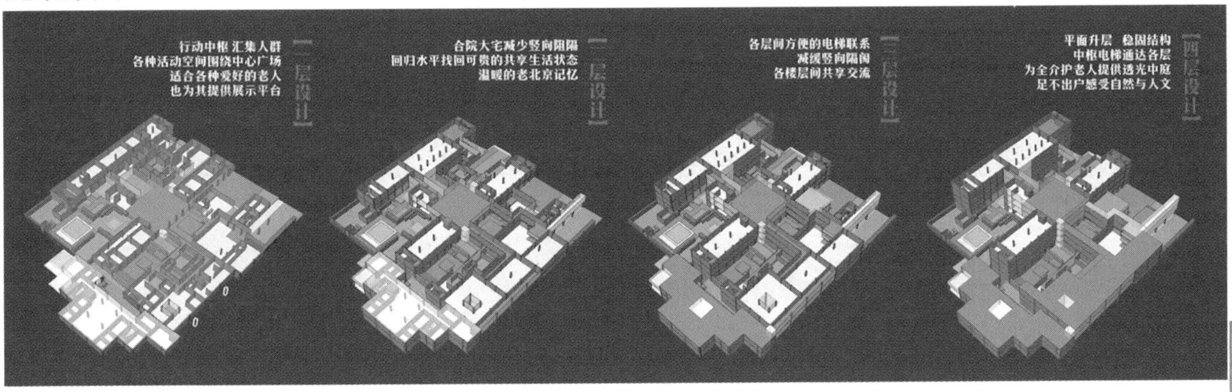

一层设计

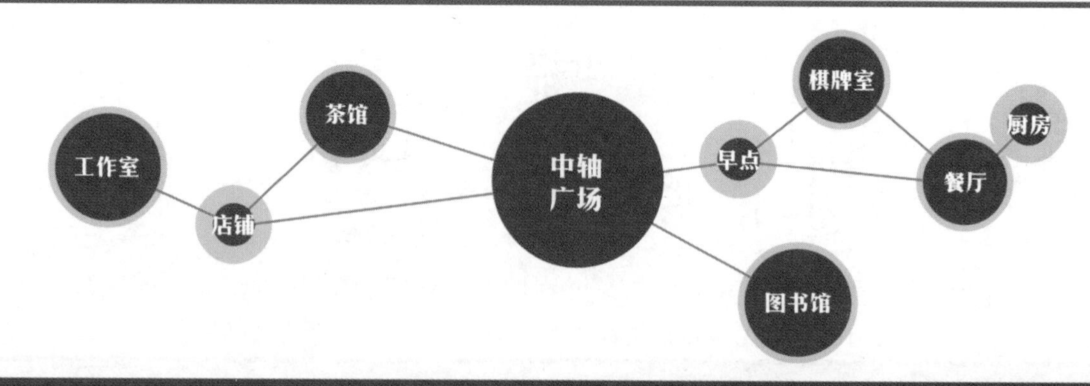

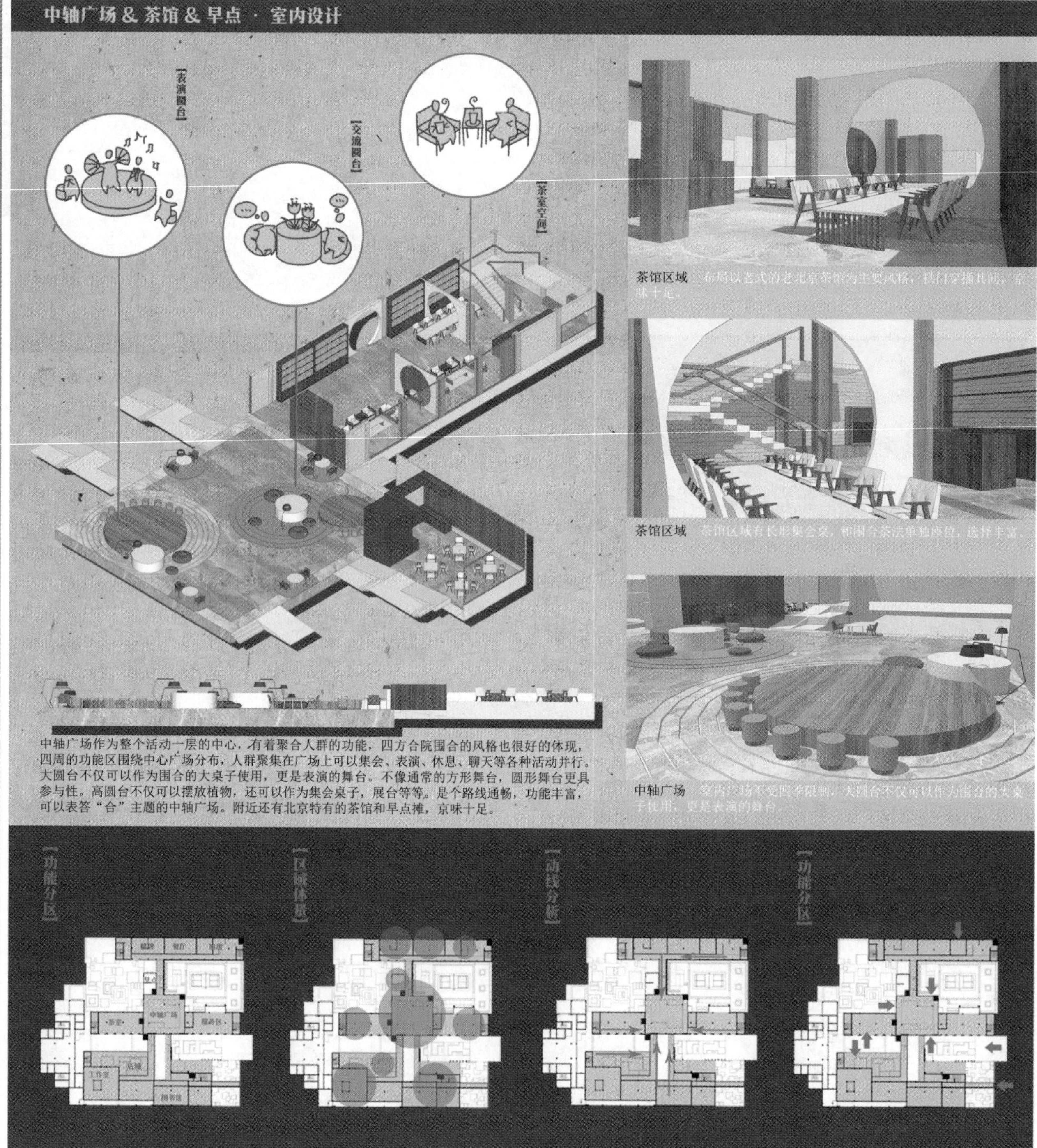

中轴广场作为整个活动一层的中心，有着聚合人群的功能，四方合院围合的风格也很好的体现，四周的功能区围绕中心广场分布，人群聚集在广场上可以集会、表演、休息、聊天等各种活动并行。大圆台不仅可以作为围合的大桌子使用，更是表演的舞台。不像通常的方形舞台，圆形舞台更具参与性。高圆台不仅可以摆放植物，还可以作为集会桌子，展台等等。是个路线通畅，功能丰富，可以表答“合”主题的中轴广场。附近还有北京特有的茶馆和早点摊，京味十足。

茶馆区域 布局以茗式的老北京茶馆为主要风格，拱门穿插其间，京味十足。

茶馆区域 茶馆区域有长形集会桌，和围合茶法单独座位，选择丰富。

中轴广场 室内广场不受四季限制，大圆台不仅可以作为围合的大桌子使用，更是表演的舞台。

认领店铺
工作室
小展示
手工艺工作室 & 店铺 · 室内设计
书籍阅读
报刊阅读
阅读自习
书籍阅读区 可供老人借阅书籍，高出的台阶可拿到更高的书籍，也有软垫可以坐下。
自习阅读区 有围合的区域可坐下阅读，也有单个的自习区可独自阅读，选择丰富。
图书馆包含三个功能，分别为报刊阅读、图
出口处安排咨询处，不仅可以办理借阅
更新书籍的同时，也为老人推荐好
椅的形式很多样化，由老人自主
书借阅以及自习区域，为老人提供更多选择。更人性化的在
手续，还可以随时为老人提供帮助。中央还有新书展示区，随时
书，丰富老人的精神生活。书架的高度适中，方便老人取阅。阅读座
选择。当然也有体现"合"主题的环形座椅，方便老人交流读后感想。
综合图书馆 · 室内设计

二层及以上设计 ——合院大宅

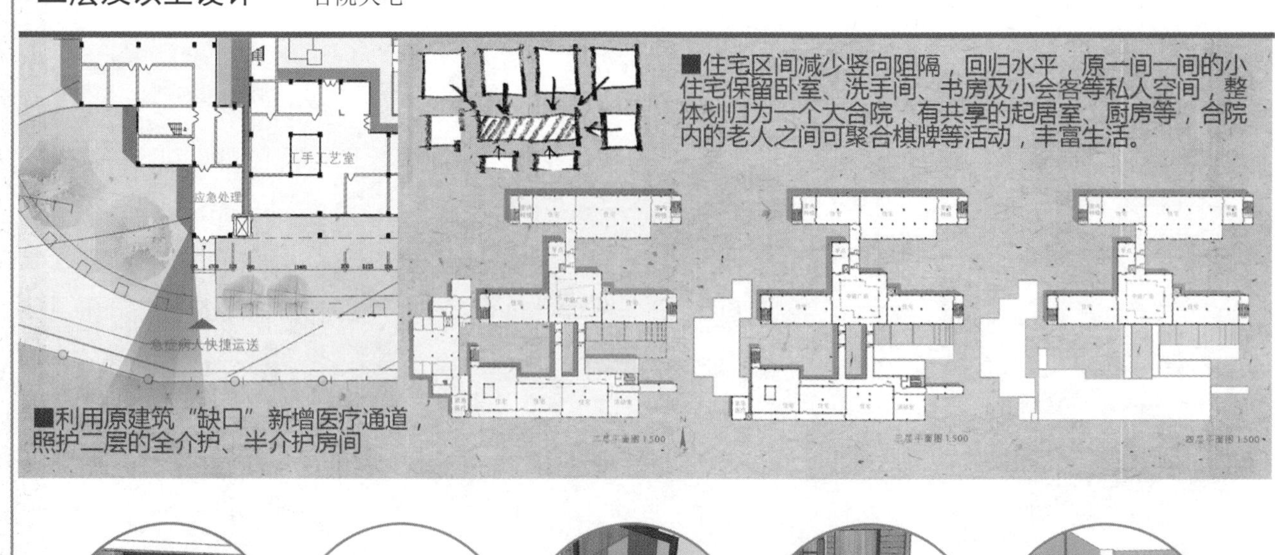

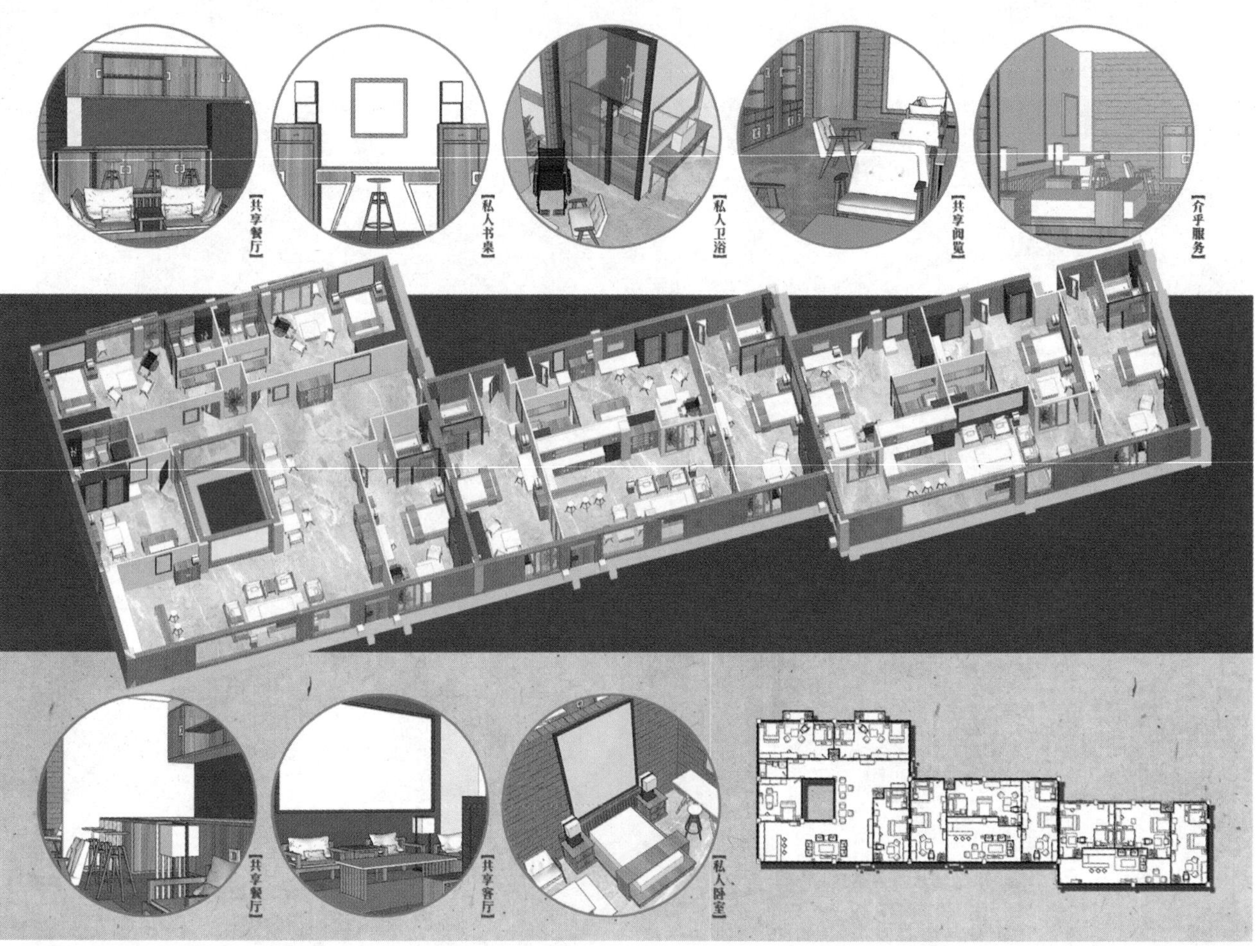

合院式集体住宅 · 室内设计

导师点评：

“四方合院”设计者以老北京文化为场所记忆，在对老年人心理行为分析的基础上对场地空间功能梳理、汇集，对传统老年疗养院进行了重新定义，在既有空间格局中植入合院、胡同式载体，形成了既融合又独立，富有趣味性内部空间的集合式疗养院。整个方案概念表达逻辑清晰，形式与功能相对统一，空间组织聚散富于变化，成果呈现系统完整。特别是在适老性设计及人性空间设计上做了较为精彩的解答，如动线组织上减少竖向阻隔，以水平流线为主；在大空间中穿插如茶馆、大舞台这样一些充满京味的公共空间等。如在概念深化，细节处理等方面进一步思考优化，则整体方案将更具说服力及合理性。

黄炎

Advisor's Comments:

Based on an analysis of seniors' psychological and behavioral characteristics, "Memories in Courtyard Houses" tries to integrate and re-divide the spaces of the project, and redefine the functions of retirement homes. Courtyard houses and hutongs are incorporated into the existing spatial layouts, creating connected but also independent interesting group living spaces. The scheme features clear and logic expression, consistency between forms and functions, flexible spatial organization, and systematic and complete results presentation. Senior-oriented spatial design is the highlight of the scheme. For example, the locations of barriers on traffic flows are changed so that they are parallel to the flows, and public spaces with the cultural characteristics of Beijing like teahouses and performance stages are arranged in large spaces. With more specific concept elaboration and further detail design, the scheme will be more convictive and rational.

Huang Yan

专家点评：

1. 梳理研究并审视养老模式的既有问题，建立以（老人）“行为类别”为基础的设计逻辑；
2. 归纳诸行为的类型，依此界定能分别承载这若干类行为的“公共子空间”群；
3. 以这样的经过“群化”的子空间为节点，重组并达成总体的养老系统；
4. 同样，以这样的子空间为节点，叠加并达成总体的营运与服务系统；
5. 以上的设计逻辑，使空间不再抽象，改建不再“想象”，界面不再“风格”，功能不再“飘浮”。
6. 从此方案中人们能感受到：“为他人”而非“我认为”的设计；“为具体人群”而非“实现我的理想”的设计；
7. 此方案较为缜密、周到；较为别致、简捷；另外，具有“健全、平衡”的时代气息。

赵健

Expert's Comments:

1. The scheme designs based on a detailed analysis of the problems existing in current aged care models, and the behavioral characteristics of "different types of seniors".
2. "Public sub-spaces" are created according to the behavior of different types of seniors.
3. An aged care system is built by taking various public sub-spaces as spatial nodes.
4. Likewise, an operating and service system is built based on such public sub-spaces as spatial nodes.
5. The above design concepts ensure formative spaces, targeted redesign, useful boundaries and practical functions.
6. The designers think and design from the perspective of and also for the users.
7. The scheme is meticulous, thoughtful, unique and compact. It's also "healthy and balanced", reflecting the characteristics of our times.

Zhao Jian

Regional Self-Sufficient Life in Groups

CIID“室内设计 6+1”2016（第四届）校企联合毕业设计

CIID“Interior Design 6+1”2016(Fourth Session)University and Enterprise Joint in Graduation Design

高　　校：西安建筑科技大学

College:　Xi`an University of Architecture and Technology

学　　生：张峰 颜强 赵凯文

Students:　Zhang Feng　Yan Qiang　Zhao Kaiwen

指导教师：刘晓军 何方瑶

Instructors:　Liu Xiaojun　He Fangyao

参赛成绩：医养建筑改造与室内设计组二等奖

Achievement: Second Prize for Aged care Building Renovation and Interior Redesign

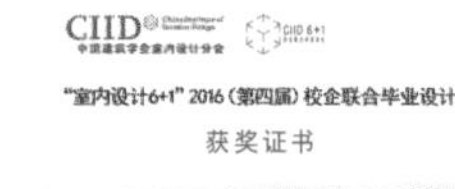

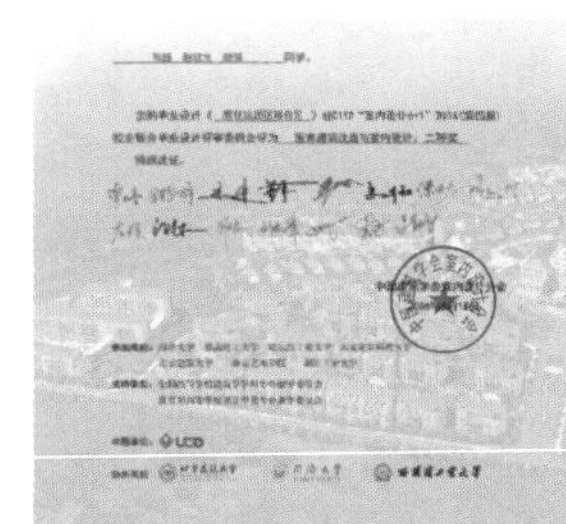

张　峰

Zhang Feng

颜　强

Yan Qiang

赵凯文

Zhao Kaiwen

学生感悟：

如果说毕业季充满了即将步入社会的紧张和师生分别的感伤，那么这次的“6+1”联合毕设之旅便让我们有一种新的情绪和感受：从 3 月的开题相聚，再到 4 月的汇报熟知，最后 6 月的答辩。我们跟 6 所学校的同学和老师共同走过了一段难忘的旅程。我们相信必定是怀揣着共同的目标和理想才让彼此有机会相聚。这次关于养老居住环境改造的命题，也让我们对设计有了新的认识——社会问题和社会需求才是设计者的设计动机和动力。此次毕业设计有遗憾但更多的是成长，是结束同样也是开始。路漫漫其修远兮，吾辈将上下而求索！

Students' Thoughts:

Usually, graduation means feeling nervous about starting a new journey and feeling sad about leaving teachers and classmates. This year's CIID “Interior Design 6+1” University-Business Graduation Projects Competition gave us new feelings. From thesis proposal presentation in March, to mid-term reporting in April, and to defense in June, we experienced an unforgettable journey with students and teachers from six other universities. We believe that it was our common goals and ideals that gathered us together. Through the retirement community renovation project, we realized something new: solving social problems and meeting social needs are the real motives and driving forces for a designer. The project recorded our regrets, but more our improvements. Its conclusion means a new start for us. The road ahead is long, and we will never stop walking forward.

建筑空间构思 || Architectural space design concepts

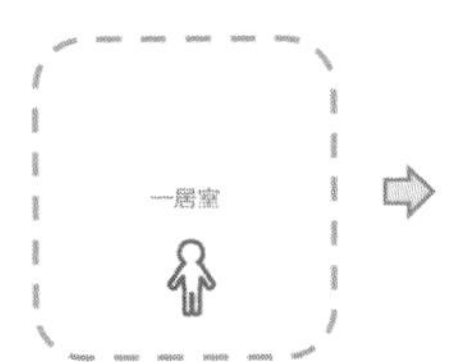

户型以一居室和单间为主
One bedroom apartments and single room apartments are the main house types.

卫浴 起居室 卧室 公共空间

将餐厅，阳台厨房作为公共空间
Turn restaurants, balconies and kitchens into public spaces.

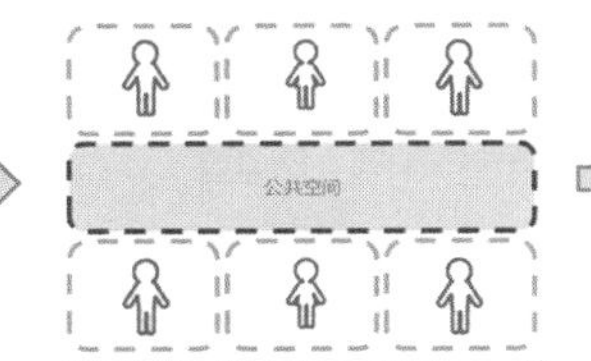

使整个楼层呈现出一种家庭式的生活模式
Create a family-like living model for occupants living on the same floor.

组团的模式，将各个住户联系起来
Groups are created to connect the occupants.

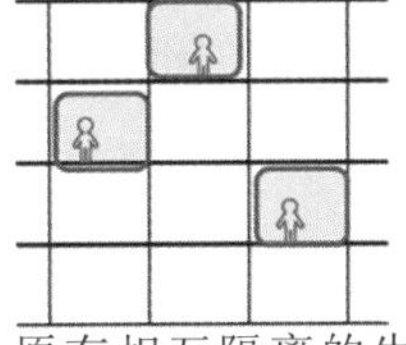

原有相互隔离的生活模式
Originally, the occupants live separately with little communication.

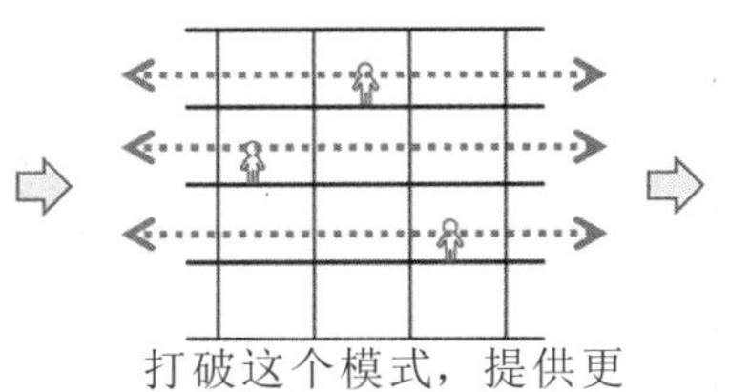

打破这个模式，提供更多的交流机会
We want remove the separation barrier, promoting the communication among the seniors.

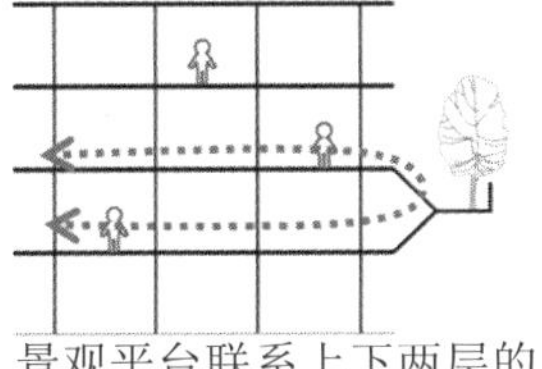

景观平台联系上下两层的沟通
The view platform creates a space of communication for the occupants living on two floors.

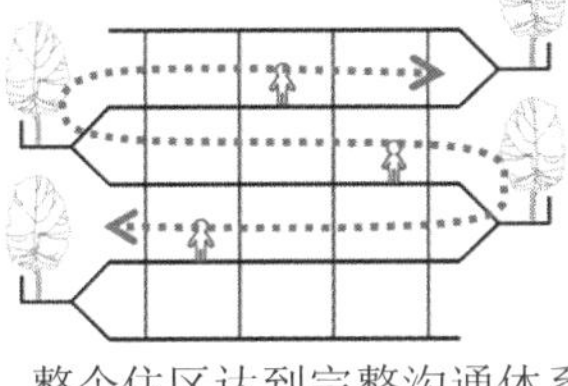

整个住区达到完整沟通体系
Our goal is building a communication network covering the whole project area.

功能分析 || Functional analysis

由于新增了两个中庭体块，所以空间布局围绕中庭展开，我们希望把厨房客餐厅等拿出来作为共享空间增大老人之间交流的可能性，使整个空间呈现出家的格局。最后三栋建筑围成的玻璃中庭活动中心，形成场地内完整的组团结构，为老人生活提供更多的可能性。

Since two atriums are newly added, the spatial layout is designed to spread out around these two atriums. The kitchen, living room and dining room are expected to serve as the shared space in order to increase the seniors' opportunities to interact with each other and to make the entire space feel like a home. The activity center in the form of the glass atrium shaped up by three buildings creates the integral structure of a group, bringing more possibilities of life to the seniors.

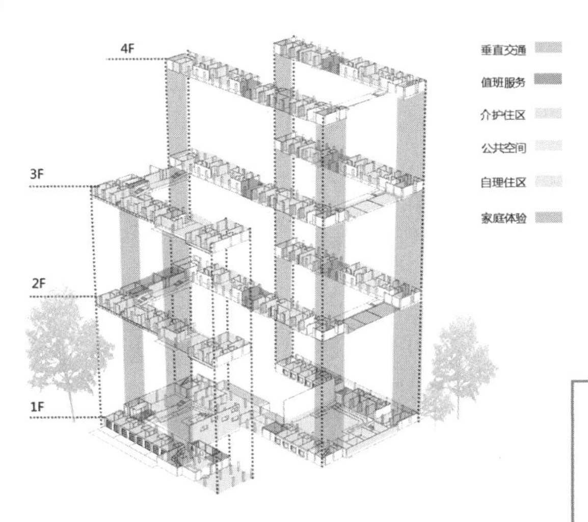

平面图 || Floor plan

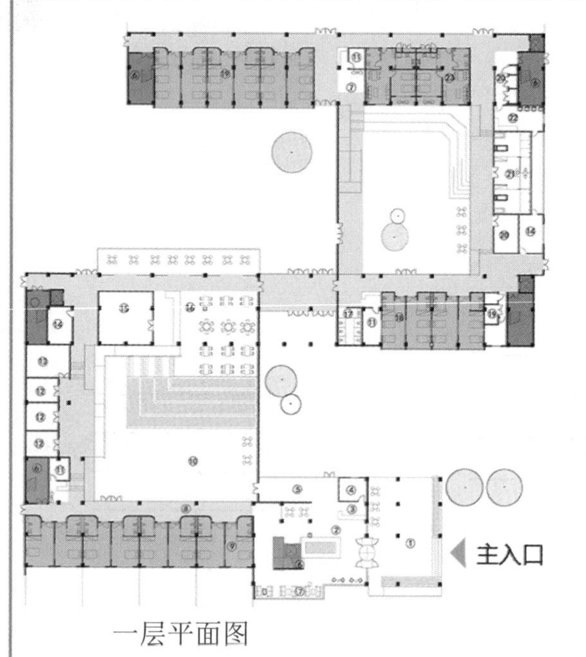

一层平面图

① 过渡空间
② 门厅
③ 服务台
④ 值班室
⑤ 玻璃房
⑥ 垂直交通
⑦ 等候区
⑧ 过道
⑨ 护理单人间
⑩ 中庭
⑪ 护士站
⑫ 商店
⑬ 手工纪念品店
⑭ 储藏室
⑮ 后厨
⑯ 餐厅
⑰ 公共卫生间
⑱ 护理双人间
⑲ 设备间
⑳ 理发室
㉑ 辅助浴室
㉒ 洗消间
㉓ 亲子体验间
㉔ 儿童娱乐
㉕ 自理单人间
㉖ 自理双人间
㉗ 组团内公共空间
㉘ 记忆书屋
㉙ 展厅
㉚ 制作间
㉛ 影音室
㉜ 冥想空间

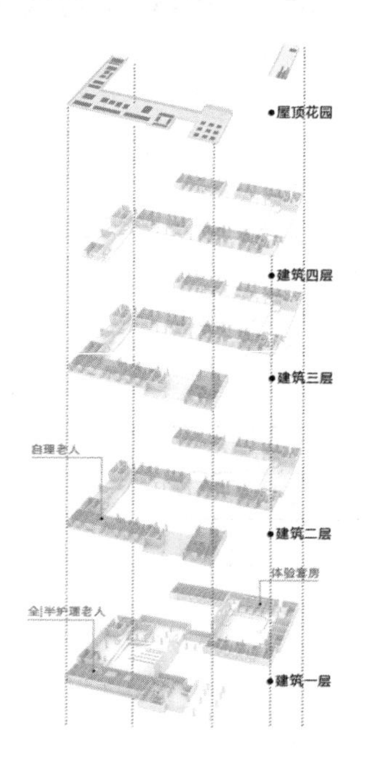

二层平面图

三层平面图

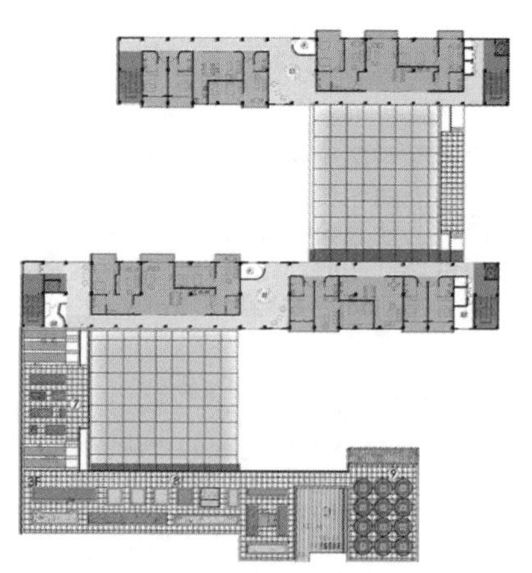

四层平面图

组团概念构思 || Concept of group living

原有的建筑居住空间以单人间为主，长廊式排列，除去过道外不再有其他公共活动空间，老人缺少停留和交流的场地。

我们把部分的单间保留供独身和喜静的老人使用，新增双人的户型满足夫妻，姐妹等两个老人相互照应。

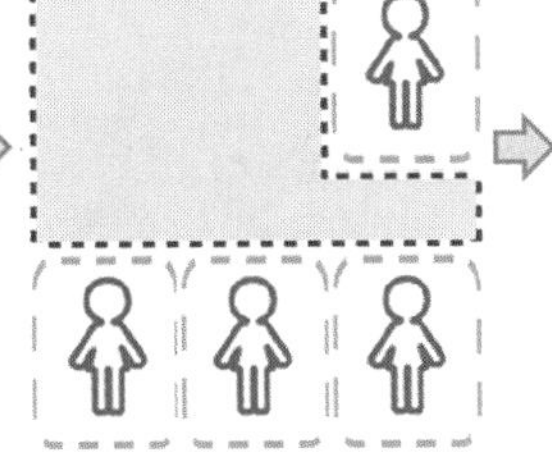

把改变后的户型单人和双人间进行重新组合，同时留出阳台，厨房，客厅，餐厅等共享空间，增加使用老人交流的可能性。

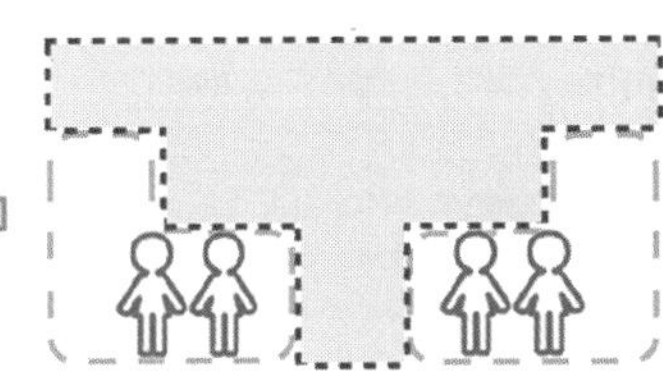

形成 3 种类型的居住组团，组团类型分为 4 户单人间组合，单人间和双人间混合组合，两户双人间组合。

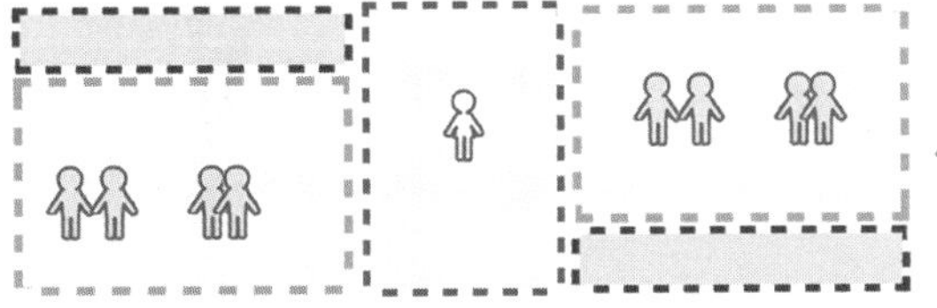

每栋一个楼层两个居住组团围绕一个护士值班服务站，便于管理和服务，同时形成通畅的交通环岛和组与组之间活动交流中心。

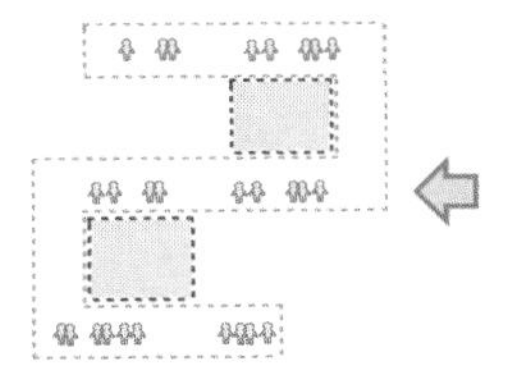

最后，三栋建筑围成两个中庭，从视线和活动联系起所有住户，也提供了一个康复，休憩活动，晒太阳空间。

室内改造构思 || Concept of interior renovation

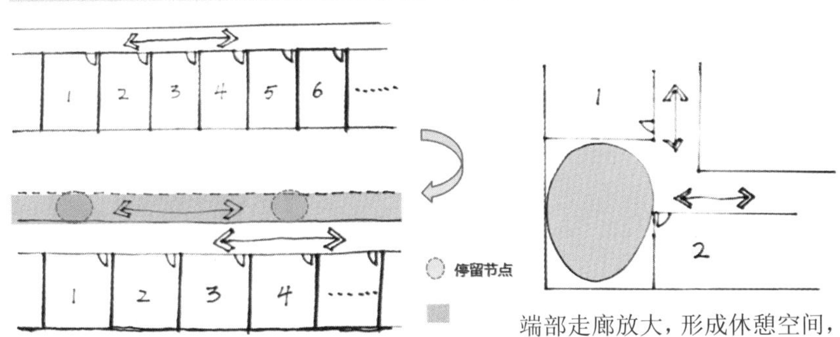

在长廊主交通不变的情况下，将走道空间放大，为走廊扩充停留节点。

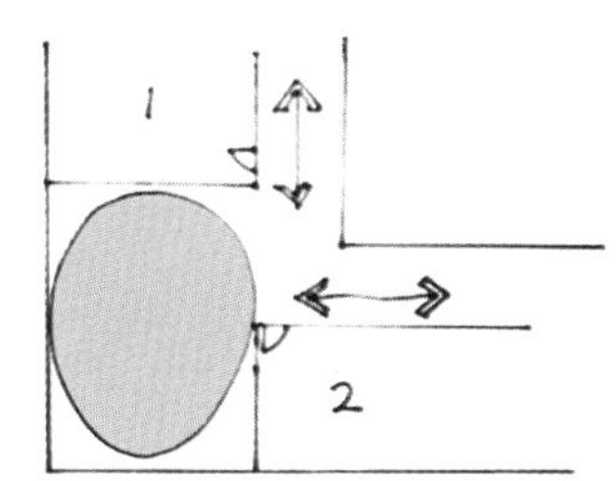

端部走廊放大，形成休憩空间，走廊与公共空间关系多样化，让其成为充满生活情趣的步道。

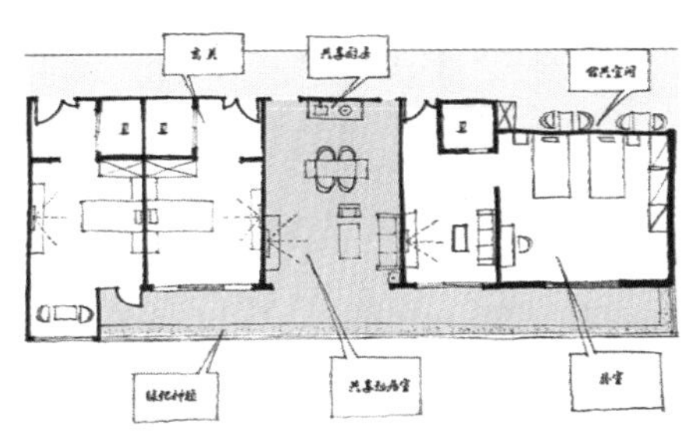

三户，四个老人共享一个公共空间和配套的服务设施，形成一个单元组团。加强邻里之间的交流，同时提高了老人生活的多样性，也更加助于住户的安全管理。

组团户型 A || Type-A apartment for group living

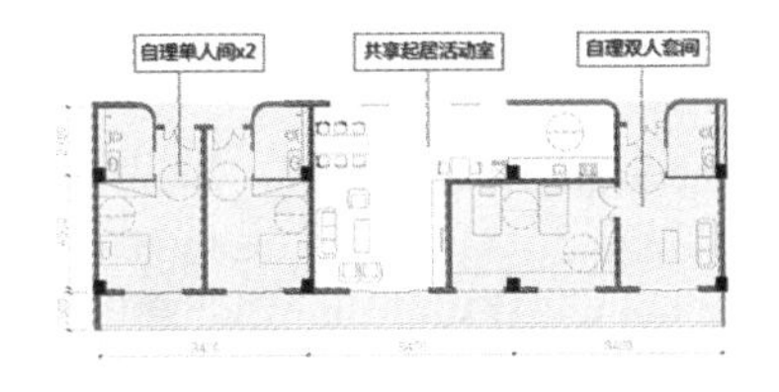

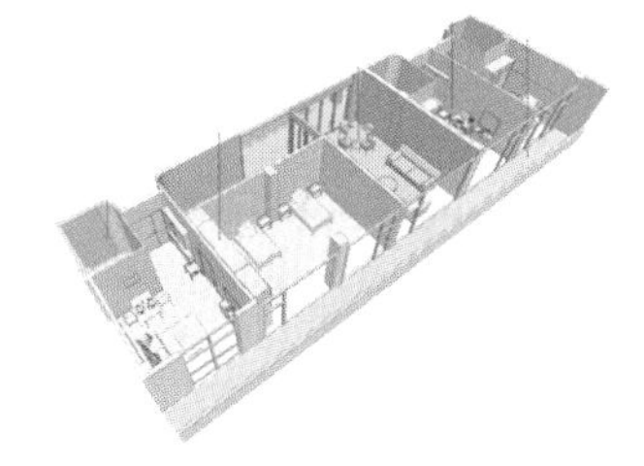

两个单人间和一个双人间的组合，即可满足独身老人又可满足夫妻或姐妹等亲密老人共同使用，双人间采用横向布局使得每个床位都有足够的采光。

The combination of two single rooms and one double room caters well not only to single seniors but also to such intimate seniors as spouses, sisters, and the like. The horizontal arrangement of the double room allows each bed to have sufficient light.

组团户型 B || Type-B apartment for group living

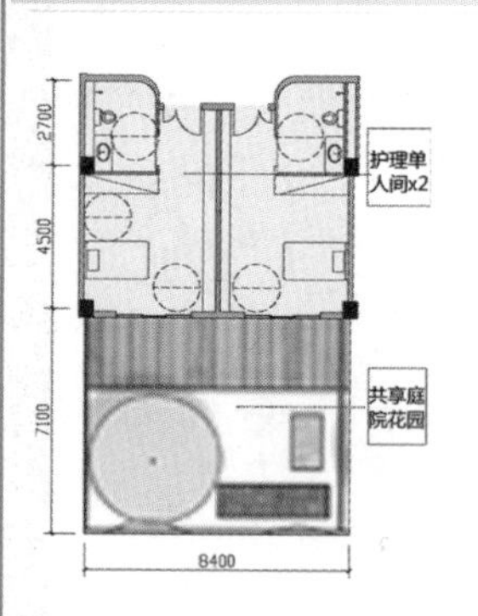

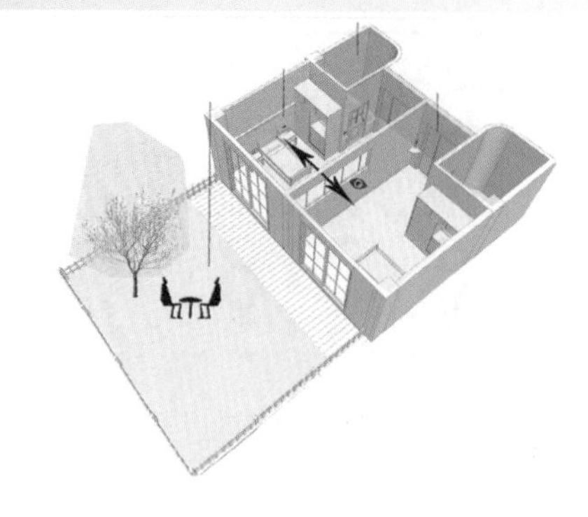

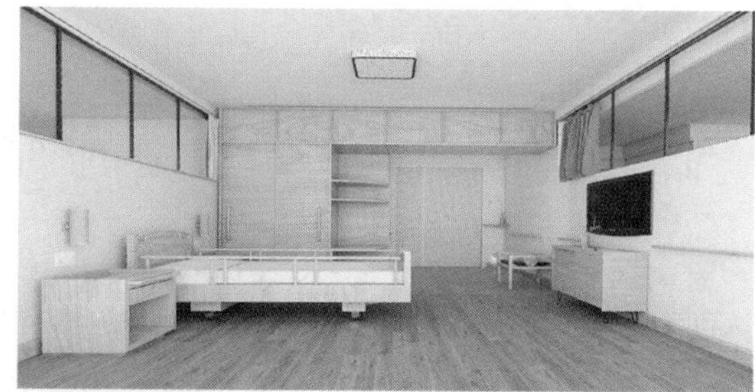

全护理房间比较单一，留下空闲空间，供特殊时期使用，并且两户共享一个小花园，方便独居老人之间的交流。

Such apartment is for fully disabled seniors. A large empty space is retained in case of special use. Occupants living in two bedrooms share a small garden which promotes communication.

组团户型 C || Type-C apartment for group living

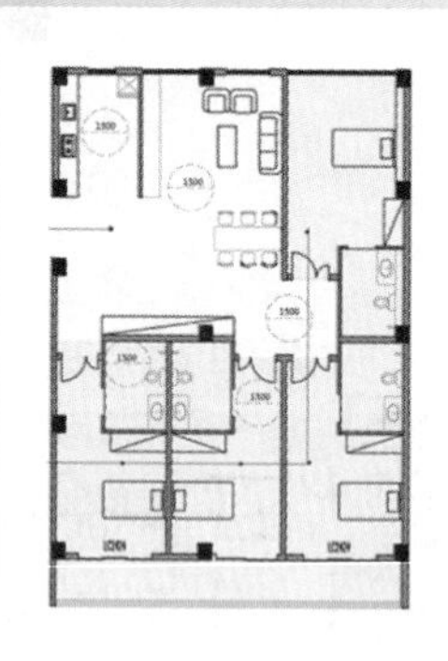

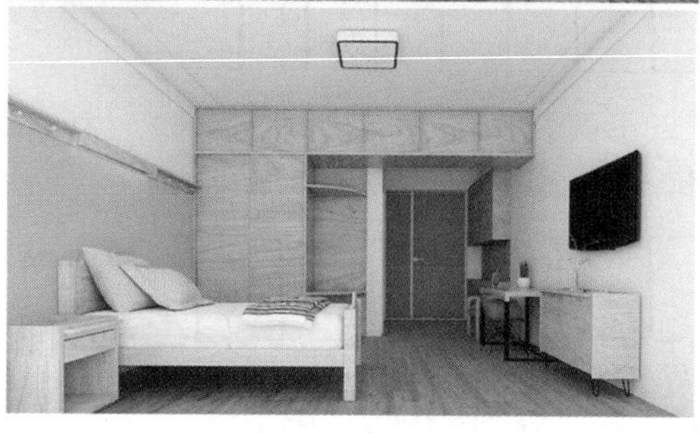

考虑独居老人的需求，所以布置了四个单人间的组团，他们共享一个客餐厅，满足他们平日里聊聊天、下下棋、交流厨艺等。

Such apartment comprises four single rooms with a shared living room to meeting the occupants' needs for living, communication (e.g. exchanging ideas about cooking) and entertainment (e.g. playing chess).

家庭体验房 || Family-style apartment

家庭体验房的布置，是为了给儿女探望老人所考虑的，增加了他们之间相处的时间，也可吸引年轻人来居住。

Such apartment is designed to meet the accommodation need of the occupant's families. It creates a space for family life, and attracts young people at the same time.

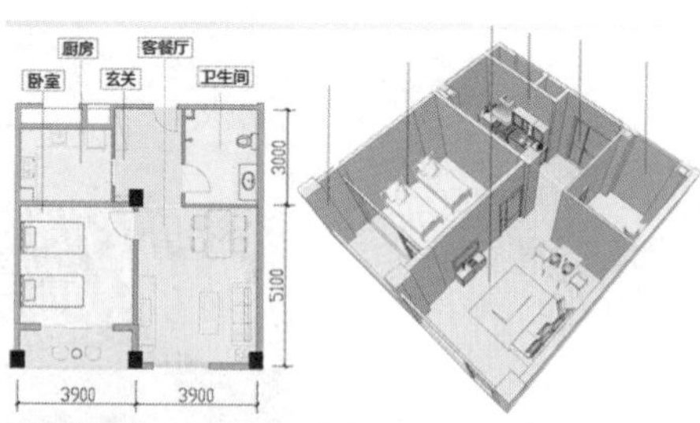

离开原来熟悉的环境，离开了身边的朋友和亲人，这是老人抗拒进入养老公寓最重要的原因。居住组团可能是维系他们之间关系最佳的居住模式。

An important reason that seniors refuse to living in retirement homes is that it means leaving their familiar environments, friends and families. Group living may be the best living mode to meet their need for communication.

1	
	2
4	3

1. 一二栋之间中庭公共空间
2. 开放式餐厅
3. 护士岛
4. 开放式图书馆

考虑到项目区位的特殊性，我们希望在项目范围内形成一个能自由运转的社区体系。在整个公寓内我们希望最终达到一种状态：由居住在此的老人主动参与社区的营造、管理和运转，赋予老人更多的话语权和操控权。而我们做的只是规划和设计一个这样的平台。

Since the project site is far away from the downtown area, we expect to build a community system running independently within the site. Our final objective about the system is that the occupants actively participate in the construction, management and operation of the community as owners with the right of speech and control. To realize the objective, what we need to do is designing the system.

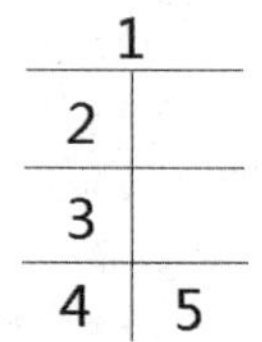

1. 一栋建筑
2. 门厅二楼回廊
3. 门厅
4. 门户网站首页
5. 门户网站推广网页

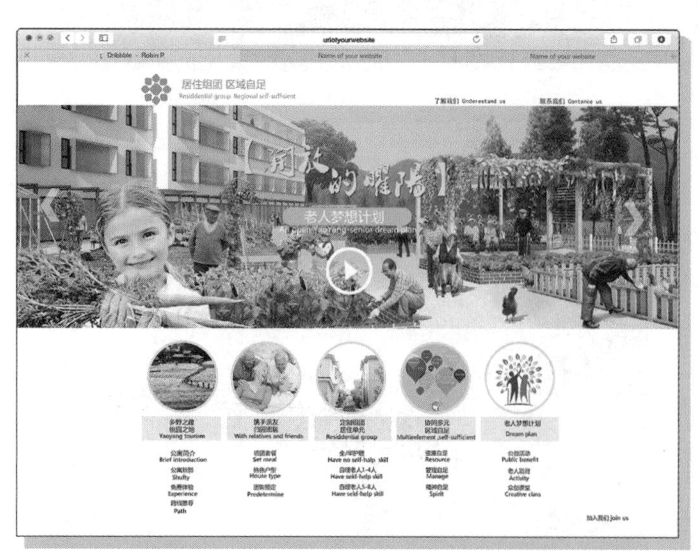

导师点评：

这个作品在思考老年人应该居住在怎样的地方，他们之间的关系怎么样，他们群体行为的独立性和相互心理的交融性的关系。学生们通过对老年人的行为和心理的探索过程来转译这个特殊体验空间的逻辑关系，强调老年人优雅、独立生存的重要性，表现了超出一般的“世外桃源”。

整个作品系统完整，表达成熟。从建筑改造、庭院景观设计、室内设计、平面导向设计到药物园植物配置、阳光利用、雨水回收等各方面的探索，无不体现出学生们对老年人的关注和爱护，以及对自然的观察和感悟，也体现了学生们扎实的专业基础和融会贯通的学习能力。

朱飞

Advisor's Comments:

The scheme tries to answer some questions: What's the living environment that the seniors really want? What's the relationship among the seniors? What's the relationship between behavior independence and the need for communication of the seniors? Through probing into the behavioral and psychological characteristics of seniors, the designers translated the logic relationship of a special experience space. Stressing on the importance of an elegant and independent lifestyle of seniors, the scheme creates an "Arcadia" for the seniors.

The scheme features high completeness and mature expression. From architectural renovation, yard landscape design, interior design, and traffic flow design, to the plant selection, sunlight exposure and rainwater harvesting of the herb garden, we can see the students' care and love for seniors, their observation and understanding of nature, their solid basic professional knowledge, and their ability to apply different kinds of knowledge in design.

Zhu Fei

专家点评：

从完成作品中可以看到，同学们深入地思考了群居老年之间生活交流的心理和行为特点。从前期调研、概念的建立、解题思路的推导、功能分析的细节表述和最终作品的绘制等各个方面，让整体方案设计的主题和要点能够逐层次地展示出来，并且创造了一种针对老年人群的具有人文关怀特色的小组团居住模式，值得大家认真品味和思考。

卓培

Expert's Comments:

From the scheme we can see that the students have an in-depth study of seniors' desire and need for communication, and their behavioral characteristics. From early-stage survey, conceptual framework formation, filling of the conceptual framework, detailed expression of functional analysis results, and the preparation of the final documents, the theme and highlights of the scheme are gradually specified and become more and more clear. The idea of group living is human-oriented, and is worth further thinking.

Zhuo Pei

Living in a Fairyland

CIID“室内设计 6+1”2016（第四届）校企联合毕业设计

CIID“Interior Design 6+1”2016(Fourth Session)University and Enterprise Joint in Graduation Design

高　　校：　同济大学
College：　Tongji University
学　　生：　胡楠　邹天格　丁思岑
Students：　Hu Nan　Zou Tiange　Ding Sicen
指导教师：　左琰
Instructors：　Zuo Yan
参赛成绩：　医养建筑改造与室内设计组二等奖
Achievement：　Second Prize for Aged care Building Renovation and Interior Redesign

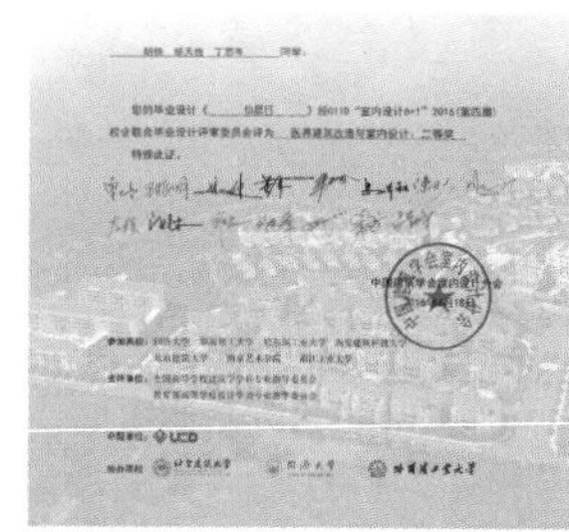

胡　楠
Hu Nan

邹天格
Zou Tiange

丁思岑
Ding Sicen

学生感悟：

在左琰老师的指导下，通过这次养老院改造与室内设计，我们学到了从老年人的角度出发去考虑设计，通过调研等各种方式了解老年人这个群体，在充分了解了社会人文背景后再去做一个设计。课题和受众群体的特殊性让我们更多地去思考建筑中的人文关怀。在老师的指导下，我们从宏观的功能定位策划、室内家具细节的设计等都尝试从老年人需求的角度出发。我从中学到了很多，十分感恩能有非常好的队友。

Students' Thoughts:

The retirement home renovation (interior design) project was conducted under the guidance of Mr. Zuo Yan. This project offered us an opportunity to design from the perspective of seniors. Through surveys and other methods, we had a better understanding of seniors, and corresponding social and cultural issues, which act as the basis of our design. Given the requirements of the project topic, and the fact that seniors are a vulnerable group, we focused on incorporating care for seniors into architectural design. Under the direction of Mr. Zuo, we determined the macroscopic functions of the buildings and completed interior design (furniture and details) to meet the real needs of the seniors. I've learned a lot in the process, and I want to express my sincere thanks to my team members.

生成分析 || Formation analysis

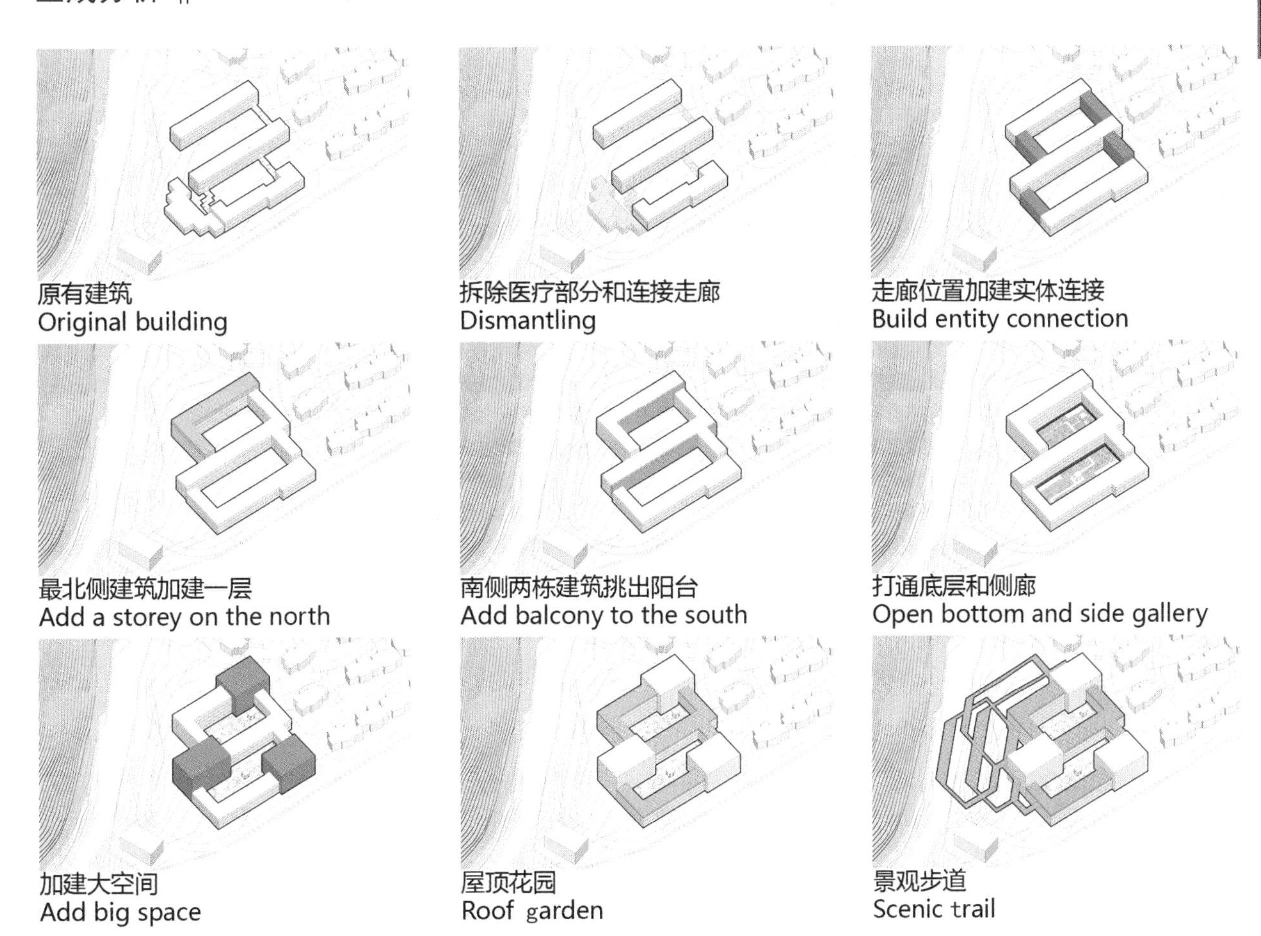

在实体中布置护士站和公共空间，在三个方盒子中解决医疗、集中公共空间和无障碍的问题。打通两个封闭的院子建立联系，打通西侧底层改善风环境。设计景观步道，联系公共空间和屋顶花园，形成环路。

A nursing station and public area are arranged in the project site, so that medical service, concentrated public areas and wheelchair friendly facilities are all accessible in three "square boxes". The two independent yards are connected, and a hall stretching through the 1st story of the building in the west is built to promote ventilation. Besides, a garden path is designed to link public areas with roof gardens, forming a loop.

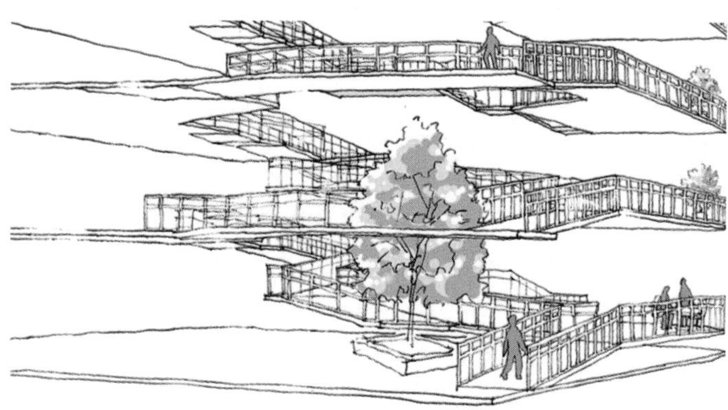

庭院景观 || Yard landscape

植物研究
Vegetation study

视线分析
View analysis

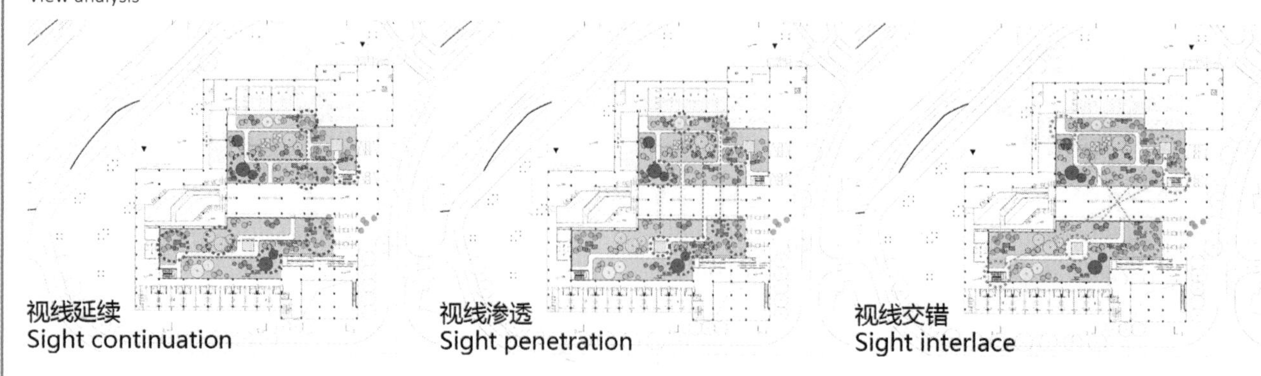

在庭院道路的行走过程中，有不同的景观和植物出现，透过中间一栋建筑打通的一层，视线通过廊道可看到不同景物。视线穿过近景、中景、远景到达最远处的标志物，能够加深景深。

The path running across the yards is landscaped with different types of plants. Through the hall stretching through the 1st story of the middle building, landscape on the other side of the building can be seen along the path into the distance where the furthest sign is.

各层平面图 || Plans of various floors

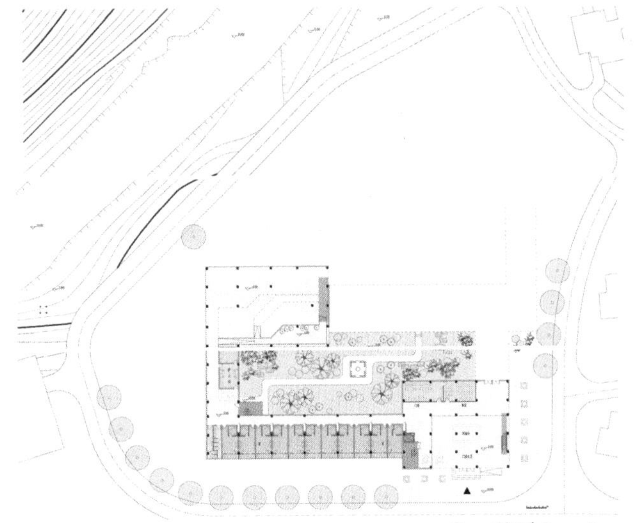
0m 平面 0m plan

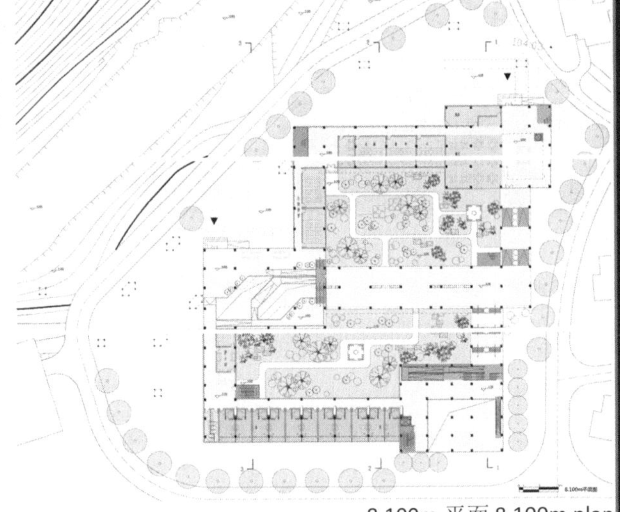
8.100m 平面 8.100m plan

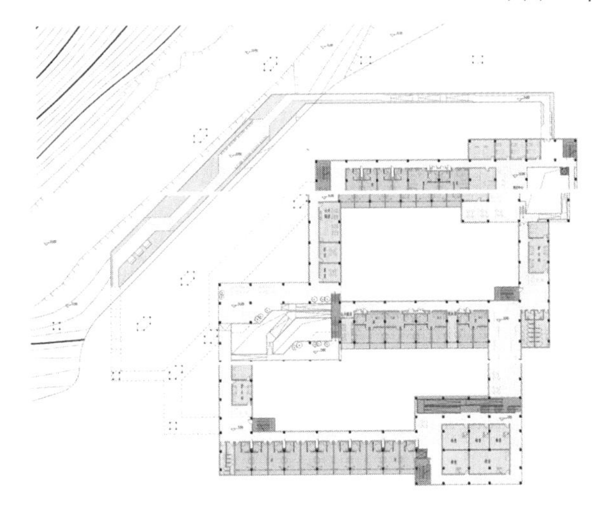
11.400m 平面 11.400m plan

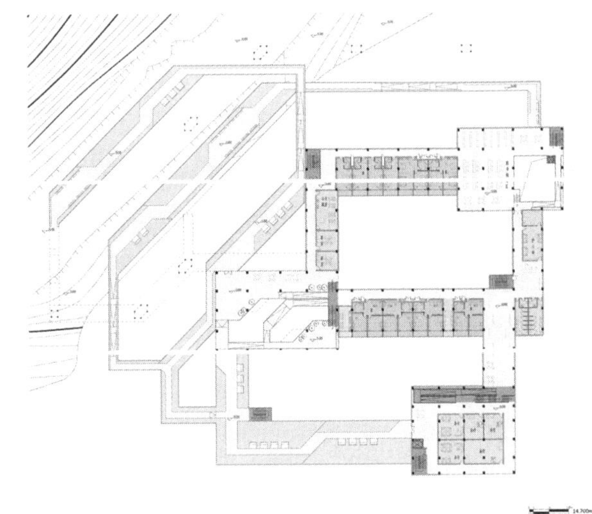
14.700m 平面 14.700m plan

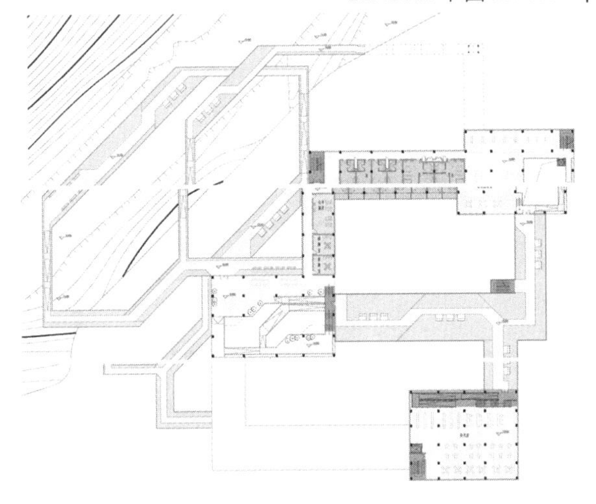
18.000m 平面 18.000m plan

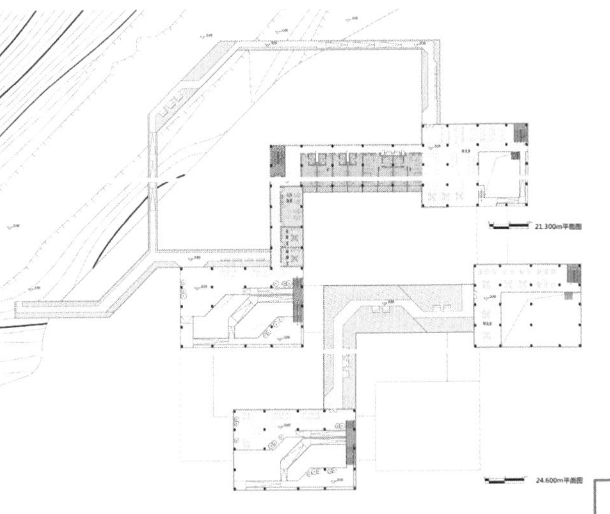
24.600m 平面 24.600m plan

景观步道 & 屋顶花园绿化 || Landscape walkway & roof gardens

为回应老年人需要更多户外锻炼的调研结果，我们结合山地设计景观步道。地面铺装和种植体系的设计呈现出软硬表面不断变化的比例关系，步道上种植灌木和地被植物，通过步道的尺度变化、节点变化设置让老年人每隔一定距离就有休憩的区域。

To meet the need of orthopedic patients for outdoor exercise, a garden path is designed according to the terrains. Different plants presenting the sense of softness and hardness, such as bushes and ground covers, are grown alternatively along the path. Besides, rest areas are arranged along the path at spatial nodes and places where the path expands.

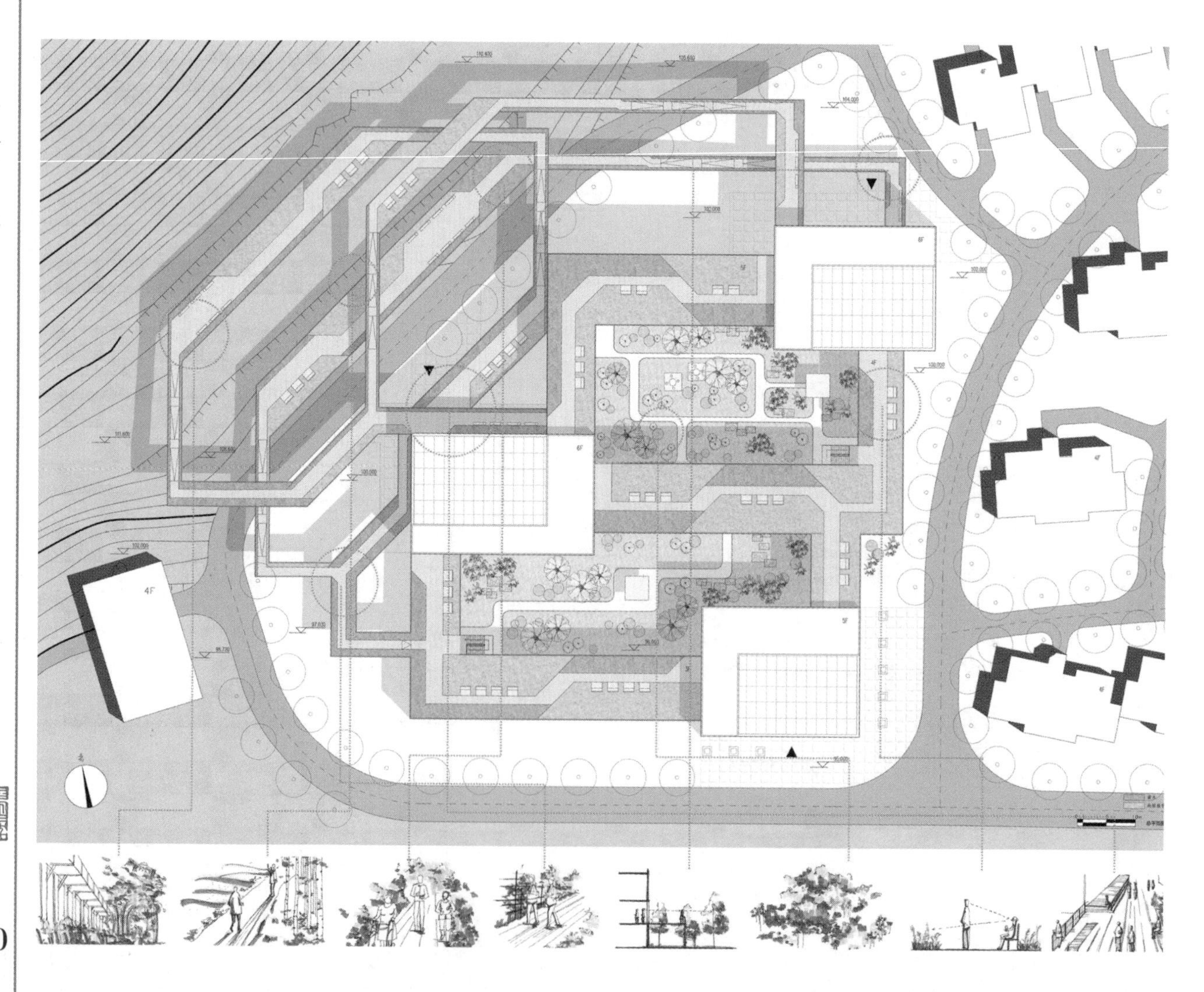

功能 & 无障碍 || Functions & wheelchair accessible facilities

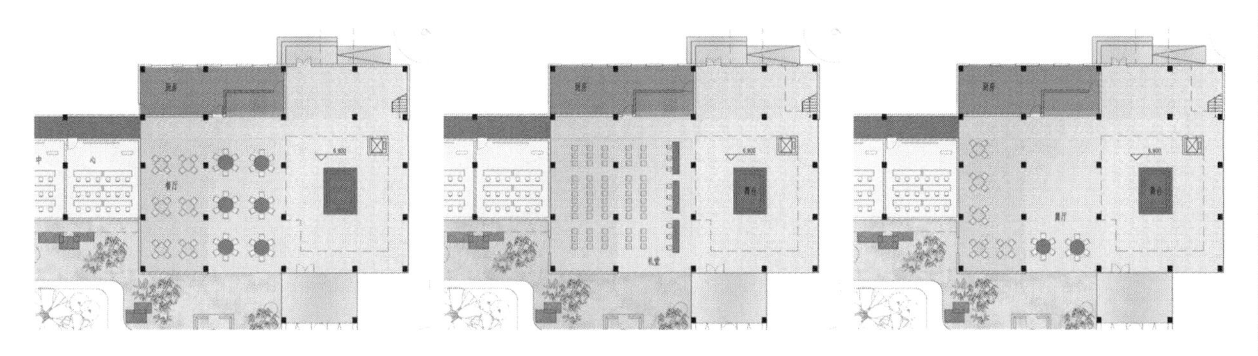

南侧建筑加建的大空间可根据空间特点，平时作为餐厅使用，如有特殊活动需求，则可进行家具的重新摆放，成为舞厅、讲堂等。
南侧通过无障碍坡道解决两栋之间同层高差问题。无障碍坡道与休息平台沿内院布置，具有良好的景观视线关系。
北侧方盒子中交通空间围绕中庭设置，无障碍坡道与电梯均环绕中庭，与建筑内部大空间有良好的视线关系与交通可能。

The large space newly added to the southern building can be used as a restaurant or a ball room, or lecture room by just changing the arrangement of the furniture.
A wheelchair ramp is arranged in the south of the project site to connect the two buildings. The ramp, together with the rest platform, is arranged along the edge of the inner yard to have a good view of the inner yard.
In the northern "square box", the transport routes, including wheelchair ramps and escalators, are arranged around the atrium, to ensure convenient transport and a good view of the large inner space of the building.

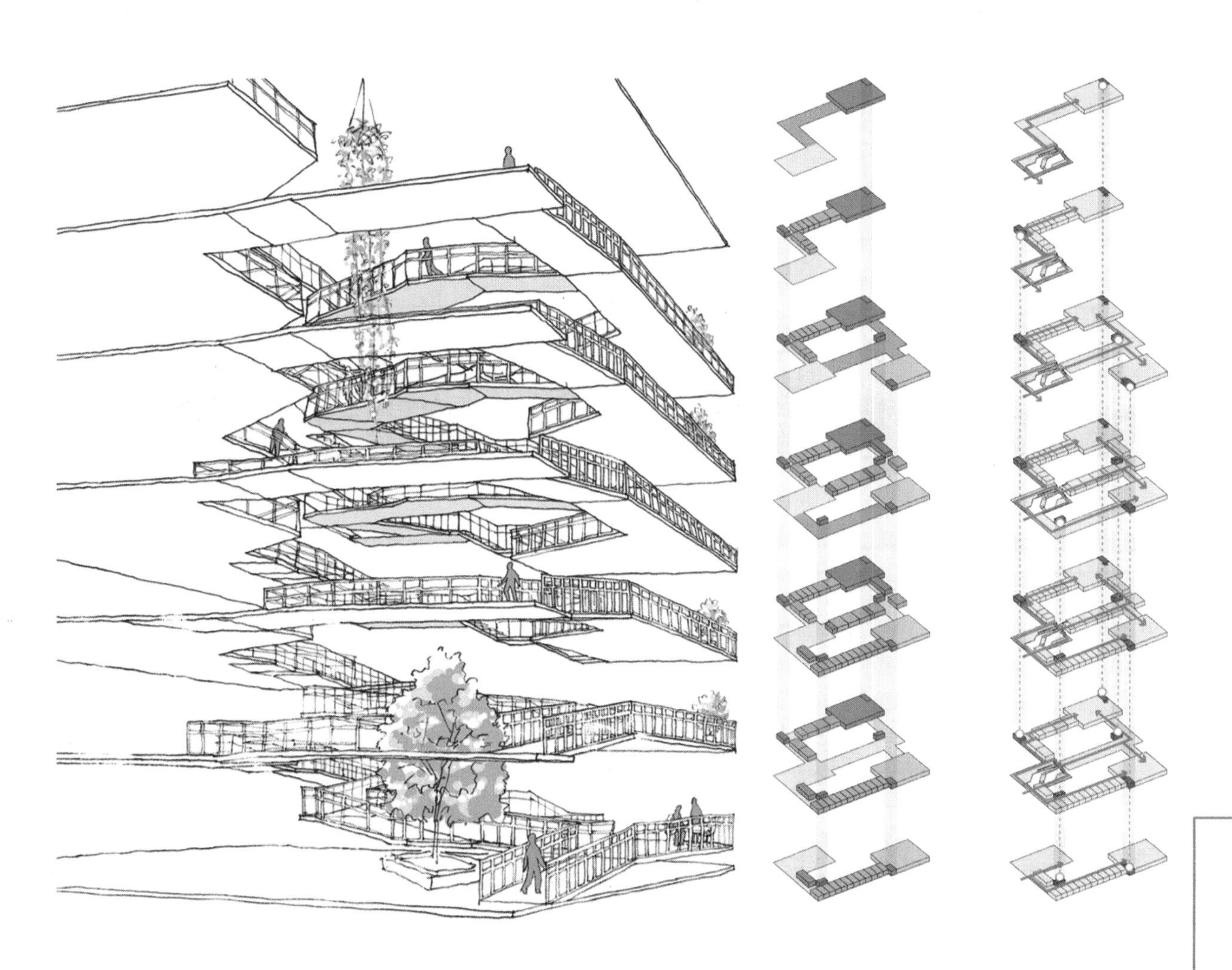

生活在其中 || Living in the retirement community

西侧三栋建筑物交会，故需集中解决两个高差问题。此时大空间单纯用作交通空间，集中利用无障碍坡道同时解决同异层交通问题。坡道围绕中庭布置，同时在经过院落景观与室外景观时，设置大休息平台，成为具有吸引力的驻留空间。

The three buildings involved in the project are of different elevations. To solve this problem, wheelchair ramps are designed to form a transport system connecting the buildings. The ramp route runs around the atrium, and expands to a large and attractive rest platform with a view of the yard landscape and outdoor landscape.

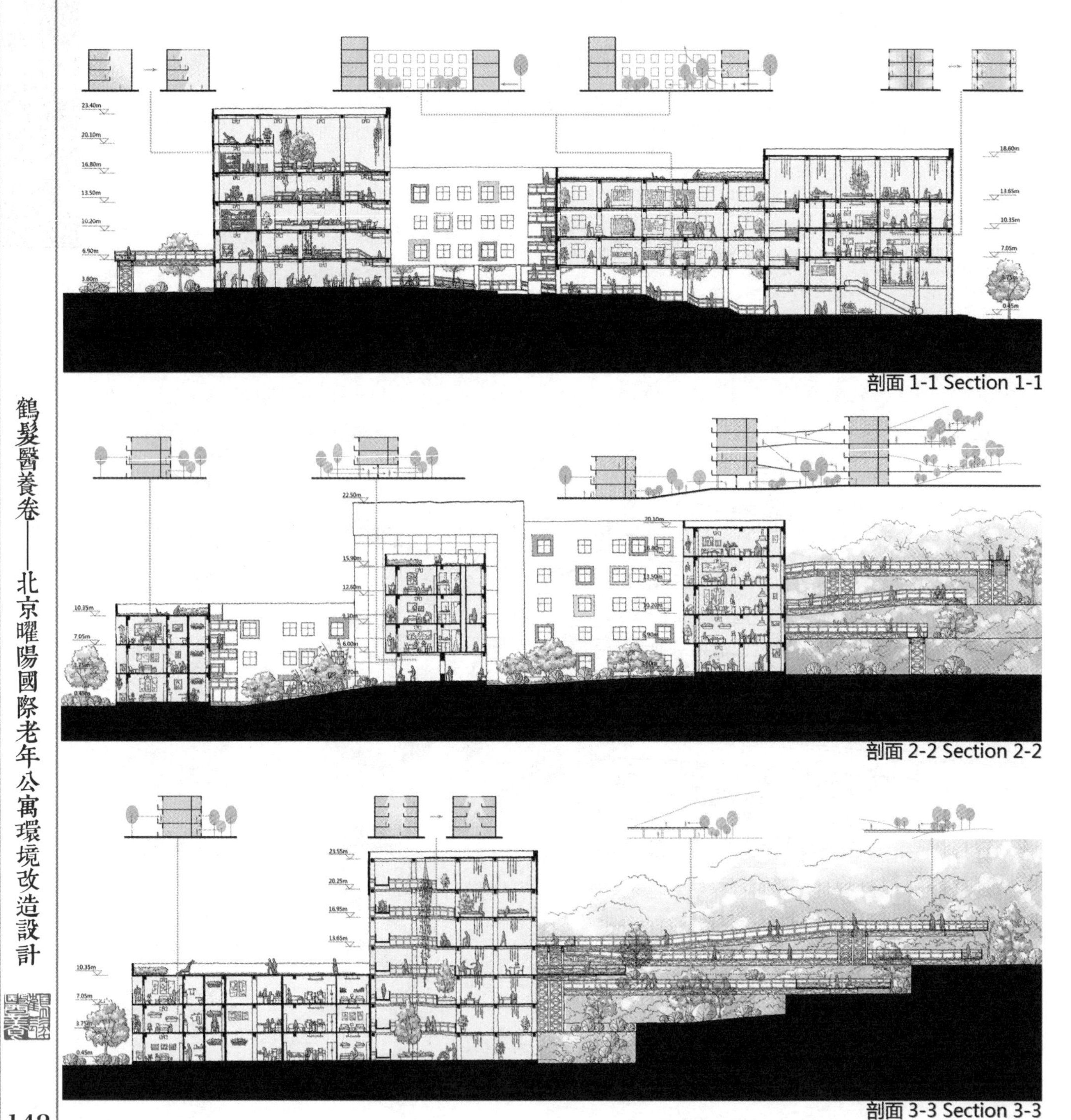

剖面 1-1 Section 1-1

剖面 2-2 Section 2-2

剖面 3-3 Section 3-3

立面设计 || Façade design

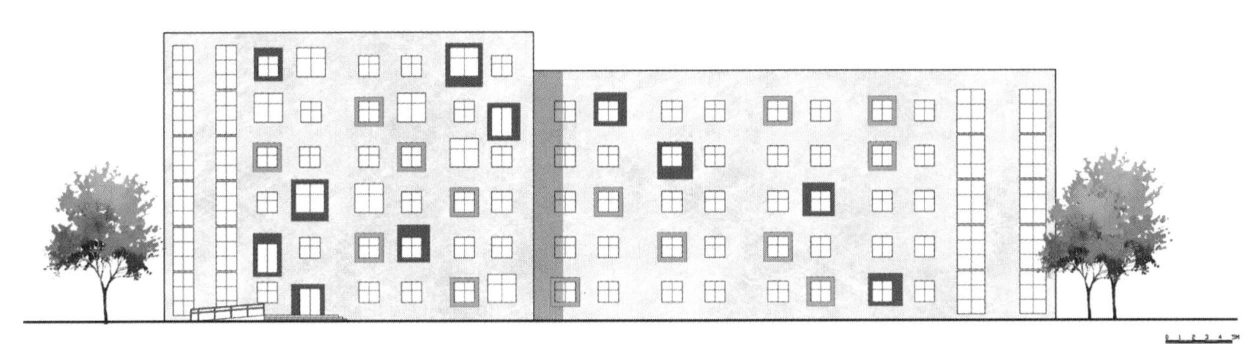

南立面 South Elecation

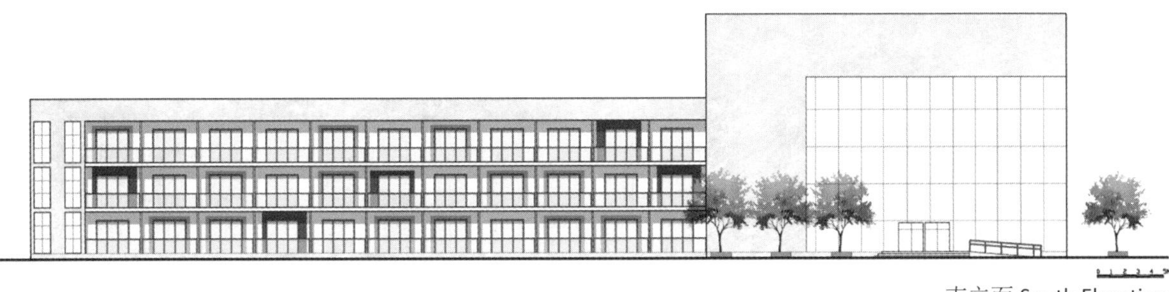

北立面 North Elecation

户型设计 || House type design

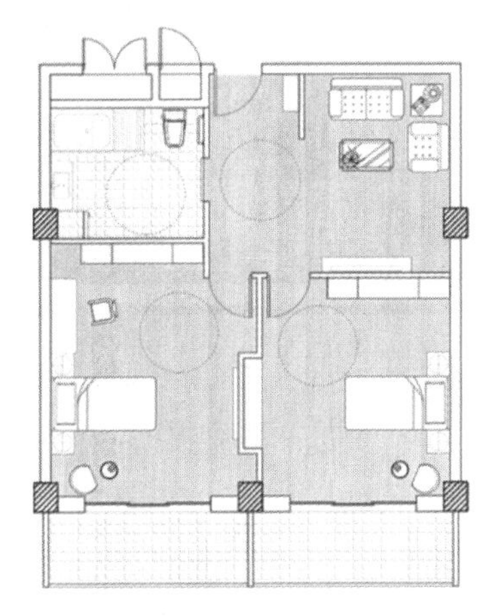

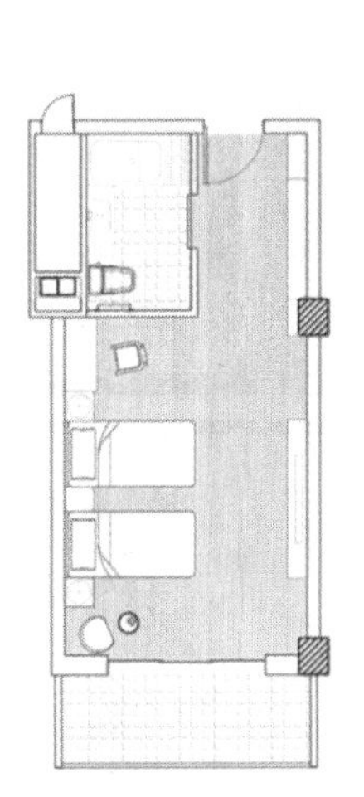

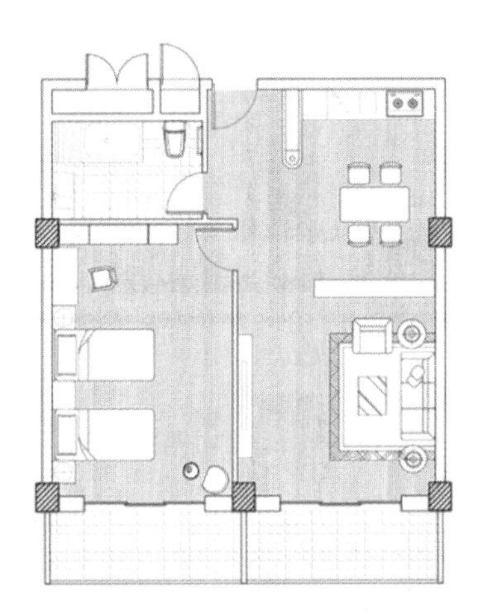

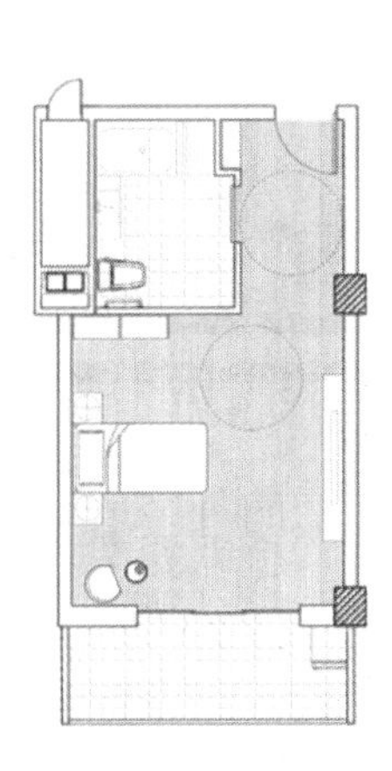

原有建筑共82套房间，改造设计后共77套。针对原平面问题，我们保留管道井，改良满足无障碍尺度，加建阳台，利用景观。

The original 82 suites are turned into 77 suites. Given the original plan design, the pipe shaft is retained, and focus is given to adding wheelchair friendly facilities and balconies, and combining buildings with landscape.

场景 || Scenes

从归园田居的画中提取传统颜色进行设计。

Colors used in the design are borrowed from the traditional Chinese painting titled "Returning to My Farm".

家居设计 || Furniture design

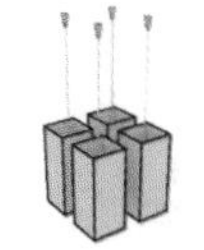

导师点评：

设计人对场地有良好的解读能力，市场调研的数据体现出严谨的工作态度，设计思路颇具条理。方案有具体的目标服务人群，欠缺的是实施方案与对骨科患者的调查结果的关联不够明显，建筑体量和空间组合的调整思路也未能与社会调研结果形成明确的对应关系。值得鼓励的是作者的视野跨越了场地的边界，将争取户外空间的意图延伸到山林之中，是值得深入探讨的思路。

方案显示出较好的专业素养和基本功，交通组织得宜，功能完整，图面内容清楚明确，素材充实，版面布置有序。但是在有限的篇幅里包含了建筑形体调整、空间重组以及景观和室内设计等大量内容，难免欠缺深度。建议目标明确、问题聚焦。

Advisor's Comments:

The designers have a good understanding of the project site's situations. The social survey data indicate their rigorous attitude, and their thoughts of design are very clear. A specific target group is included in the scheme, but the relationship between the implementation plan and the orthopedic patient survey results is not very clear, and the adjustments of building dimensions and spatial division fail to well respond to social survey results. One commendable thing is that the designers take a broader view of the project site, trying to extend exterior space to the further forests, which is an idea worth in-depth discussion.

From this scheme I see the professional qualities and solid basic knowledge of the designers. The fully functioning traffic system is a successful highlight. The drawings clearly illustrate the ideas of the designers, sufficient information is contained in the scheme, and the typesetting is neat. A problem is that a lot of information covering many aspects like building shape adjustment, space re-division, landscape design and interior design is contained in a short article, which means in-depth exploration cannot be realized. My suggestion is design focuses are needed in the scheme.

Xie Guanyi

专家点评：

该方案能够根据使用者的特殊需求，结合建筑本体进行功能分析并加以改造，完善建筑内部结构，合理划分各功能空间。

在景观设计中，根据花期、花色、株高等要素能够合理选取植物搭配。多层次的景观节点满足不同使用者的需求，形成了多角度的视觉中心。

在室内空间设计中，整体风格符合中老年人的审美要求，陈设选择符合患者人体工程学比例尺度的要求，无障碍设施的应用体现出建筑本身的价值。

整体设计能够遵循以人为本的设计理念，向我们展示了人、建筑、景观的有机共存。

张红松

Expert's Comments:

According to the special needs of the occupants, the scheme renovates the buildings by improving the interior structures of the buildings and re-dividing their functional spaces based on an analysis of spatial functions.

For landscape design, elements like the flowering periods, colors and stem heights of plants are considered during selecting plants. Spatial nodes are arranged to create diverse landscapes which meet the needs of different people, and can be appreciated from various angles.

Interior spaces are designed to meet the aesthetic requirements of middle aged and elderly people. All the furnishings are selected according to the ergonomic characteristics of occupants with disability, and the application of wheelchair friendly facilities adds value to the buildings.

Generally speaking, the scheme successfully follows the principle of human-centered design, and gives an example of harmonious coexistence of human, architecture and landscape.

Zhang Hongsong

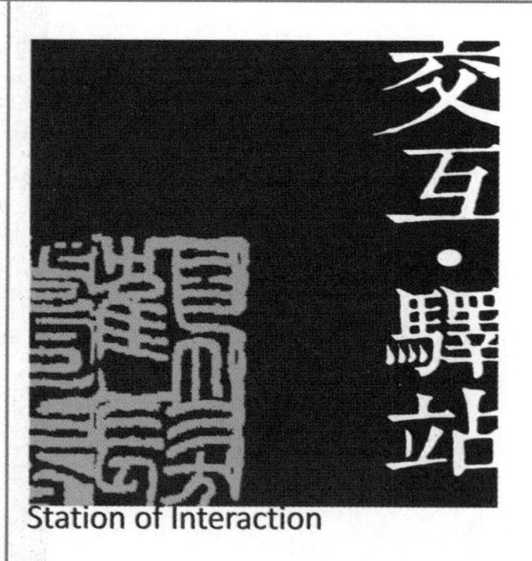

Station of Interaction

CIID“室内设计 6+1”2016（第四届）校企联合毕业设计

CIID“Interior Design 6+1”2016(Fourth Session)University and Enterprise Joint in Graduation Design

高　　校：浙江工业大学

College: Zhejiang University of Technology

学　　生：卢倩雯　邱丽珉

Students: Lu Qianwen　Qiu Limin

指导教师：吕勤智　黄焱

Instructors: Lv Qinzhi　Huang Yan

参赛成绩：医养建筑改造与室内设计组二等奖

Achievement: Second Prize for Aged care Building Renovation and Interior Redesign

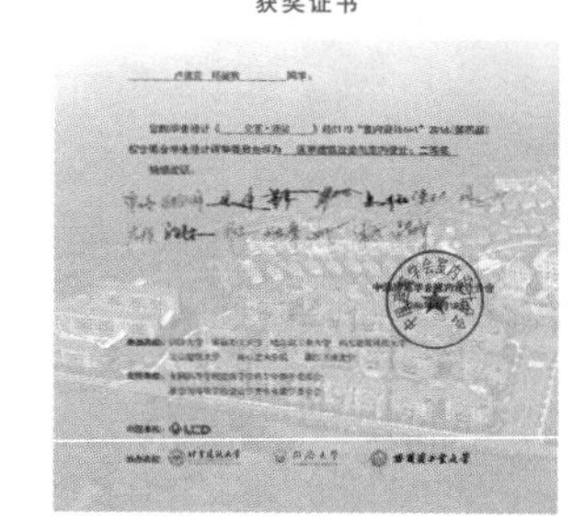

卢倩雯
Lu Qianwen

邱丽珉
Qiu Limin

学生感悟：

通过本次毕业设计的交流学习，我们慢慢学会有逻辑性、条理性地分析和解决问题，从老年人的角度去思考问题。探讨老年人真正需要的设计，不单单在表面，而是探求现代社会和人文精神的共鸣点。对于老年人来说，精神需求远远大于物质，心理的障碍远远比生理的毁灭性更大。针对这点，我们利用交互·驿站的主题思想开展了一系列方案设计。通过老师的耐心指导与点评，以及竞赛带来的一次次视觉和精神上的冲击，我们体会到的远远不止于表皮，用心感悟追求，为人们的行为心理需求而设计，是这次毕业设计最大的收获。

Students' Thoughts:

The process of completing the graduation project is a process of communication and learning. What we've learned is how to analyze and solve problems in a logic and rational way, and how to view thing from the perspective of seniors. We realized that a design discussing the real needs of seniors should not only solve superficial issues, but more importantly, probe into relevant social concerns. For seniors, spiritual needs are more important than physical needs, and psychological problems are much more destructive than physiological problems. Given this, we developed a design scheme themed on “Station of Interaction”. Thanks to the patient guidance and comments from our teachers, and the visual impacts that we experienced during the competition, we gained significant understanding of many things. A real designer should design with his heart to meet the psychological and physiological needs of the target group – this is what I've learned the most.

设计说明 || Design description

人的一生从幼年、青年、壮年直到老年，其生活的每一个地方都是一个站点，而这个站点，有一个统一的名字叫“驿站”。当下的老年公寓，或许就是老年人在生命中最后停留的站点，如何让这类常被世界所遗忘的弱势群体在有生之年仍可以享受生活，老有所享、老有所乐、老有所用？结合社会背景，了解老年人的自身需求，利用交互空间的设计，使老年人重新焕发生命的热情。

北京曜阳老年公寓的设计中，保留建筑原有的柱网及排污管道，在减少资金成本的前提下，将空间有序结合重组，形成有机的交互空间为老年人提供一个交流互动的场所。

Life is a journey, in which we will stop at different places, and each place is called a station. Retirement homes are the last station of seniors. How to help seniors, a disadvantaged group that tends to be ignored, enjoy their life and create value is a problem worth thinking. In this project, we try to arouse the enthusiasm of seniors for life using the idea of interactive space, according to the current social context and the actual needs of seniors.

We try to provide the seniors with a place of communication and interaction through systematic spatial re-division to form organic interactive spaces at a cost as low as possible. The original column grid and sewage piping remain unchanged.

设计理念 || Design philosophy

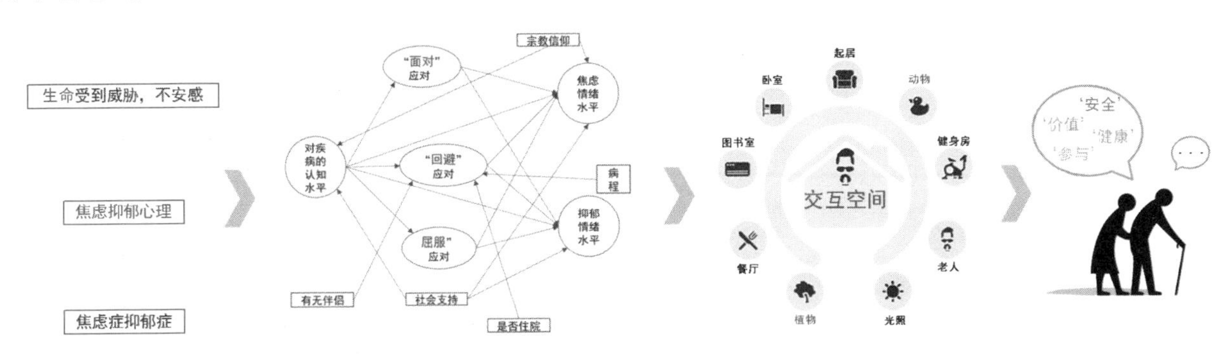

结合老年人的生理和心理特点，重点针对身心疾病老人焦虑抑郁的心理特点，了解其心理问题带来的一系列行为特点，借用交互空间的设计，使老年人在与空间元素交流互动的过程中产生情感体验，满足老年人真正的心理需求，促进身心疾病老人的身心健康。

According to the physiological, psychological and behavioral characteristics of seniors, especially their proneness to anxiety and depression, interactive spaces are designed to help the occupants have emotional experiences, satisfy their psychological needs, and ultimately obtain physical and mental health in the process of communicating and interacting with spatial elements.

设计思考 || Design thinking

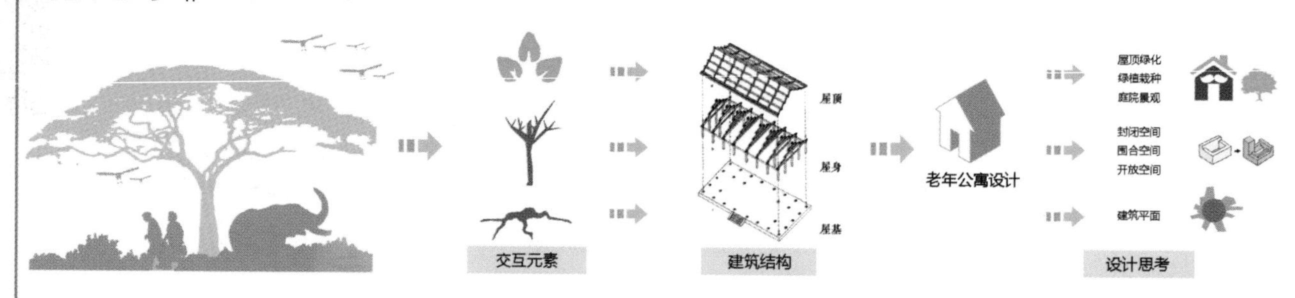

通过对树这一交互元素的思考，将生命注入建筑空间，生成方案设计。
Trees, an interactive element representing "life" are incorporated into the architectural design.

设计总平面 || Site plan

在保留原有建筑形式的同时，增加了老年人活动的公共娱乐空间，借用树根的生长形状，组织空间与空间的关系，用道路延续生长空间，给予公寓空间顽强的生命力。

On the basis of retaining the original architectural form, public entertainment spaces are added. Spaces are connected by the "roots of trees", and paths act as the extension of spaces to present the strength of life.

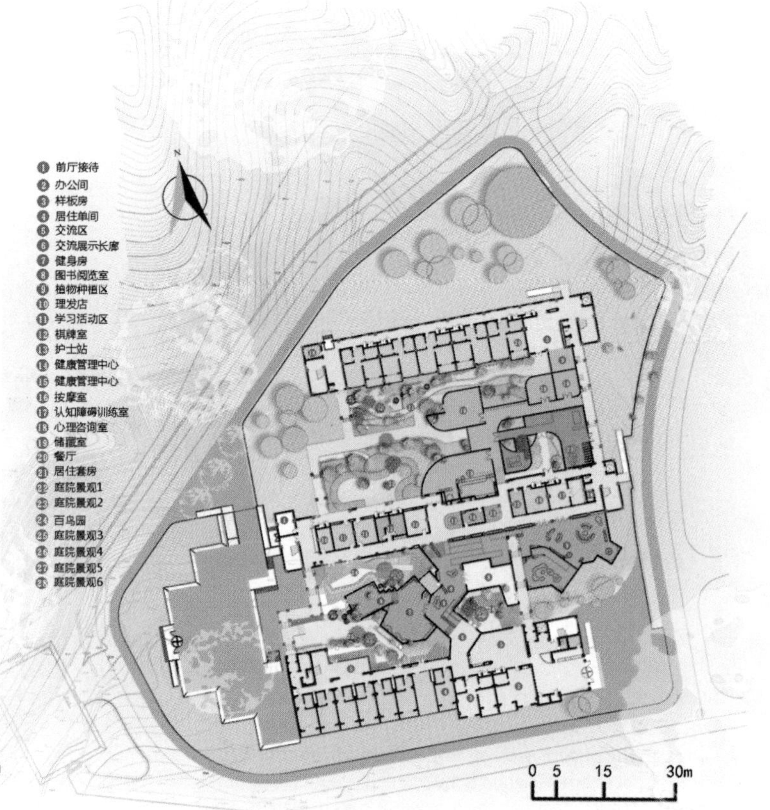

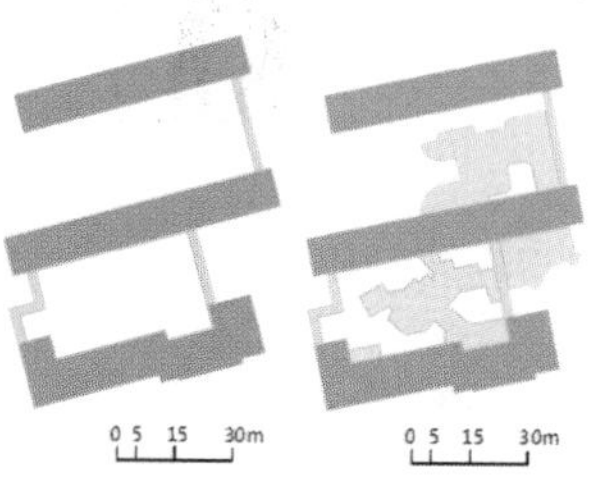

原建筑平面图
Original floor plan

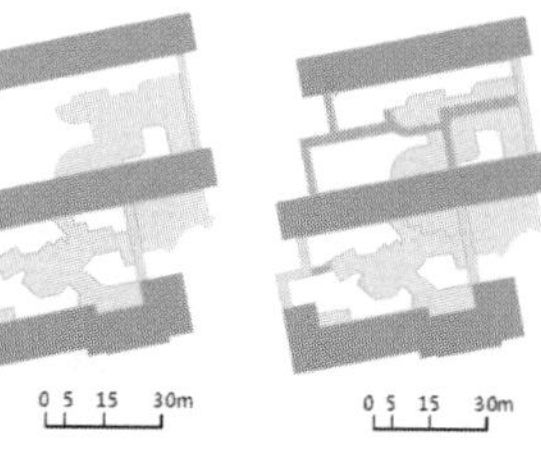

新增平面图
Plan of newly added structures

改造后平面图
Redesigned floor plan

设计分析 || Design analysis

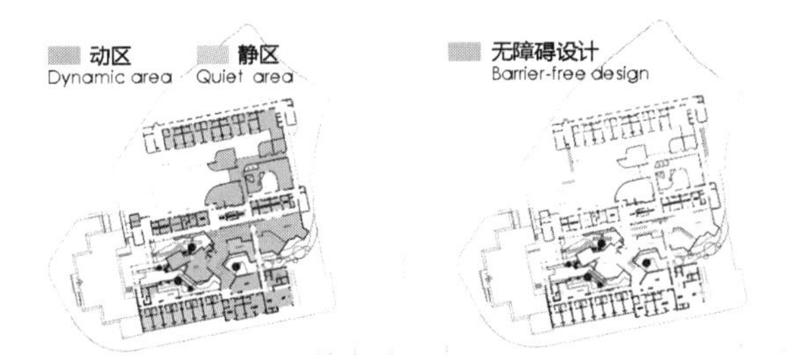

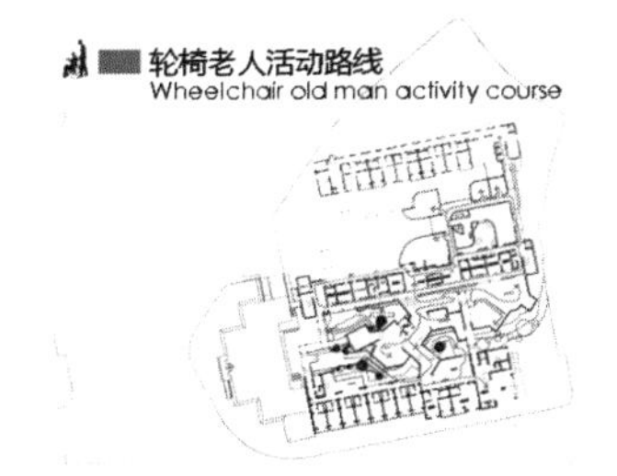

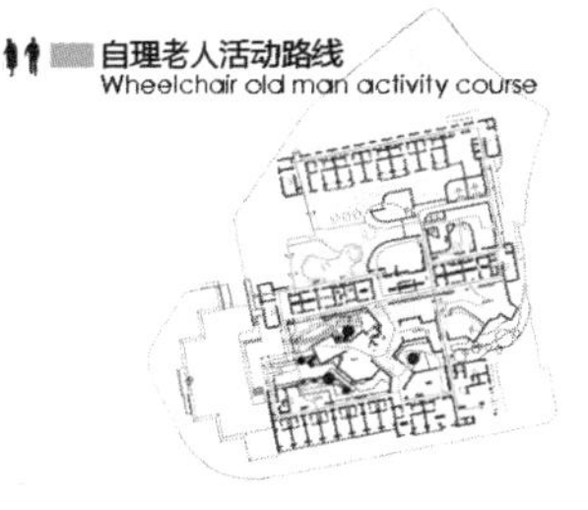

公寓的动静分区。大部分开放活动的公共空间设置于B区首层，并向南北扩散开，满足半自理老人的活动需求。可以根据地形剖面得知建筑楼之间存在1～2米的地形抬高。通过无障碍设计，满足更多老年人进入空间的便利性和可达性。

人群流线分为自理老人的活动流线和轮椅老人的活动流线，贯穿于整个设计空间，设计功能的交织使得空间变得有序合理，给老年人带来更好的体验感受。

Living and activity areas are separated. Most open spaces are arranged on the 1st floor of District B, and extend southward and northward, meeting the need for activity of partially disabled seniors.

From the terrain section, it can be seen that there is a 1-2 m elevation difference between adjacent buildings. Given this situation, wheelchair ramps are arranged to create convenient transport.

The traffic flows of able-bodied seniors and seniors depending on wheelchairs are considered during spatial design, for the purpose of bringing better experiences through systematic and rational spatial re-division.

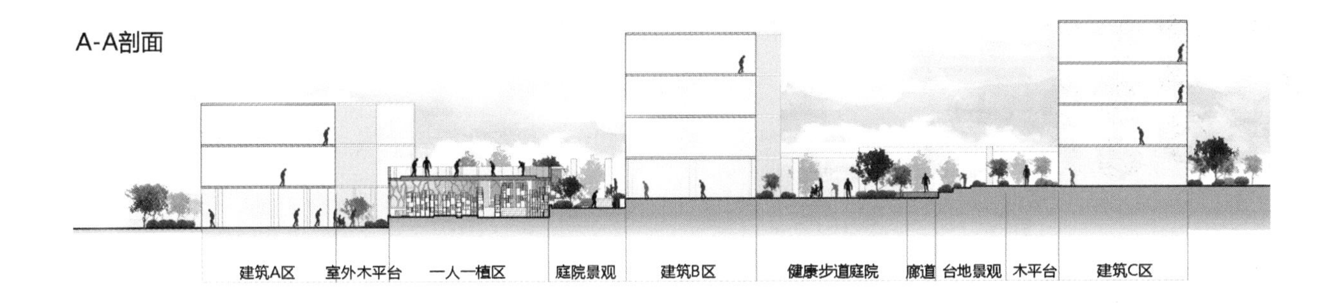

节点设计 || Node design

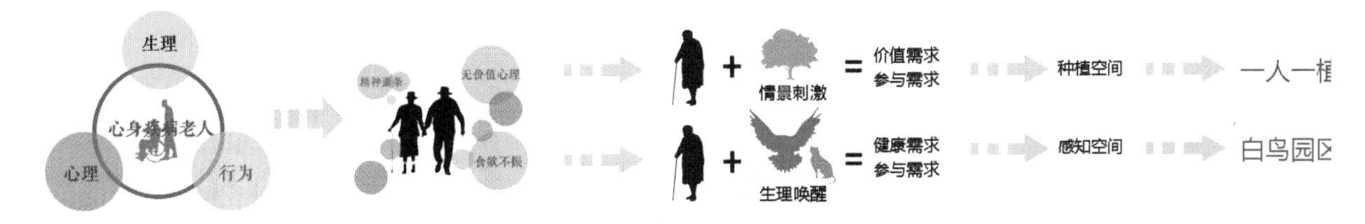

在节点的选择上，交互元素是首要思考的设计因素。因此，结合老年人的生理、心理及行为特点和心理需求，提炼出了植物和动物这两大交互元素，得到一人一植和白鸟园区两个交互节点空间。

Interactive elements are important for the selection of spatial nodes. Given the physical, physiological and behavioral characteristics of seniors, and their psychological needs, plants and animals are adopted as main interactive elements. Then the interactive spatial nodes of Seniors & Plants and Birds Garden are determined.

设计分析 || Design analysis

植物本身就是生命，并千变万化，它的变化让人们感受到流动的色彩旋律，体会到生命的节奏，促进身心的康复。
通过老年人与植物的共成长，形成一对一的视觉对话，以植物作为生命寄托，创造人与环境持续性的情景交互，从而产生新的人与人的交互形式，分享植物成果，交流种植心得，使他们的自我认知得到重建，拾起追求生命的本真动力。

Plants are forms of life. The changes of plants in shape and color present the rhythms of life, helping seniors to recover, obtaining physical and mental health.
When seniors meet plants, they can "dialogue with" plants through eye contact. When they inject emotions into plants, constant interaction between them and the environment occurs. When they exchange ideas of planting and share their harvests, interaction among seniors happen, and in this process, their self-cognition is reconstructed, and their enthusiasm for pursuing a better life is aroused.

空间元素 || Spatial elements

种植结构上，满足植物根系和株高的生长空间尺度，进行有序的组合变化形成私密、半私密的活动空间。

Plants are grown in such a way that sufficient space for the growth of their roots and stems is provided. Through skillful design, private and partially private activity spaces are created.

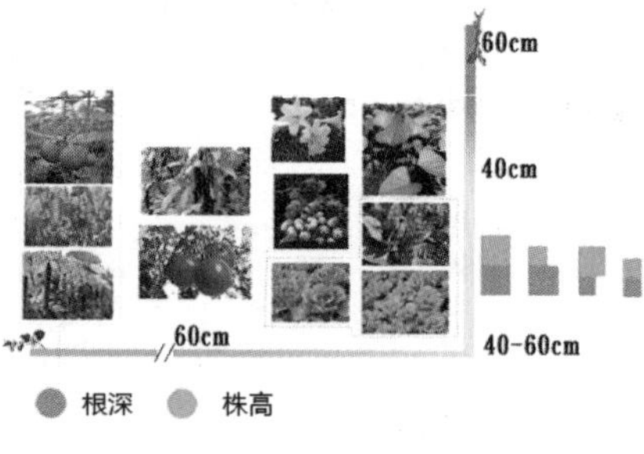

考虑到老年人本身多病易病的体质，有利于老年人身心健康的草本植物及瓜果蔬菜成为了种植首选。

Given that seniors are physically weaker, herbs, melons and vegetables which are good for their health are grown.

节点区位 || Spatial node locations

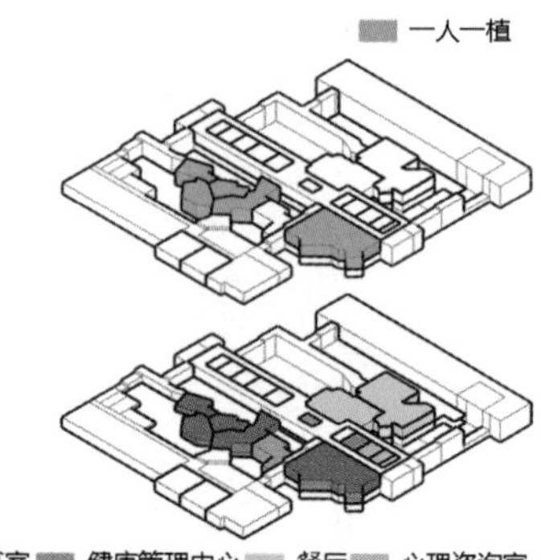

建立老人自己发现病源、寻求解决方法并及时做好防病愈病的意识，实现老人身心健康。

Help the seniors to develop the awareness of preventing, discovering and seeking solutions to treat diseases by themselves, promoting their health.

a 健身空间
b 展示休闲空间
c 种植交流空间
d 无障碍设计

功能空间

种植区
交流休闲区
活动区
健身区
花艺区
棋牌区

功能分区

灰色橡胶地板 黄色橡胶地板 黄色橡胶地板 草地 防腐木地板

材质分析

设计分析 || Design analysis

建筑及立面结构上，通过对树这一基本物体的分解，将树干和树枝通过二维和三维的方式表现在建筑立面上。

2D and 3D images of trunks and branches are presented on facades based on architectural and façade structure analysis.

建筑结构及立面分析

Architectural structure and façade analysis

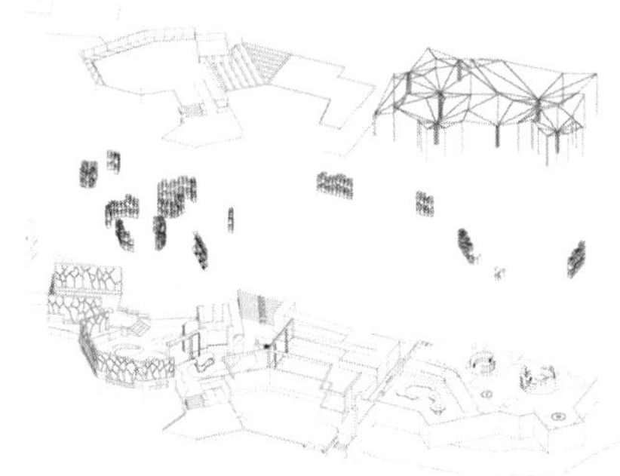

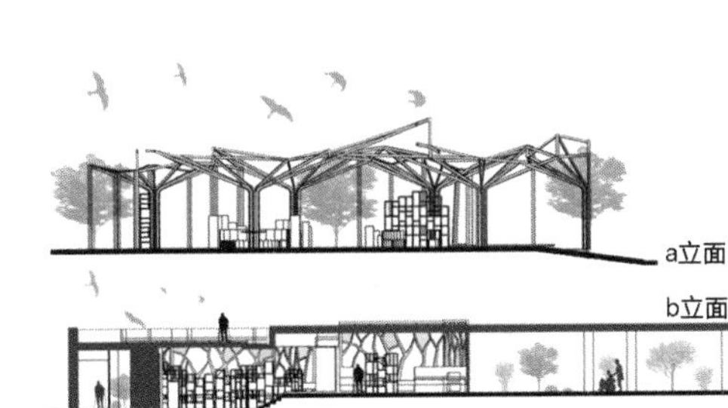

设计分析 || Design analysis

放大变形

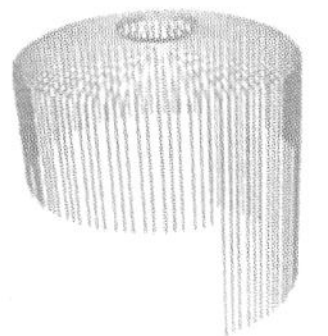

空间组合

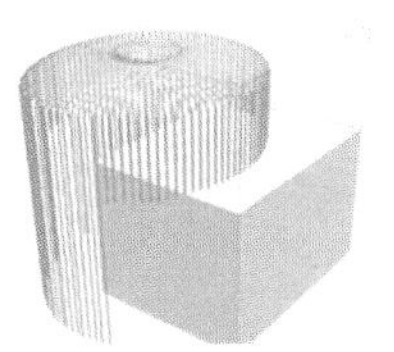

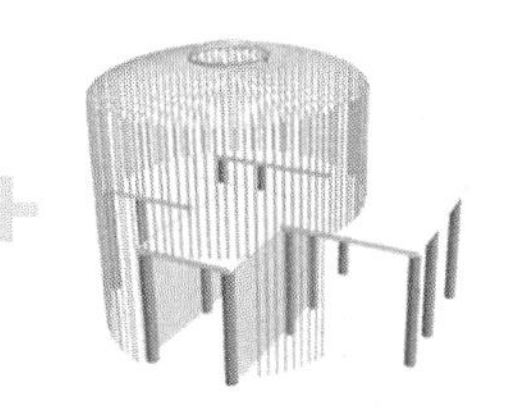

鸟笼单体　鸟笼空间　鸟笼空间组合 1　鸟笼空间组合 2

从北京当地文化特色中的鸟文化入手，提取鸟笼元素，将鸟笼放大变型，形成建筑空间。

Keeping birds is a characteristic local culture of Beijing. In the design, large birdcages are built to form architectural spaces.

结构分析 ‖ Structural analysis

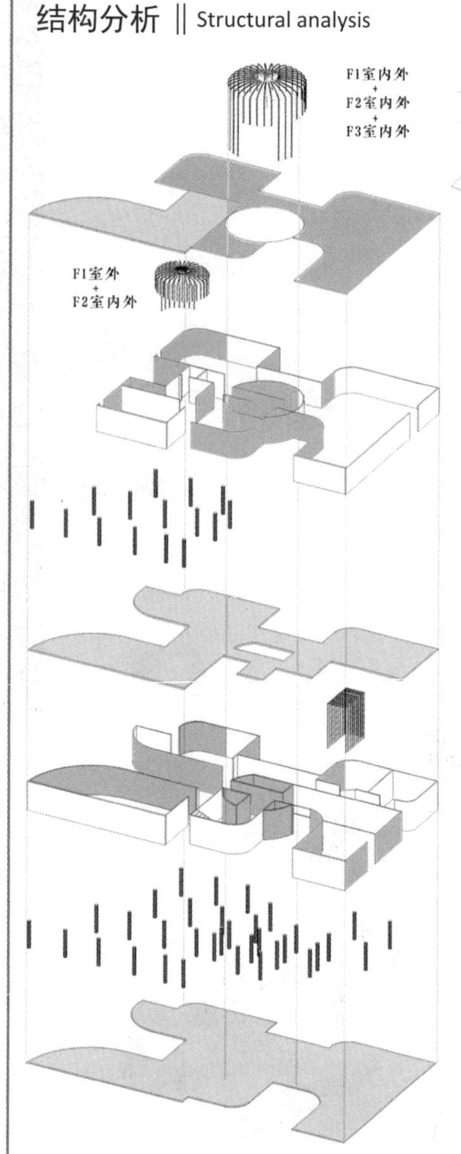

平面设计 ‖ Plan design

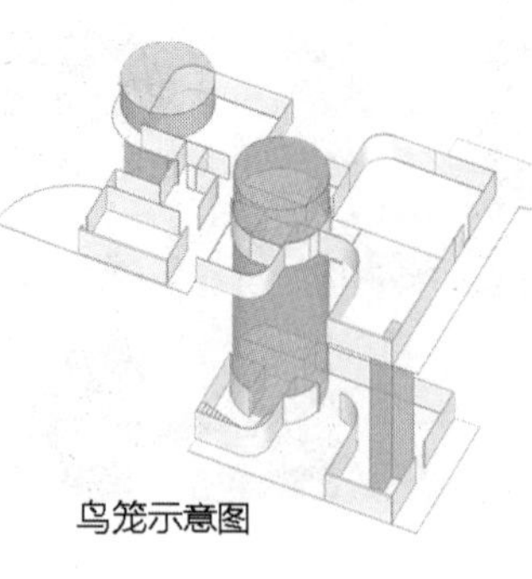

鸟笼示意图

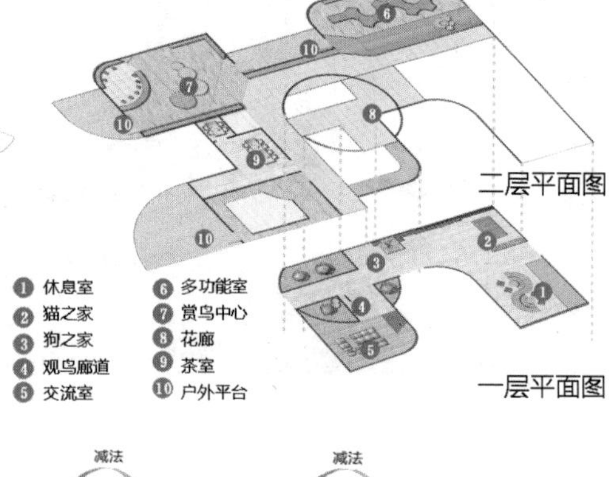

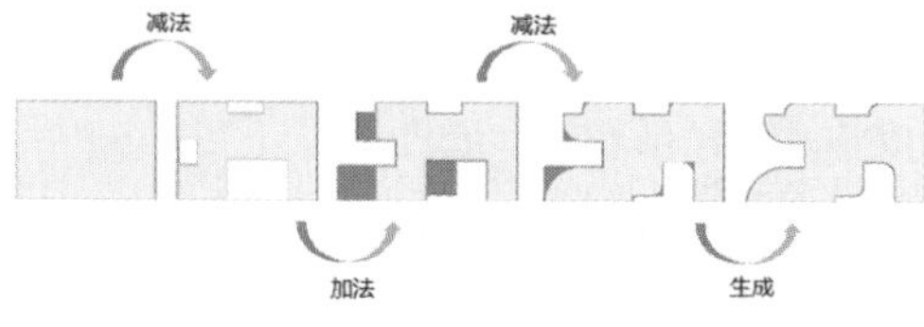

百鸟园的平面轮廓由一个长方形减去部分空间，形成外向庭院；根据功能需求，增加部分空间。又因直角不适宜轮椅老人转弯，于是，将直角进行倒角。

鸟之家，作为人与鸟的自由交往空间，贯穿整个百鸟园。不仅连接一、二层功能空间与室外庭院，更在空间的竖向关系上，冲破屋顶界限，将室内空间与天空融为一体，使整个百鸟园从各个角度看都生机盎然。

功能空间多样，设有喝茶、聊天、种植、展示、遛鸟、逗猫、遛狗、健身、打牌、下棋等不同功能。

Seniors & Birds is a quasi-rectangle outward yard. Some spaces are added to meet the needs of the seniors. All right angles are chamfered for the movement of wheelchairs.

"Home of Birds", as a space for the communication between man and birds, stretches through the whole Birds Garden. Horizontally, it stretches to the functional spaces on the 1st and 2nd floors, and the outdoor yards, and vertically, it reaches the roofs, removing the boundaries between interior spaces and the sky, bringing vitality to every corner of the Birds Garden.

Diverse functional spaces are created for the seniors to drink tea, chat, grow plants, exhibit, walk with caged birds, play with cats, walk dogs, do exercises, play cards, play chess, etc.

流线分析 ‖ Traffic flow analysis

交织流线　鸟·流线　人·流线　猫·流线　狗·流线

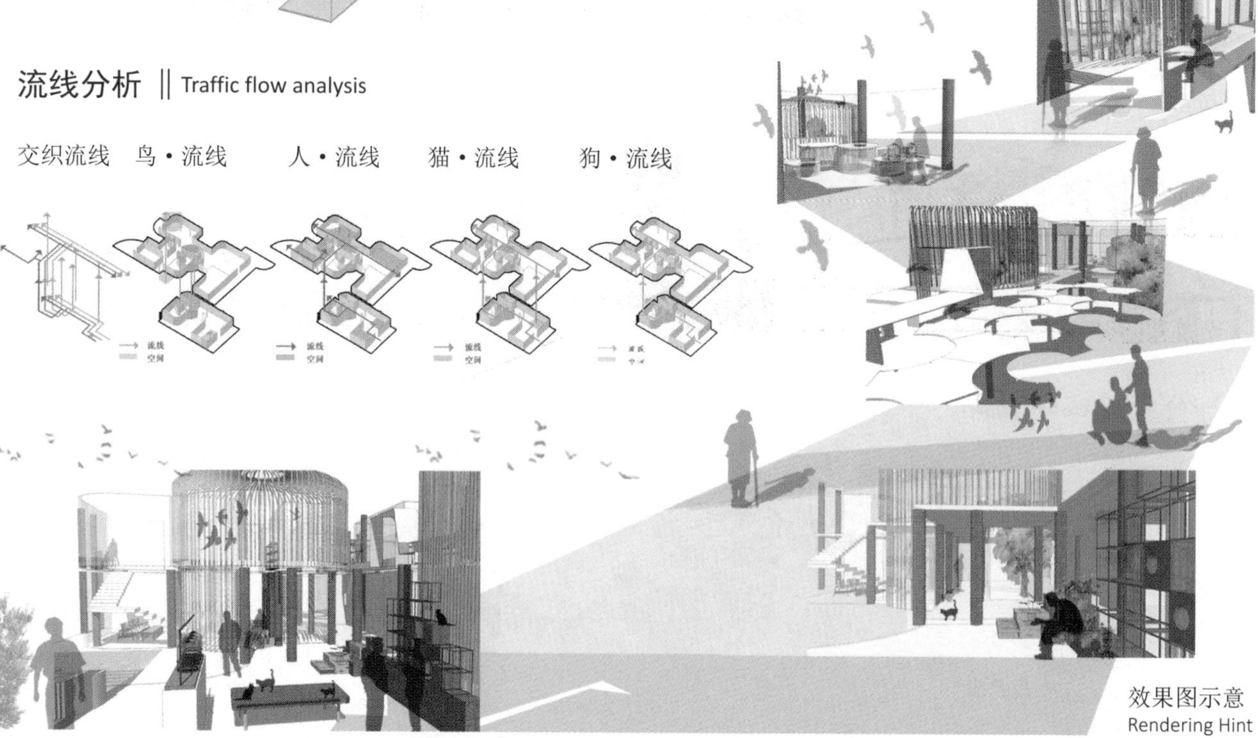

效果图示意
Rendering Hint

导师点评：

该方案对原有基地进行了全面细致的调研及分析，对适应人群、外环境条件、原方案的不足之处做了深入的解析。

方案优点：对环境的改造非常富有变化，功能分区清晰明了，交通流线层次清楚，将环境空间进行了有序的梯级化，给人以更多的想象空间。

方案不足：对外环境的改造有些琐碎，空间变化过多，适老性上考虑不足；对内部空间的改造欠缺，应该更多从整体上分析考虑。

兆翚

Advisor's Comments:

The scheme is developed based on a comprehensive and detailed survey of the project site, and an in-depth analysis of the target group, exterior environment, and the shortcomings of the original design scheme.

Advantages: The scheme features flexible environmental redesign, clear functional zoning and traffic flows, and layered landscape which creates more possibilities of imagination.

Deficiencies: The external environmental design is too flexible without a focus and full consideration of the seniors' needs. Interior spatial design should be conducted from an overall perspective.

Zhao Hui

专家点评：

该方案试图通过老年人心理需求的场所回应，来影响其行为健康的空间设计理念，利用不同的交互形式与空间元素的有机互动，产生唤醒认知的心理因素，从而获得情感与空间的多重体验。

对建筑室外场地的重新规划与设计，增加了有趣的空间体验，整个室外空间的布局流动性较好、动线清晰、植物配比合理。然而在垂直交通方面还需进一步完善无障碍设施。

该方案若能将室外空间与建筑内部空间有机联系起来，方案的立意就会更有说服力，同时也会更好地展示独特的创意。

马本和

Expert's Comments:

The scheme tries to, through interactive spaces, satisfy the psychological needs of the seniors, affect their behavior, and ultimately, improve their health. In the process of interacting with spatial elements through different ways of interaction, the seniors' cognition is awakened, and consequently, they can obtain both emotional and spatial experiences.

External spaces are redesigned by introducing interesting spatial experiences. The overall exterior space layout presents good fluidity, clear traffic flows and rational plant combination. However, wheelchair friendly facilities still need to be improved in the vertical transport system. If the interior and exterior spaces of the buildings can be connected harmoniously, the scheme will be more convincible, and its unique innovations can be better embodied.

Ma Benhe

To Live, To Learn

CIID“室内设计 6+1”2016（第四届）校企联合毕业设计

CIID“Interior Design 6+1”2016(Fourth Session)University and Enterprise Joint in Graduation Design

高　　校：北京建筑大学

College: Beijing University of Civil Engineering and Architecture

学　　生：陆昊 魏卿 曾宁

Students: Lu Hao Wei Qing Zeng Ning

指导教师：杨琳 朱宁克

Instructors: Yang Lin Zhu Ningke

参赛成绩：医养建筑改造与室内设计组三等奖

Achievement: Third Prize for Aged care Building Renovation and Interior Redesign

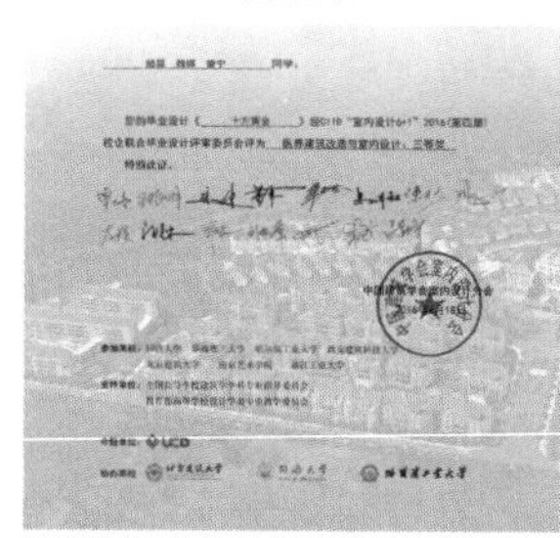

陆　昊 Lu Hao　　魏　卿 Wei Qing　　曾　宁 Zeng Ning

学生感悟：

本次 CIID“室内设计 6+1”2016（第四届）校企联合毕业设计让我收获颇丰。首先是知识面的扩展和深化。以前很少涉及老年建筑、室内和景观的调研和设计，这次不但在设计思维的转化上学习到了很多，也在各方面的标准、规范上了解和掌握了很多。其次是在设计过程的思路上得到了很大启发，不同于平时课业所涉及的虚拟题目，本次毕业设计更实际、更理性，对于前期的工作量，比如实地调研、访谈、总结、文献综述等，有很大的需求。这也让我在面对一个新的设计题目时，对于适合此题目的设计方法逐渐有了更加明晰的掌控。

Students' Thoughts:

I had a lot of gains from the 2016 fourth "Interior Design 6+1" joint graduation design of colleges and enterprises, sponsored by China Institute of Interior Design (CIID). First of all, my scope of knowledge was extended and deepened. I was seldom involved in research and design of architecture, interior and landscape for the elderly. This time I not only learned more about transformation in design thinking, also understood and mastered many standards and specifications in all aspects for this. Furthermore, I had much inspiration on the ideas of design process. Different from virtual topics involved in school, the graduation design is more practical and rational, and also has a lot of demand for the early work such as field survey, interviews, summary, and literature review. This also allowed me to gradually have clearer control over appropriate design methods in the face of a new design topic.

建筑平面分析

C1：学堂功能，根据不同空间尺寸和位置灵活安排具体功能。
C2：老年保健品牌入驻，可分为展示区、售卖区、体验区。

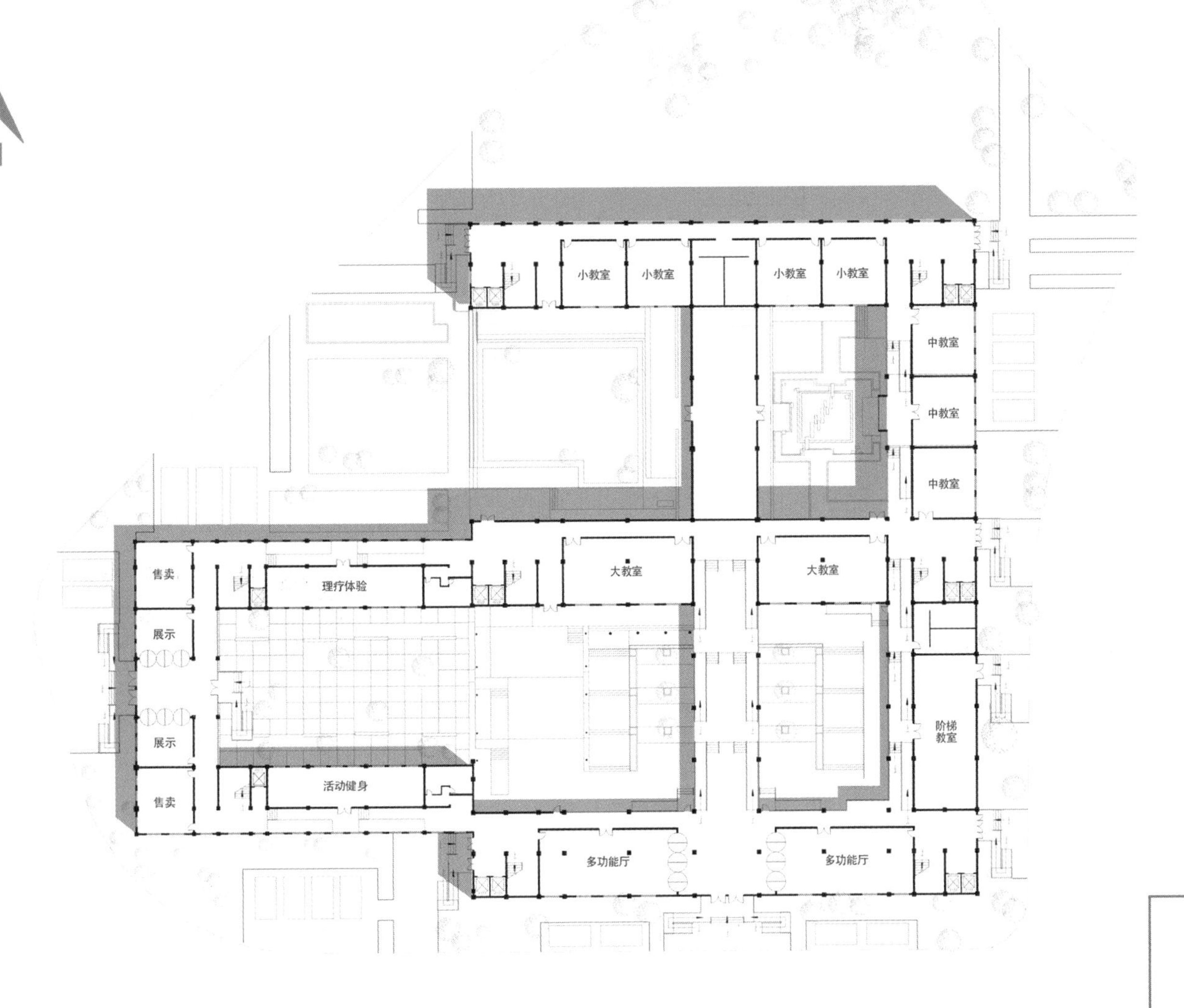

室内分析

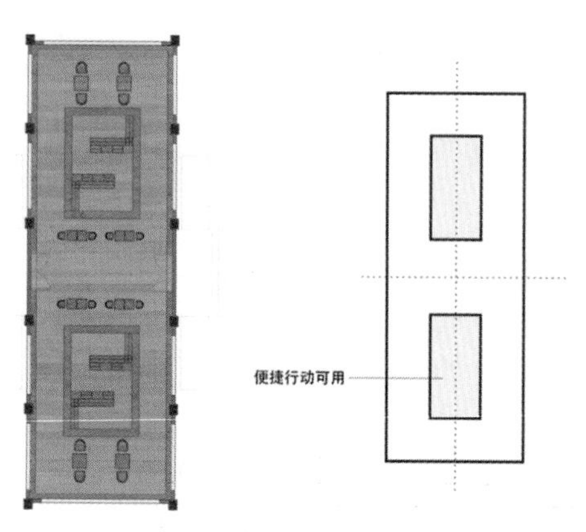

• 下沉的阅览空间增加了藏书的数量，增添了空间的趣味。沿建筑周边做木台，不仅增加了藏书空间，而且增加了座位，设计更为人性化，老人想坐随时有位置可坐。

• 一层有桌椅可供腿脚不便的老人阅读、写字、休息，不用进入下沉空间。下沉空间桌椅可供老人阅读休息，增加了空间使用面积。

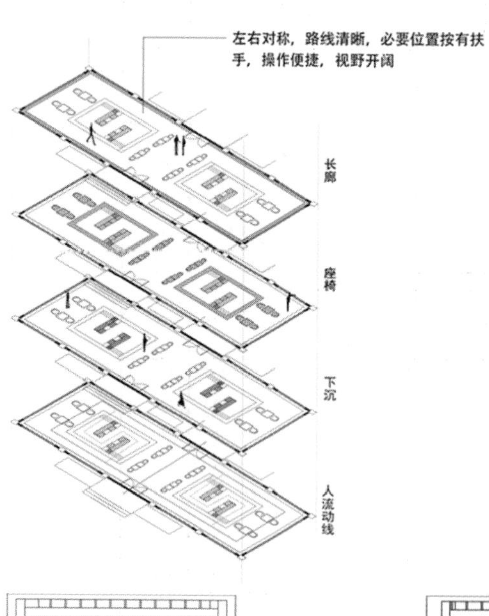

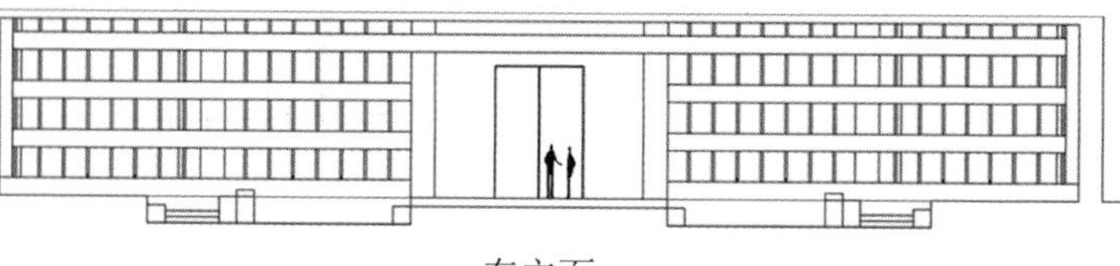

北立面　　东立面

建筑平面分析

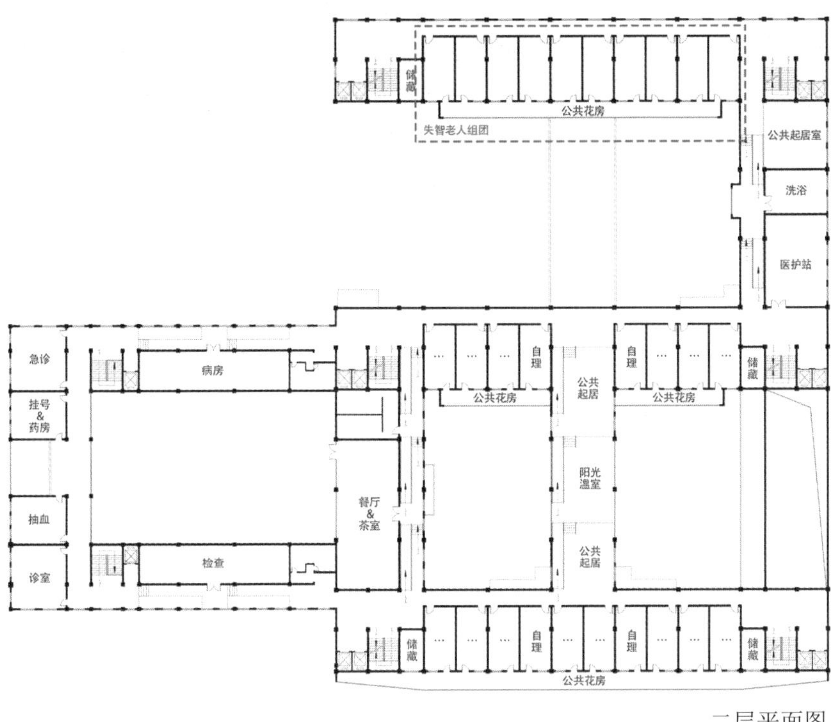

二层平面图

C1：失智老人组团、自理老人居室、公共服务空间。
C2：医疗功能空间。

• 二层及以上设置为不同状态老年人的户型标准。各层中设置有公共服务空间，面向该层的所有老人。

• 二层最北设置失智老人组团，将所有失智老人集中照护，并配置有洗浴室和公共起居室。

• 南侧设置自理老人户型，连接处设置阳光温室，每层的户型南侧有公共花房。

• C1 的西侧设置有自助餐厅和茶室。

室内分析

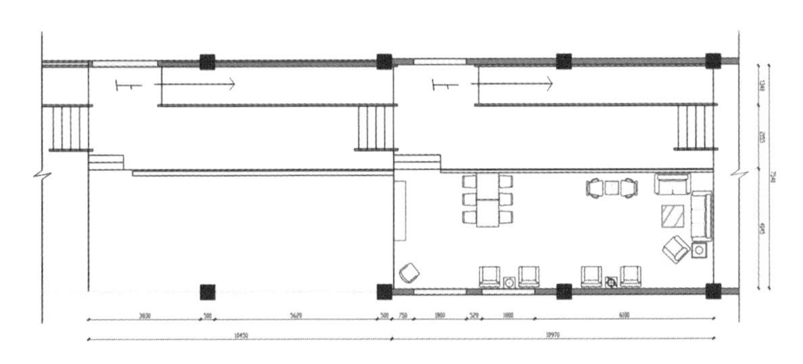

阳光温室 & 公共起居室平面图

• 阳光温室——旨在为老人提供一种学习、劳动、自然相互关联的生活态度和生活方式，在公共起居室中，老人通过阳光温室中的活动产生共同话题，讨论、交流、学习。

建筑平面分析

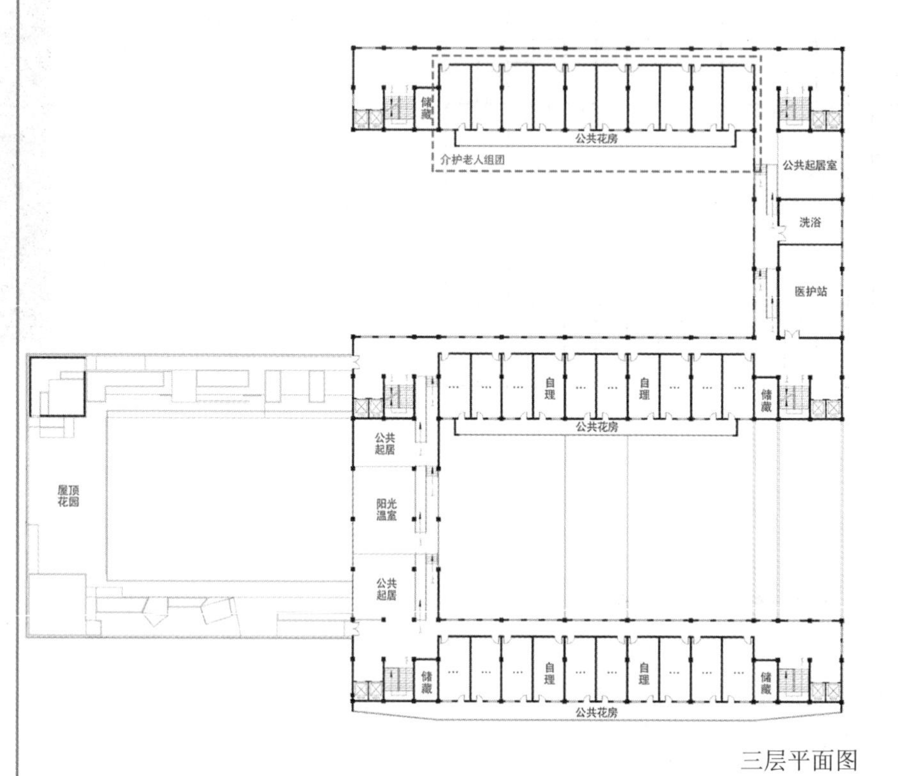

三层平面图

C1：全介护老人组团、自理老人居室、公共服务空间。
C2：屋顶花园。

• 北侧为全介护老人组团，配置公共服务空间。

• 南侧与二层相同，为自理老人户型，设置公共服务空间和公共花房。

•C2屋顶有面向C区的屋顶花园，有较为开放的长椅和长桌，也有较为封闭的围合式私密空间。屋顶南侧有视野很好的观景平台。

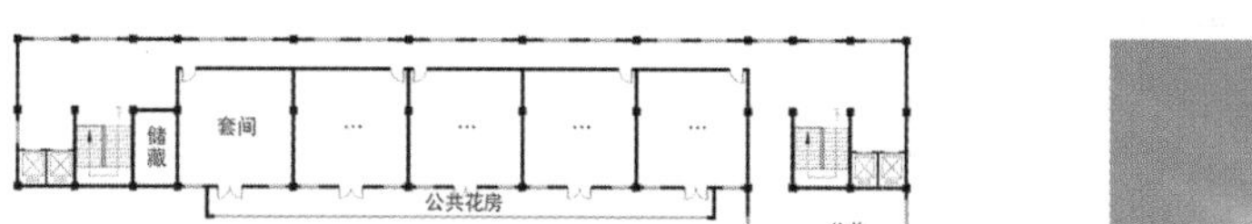
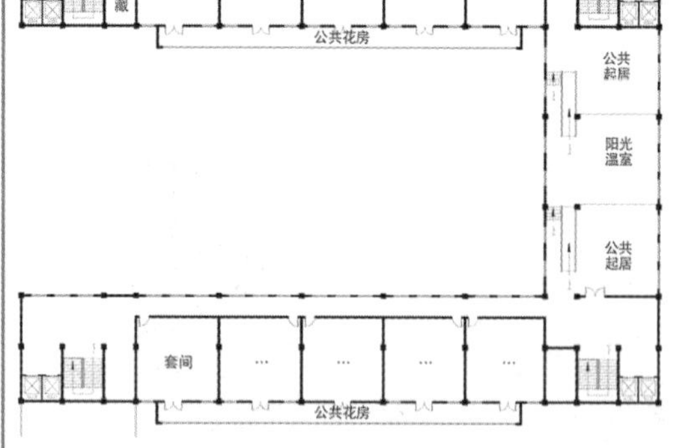

四层平面图

导师点评：

“十方黉舍”，题目别致，立意较好。黉舍有学校、学堂之意，点明设计场所目标为老年大学，宗旨即是设计一处能满足老年人精神需求的场所。设计前期调研扎实，对老年人的生理、心理需求进行了深入解析以指导设计。在建筑一层部分压缩原有居住面积，为老人提供了多个公共活动空间，如活动健身、理疗体验、多功能厅以及图书馆等。

值得一提的是，图书馆与花房的设计。图书馆的空间设计层次丰富，安逸宁静的氛围符合功能要求。最温暖关怀的设计部分在于三栋建筑的每层随处可见公共花房空间，让自理与失智老人都能感受到大自然的气息，调节身心健康。提出几点建议：第一，三栋建筑一层整体作为公共活动功能使用，活动面积过多，是否存在利用率低的问题？可以考虑资源有效整合，一个活动室可分时段供多种活动使用。节省出一些一层阳面的空间给失智老人作为居住使用。第二，图书馆的下沉空间设计虽有趣别致，却忽略考虑腿脚不力的自理、半自理老人使用问题。除此之外，一楼的阶梯教室等多个公共区域设计，除了一般性的功能家具、设备布置外。同样重要的应是更多地考虑老年人的特殊关怀设计，如轮椅坡道。第三，前期调研分析中提到的禅学空间让人意犹未尽，但在最终的设计中没有体现。同时，对居住部分室内空间的设计考虑较少，这是比较遗憾的地方。

总的来说，设计作品对基地三栋楼的整体设计，功能布局明确，交通流线清晰，做到了题目的内涵追求——精神至上的老年家园。

何方瑶

Advisor's Comments:

"To Live, To Learn" is a very special and meaningful theme. It indicates that the objective of the scheme is turning the retirement community into a school for the elderly where the mental needs of the seniors can be met. In early survey stage, the students carefully studied the physiological and psychological needs of seniors as the basis for later design. On the 1st floors of the buildings, some living spaces are turned to public spaces such as gym, physiotherapy treatment room, multifunctional hall and library.

The laboratory and indoor gardens are the highlights of the design. Diverse functional spaces of the laboratory are rationally designed, creating a comfortable and quiet environment. Indoor gardens on every floor of the three buildings representatively show the designers' care for the seniors. Such gardens create a natural environment for both the able-bodied seniors and seniors with dementia, promoting their health.

Lastly, I'd like to give some suggestions. Firstly, turning the 1st floors of all the three buildings into public areas will actually lower the utilization rate of the spaces. In my view, spatial resources should be effectively integrated. For example, an activity room can be used to hold different activities at different times, saving some sunny spaces for seniors with dementia to live in. Secondly, the idea of designing a sinking space in the laboratory is interesting, but it ignores the access of seniors with disability. Likewise, in several public areas, such as the lecture hall on the 1st floor, in addition to functional furniture and facilities, more importantly, wheelchair friendly facilities like wheelchair ramps should also be designed. Thirdly, the idea of creating a Zen space is mentioned in the early-stage survey. It's a very interesting idea, but it doesn't appear in the final report. Another unsatisfactory thing is that interior design of the living areas is not described in detail.

Generally speaking, the functional layouts and traffic flows of the three buildings are clearly and thoughtfully designed for the purpose of building a spiritual home for the senio.

He Fangyao

专家点评：

同学们解决了中期评审时对佛教中关于禅修的理解太空泛和思路太分散的不足，把握住了对禅修文化的诠释方法，使设计方案具备了一定的思想深度，但细微之处没有能够全面展现出来。在建筑规划和功能布局上，采用拆改结合、部分增建的改造手法，改小了院落的尺度、增加了院落的层次，使得建筑的内部空间感与佛教建筑中庙宇的布局有共通之处，让居住人群在精神层面上有所感悟和升华。

卓培

Expert's Comments:

Compared with the mid-term report which lacks a substantial understanding of Zen meditation and a focus of related design, the final scheme profoundly interprets the Zen meditation culture which makes the scheme more insightful. However, detailed interpretation is still not given. In terms of architectural planning and functional layout, some spaces are removed, some are transformed, and some are newly added. The yards are smaller, but are given more functions and design elements. Interior design borrows some elements of Buddhist temples, promoting the self-cultivation of the occupants.

Zhuo Pei

Six Harmonies

CIID“室内设计 6+1”2016（第四届）校企联合毕业设计

CIID“Interior Design 6+1”2016(Fourth Session)University and Enterprise Joint in Graduation Design

CIID 中国建筑学会室内设计分会

“室内设计6+1”2016（第四届）校企联合毕业设计

获奖证书

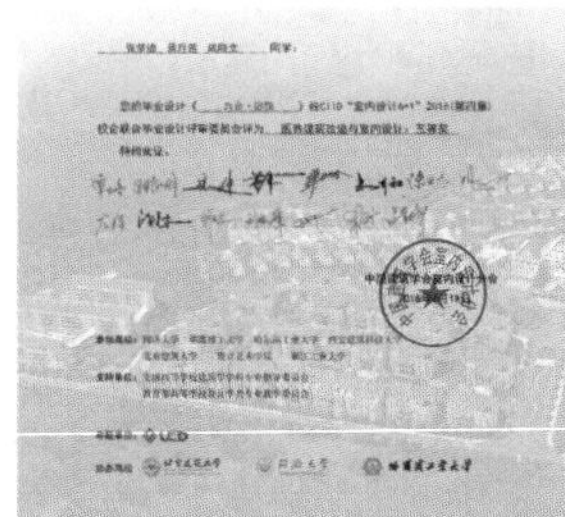

高　　校：南京艺术学院

College: Nanjing University of the Arts

学　　生：张梦迪　龚月茜　戚晓文

Students: Zhang Mengdi　Gong Yuexi　Qi Xiaowen

指导教师：朱飞

Instructors: Zhu Fei

参赛成绩：医养建筑改造与室内设计组三等奖

Achievement: Third Prize for Aged care Building Renovation and Interior Redesign

张梦迪
Zhang Mengdi

龚月茜
Gong Yuexi

戚晓文
Qi Xiaowen

学生感悟：

经过大半年的奋战，从北京到上海再到哈尔滨，我们经历了很多，也收获了很多。在没有参加“6＋1”之前，我们觉得毕业设计只是对这几年来所学知识的单纯总结，但是通过这次锻炼，我们发现我们的认识过于片面。毕业设计不仅仅是对前面所学知识的一种检验，更是对自己能力的一种提高。通过这次互动我们明白了自身的知识和经验还是缺乏的。虽然在设计的过程中，我们遇到了很多的困难，方案改了一次又一次，也被批评了很多次，但是在方案完成的那一刻，我们是快乐的。只有坚持，才会有真正的收获。

Students' Thoughts:

From Beijing to Shanghai to Harbin, during the competition which lasted for more than half a year, we experienced a lot and also learned a lot. Before participating in the competition, we thought that a gradation project is just a test of the knowledge that we'd learned in the university years. Later we realized that our understanding is too partial – a graduation project is not only a test, but also a process of improvement which helps us realize our lack of experience and knowledge. It's true that the process of design was hard. We met a lot of difficulties, changed our scheme again and again, and were criticized for many times. However, at the moment when the scheme was finally completed, we were very happy and excited. Success cannot be made without insistence.

病房区域 || Ward area

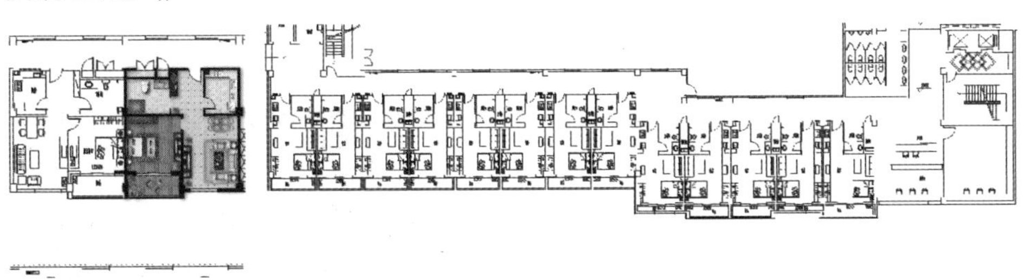

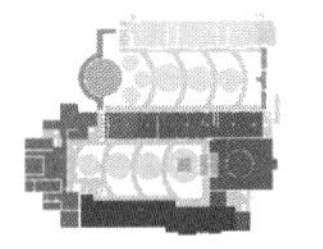

残障老人住房区
Living area of disabled seniors

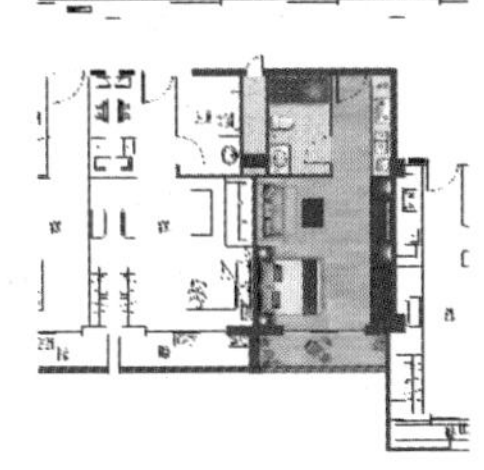

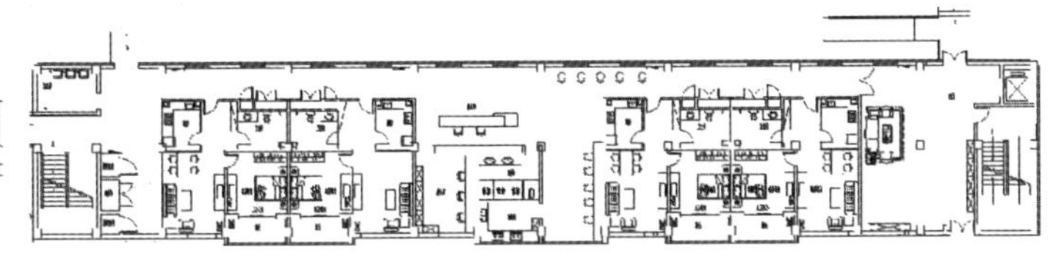

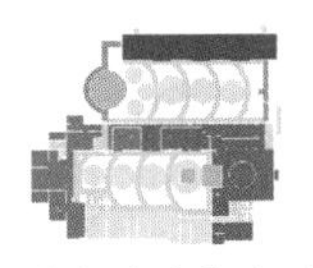

高级老人住房区
VIP three-room suite area

庭院区域 || Yards

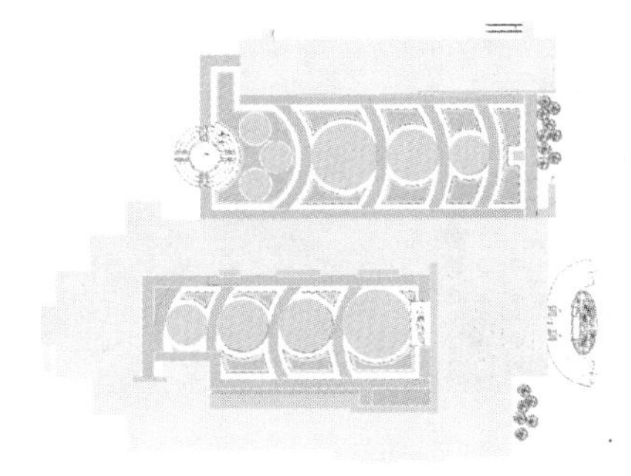

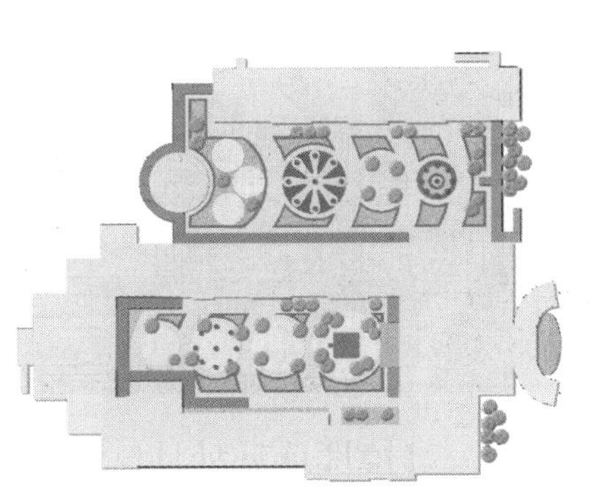

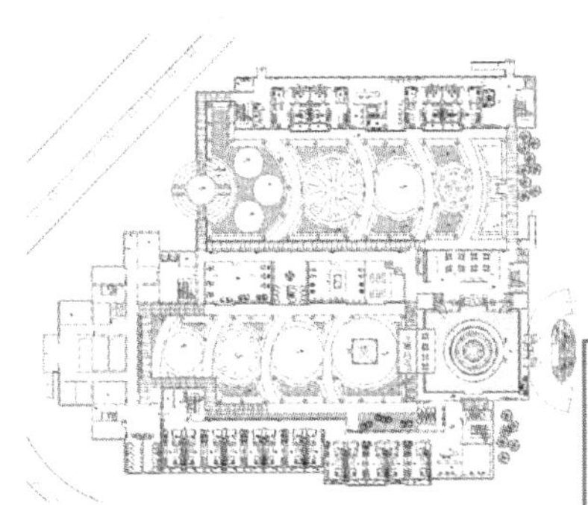

调研分析 || Survey analysis

类型	研究分析
展示型和观赏型老人	现代老年人都有一定的特长，能歌善舞、能画会写，有一定的艺术表现能力,他们往往喜欢参加展示活动，如老年合唱团、老年时装模特表演、老年书画展、老年舞蹈队、乐队、街头广告宣传队等活动。而没有表现能力的老年人则喜欢“看热闹”，他们成群结队地看表演、听故事、看展览，相互交流观看心得，不亦乐乎。展示型和观赏型的两个群体在老年人中形成了良好的互补互动。
根据受众，搭建展示公共平台方便老年人交流与互动	人各有所长，很多老年人往往有很多艺术才能，生活技能等，可因为居住空间，居住群体，而没有使其得到良好的展示与交流互动，因此其兴趣也随之减弱，从而影响身心发展，设计者的我们如果能将空间构造成一个展示平台,会促进老人们的交流与互动，增进其各方面的兴趣，从而老有所爱,老有所乐。

元素分析 || Element analysis

导师点评：

“六合·宓馆”命题新颖，充满东方意味，设计方案从中国传统中医养生切入，“顺气一日分为四时，中医养生、养气，气和而神形、阴阳平衡”，并提出宅物、宅己、宅生、合形、合气、合心的设计思想，方案带入感极强。

方案充分考虑了普通健全老人、残障老人、旅游人群、家属子女、医疗服务人员、物业管理人员六类受众人群，并创新性地将老人分为展示型老人和观赏型老人，注重搭建展示公共平台、方便老年人交流互动，采用竹、木、石、土、水的设计元素，使得空间充满淡雅、传统的中式味道。平面方案规划合理，注重阴阳结合，病房区域分类设置，合理布局，入口门厅重点设计，效果尤为突出。庭院设计有机结合太极八卦的意向，设置2处凉亭，体现中式建筑的特征。

方案设计新颖，紧扣中式建筑、园林、室内的特征，有一定创新性。但方案图纸部分有所欠缺，制图规范性不足，需进一步提升。

朱宁克

Advisor's Comments :

"Six Harmonies" is a typical oriental theme. Adopting TCM as the highlight, the scheme incorporates such TCM concepts as "variation of qi in a day", "health maintenance", "cultivation of qi", "qi-spirit harmony", and "yin-yang balance" into the design. It also discusses the relationship between homes and objects, man, and life, and the coexistence of different forms, types of qi, and ideas in the context of design. The scheme is very attractive.

The needs of six groups including able-bodied seniors, seniors with disability, tourists, occupants' families, medical personnel, and property management are discussed in the scheme. The scheme innovatively divides seniors into seniors preferring to show and seniors preferring to watch, and corresponding spaces for performance and communication are designed. Besides, elements like bamboo, wood, stone, earth and water are employed to create an elegant environment filled with traditional Chinese culture. The plan design features yin-yang integration, ward classification, and rational spatial division. The entrance hall is a highlight of plan design. The design of the yards applies the concepts of Taiji and Eight Diagrams, with two pavilions presenting yin and yang in each yard.

The scheme is creative, incorporating traditional Chinese culture elements into architectural, landscape and interior design. The drawings should be more normative, and relevant improvements are needed.

Zhu Ningke

专家点评：

该方案通过对原有建筑的门厅、公共活动区域，各层病房和室外庭院进行改造，试图改善目前的居住环境。门厅的改造，引入视觉冲击力较强的体块，其实质的功能、空间感受尚待商榷；公共活动区域的引入能增加使用人群的交流，但应做到动静分区，设置更符合老年人生理及心理需求的内容；病房的分区规划有一定的合理性，建议根据各种设定人群的需求推导病房护理单元可提供的服务，从而进行更为细致的空间调整和划分；室外庭院的形式及材料的选择等，可紧扣提出的“太极”“中医”“道家”这几个主题进行深入设计。方案思路明确，各图纸在规范表达及突出表现方面欠佳，版面略松散。

李莉

Expert's Comments:

In the scheme, the halls and wards of the buildings, public areas, and yards are redesigned, for the purpose of creating a better living environment. The hall design, including functional and spatial design, is impressed, but its actual effects still need further discussion. More public areas are created to promote the communication among the seniors, but areas for active and quiet activities should be separated to meeting the physiological and psychological needs of different occupants. The zoning of wards is kind of rational. My suggestion is that the ward area can be divided in a more specific way that seniors needing different medical services are arranged in different zones correspondingly. As for the design of the yards, forms and the selection of materials should be themed on Taiji, TCM and Taoism. The scheme expresses design ideas clearly and logically, but the drawings are less normative, impressive, and compact.

Li Li

Lotus Garden

CIID“室内设计 6+1”2016（第四届）校企联合毕业设计
CIID"Interior Design 6+1"2016(Fourth Session)University and Enterprise Joint in Graduation Design

高　　校：同济大学
College: Tongji University
学　　生：王雨林 王舟童 张晗婧
Students: Wang Yulin Wang Zhoutong Zhang Hanjing
指导教师：左琰
Instructors: Zuo Yan
参赛成绩：医养建筑改造与室内设计组三等奖
Achievement: Third Prize for Aged care Building Renovation and Interior Redesign

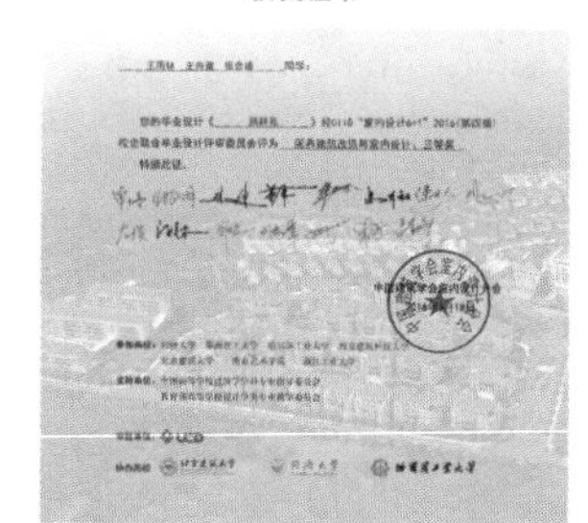

王雨林
Wang Yulin

王舟童
Wang Zhoutong

张晗婧
Zhang Hanjing

学生感悟：

本次毕业设计的特殊性一方面在于基地位于市区之外的密云水库附近，脱离了我们熟悉的既有城市环境，进行建筑设计时既自由又有挑战性。另一方面则是因为设计是基于对并不成功的已有项目的改造，因此需要考虑经济因素。这两个特点决定了这个设计不会像其他老年公寓那样简单。设计中强调中国传统医学与孝道的概念，随着对于这两个母题的探索，我们逐渐明白了自然环境在老年人医养中的重要程度，以及如何以传统的医疗手段和合理的功能配置吸引老年人入住，进而找到了回应基地特殊性的方法。

Students' Thoughts:

The graduation project has two features. Firstly, the project is located near the Miyun Reservoir – such an unfamiliar suburban environment gives us the freedom to design but also a challenge. Secondly, the project is about the renovation of an unsuccessful project, which makes cost control an important matter. These two features make the project a tougher task. TCM and filial piety are the highlights of our design. In the process of probing into the two fields, we gradually understood the importance of natural environment in the health care of seniors, and found the way to rationally incorporate TCM into the design, attracting seniors. In other words, TCM is our solution to the above-mentioned challenges.

概念分析 || Conceptual analysis

—大环境— Environment

[燕云]

燕指燕山；
云染空气。
取燕山云气之意。
借此地清新空气与优美山色。
调养康复、度假休闲。
亦点出闲云野鹤不挂尘世之境。

[愚谷]

典出王维《田家》
“住处名愚谷，何烦问是非”
意忘忧解愁。

又配合地形上从山脉大谷地到山脊小谷地到山丘围合谷地。取谷的地理意向。

—设计— Design

[耦耕]

典出《论语·微子篇》
“长沮桀溺耦而耕”
良家合作的农耕形式
隐居/志同道合之友/独乐乐众乐乐/
单人单家庭到多人多家庭的社交关系。

又典苏轼《浣溪沙》
“何时收拾耦耕身”

[苑]

值林为苑。
——左思《吴都赋》。
刘注：“有木曰苑。”
指示设计环境。
亦取文艺荟萃之处意。

燕雲愚谷
耦耕苑

整体意向：【白屋花开里 孤城麦秀边】（杜甫）活动意向：【软草平莎 轻沙走马 日暖桑麻 风来蒿气】（苏轼）

【短期】Short Term
- 酒店经营
- 专项病专业疗养
- 企业品牌打造
- 实现第一批人的吸引

C区自我价值实现
The Uplift of Value of Section C

↓

【中期】Mid Term
- 园区商品房租赁售卖
- 主打候鸟养老
- 周边旅游资源共游组合
- 园区附近村落有集农业组织互动

园区整体价值复苏
The Uplift of Value of the Area

↓

【长期】Long Term
- 酒店式经营高端疗养的模式推广
- 园区住户世代更替与混合
- 社区组织与复合家庭的凝聚

美好愿景
Future Development

燕雲愚谷
耦耕苑

屋顶种植设计 ‖ Roof gardens

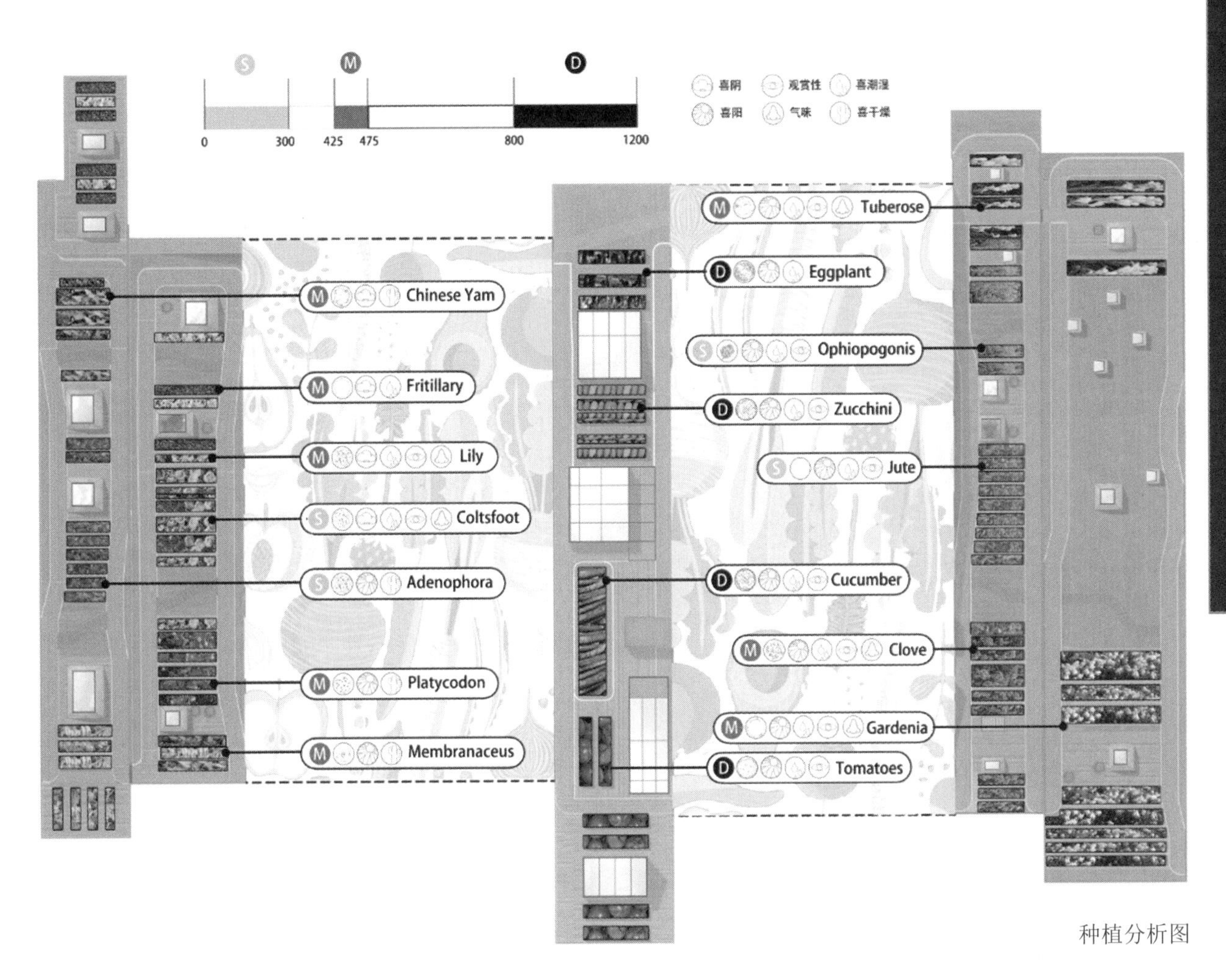

种植分析图

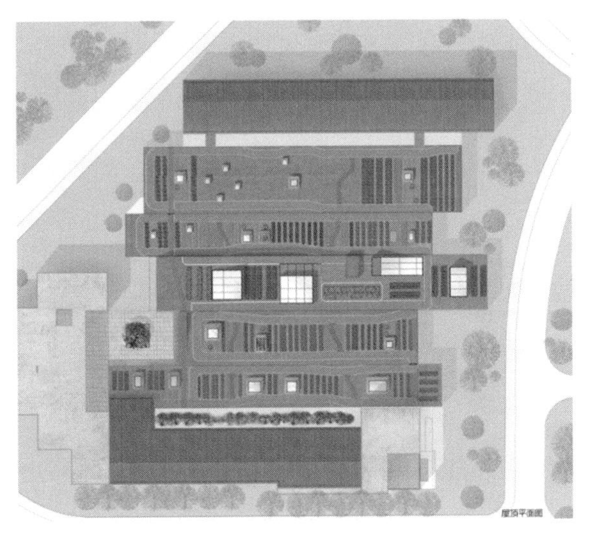

我们又根据形体大小分为中型灌木、小型灌木和草本，依次对应不同土壤深度。由于坡道上覆土厚度最薄接近 100 毫米，所以并不适合种植相对大型的植物，所以我们尽量将小型植物种植在中间的坡面上，将大型植物种植在东侧和西侧水平的平台上。另外南侧的屋顶平台由于建筑的遮挡，日照时间较短，北侧没有遮挡，日照时间较长。这种日照情况的不同我们也有相应的对策。对于川贝和款冬花这些喜阴耐寒的植物，我们尽量把它们安置在南边，其他一些喜温暖且适合阳光充足环境的植物则大多数被安排在北侧。

Middle-sized bushes, small-sized bushes and grass are grown according to the thickness of soil. Since the slope is covered by the thinnest soil of only nearly 100 mm, small-sized plants are grown on the slope, and large-sized plants are grown on the flat platforms to the east and west of the slope. The roof terraces in the south cannot get enough sunlight in the shade of the building nearby, and the roof terraces in the north can enjoy a long sunbath. Given this situation, plants preferring a shadily and cold environment like fritillaria cirrhosa and flos farfarae are grown in the south, and plants preferring a warm and sunny environment are grown in the north.

活动分析 || Activity analysis

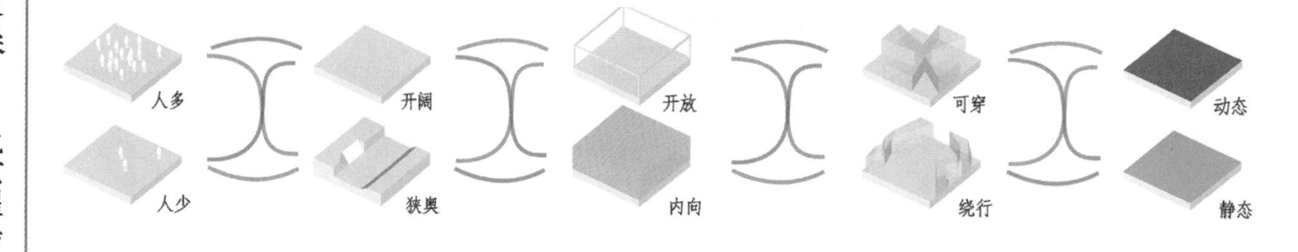

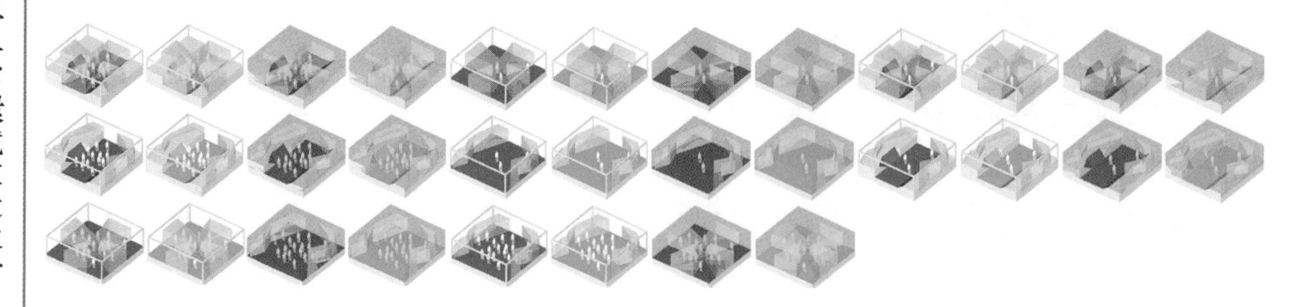

我们想通过分析并抽象出公园中的活动来归纳每一种活动的特性，例如静与动。最终以公园的形式应用到大空间当中来。

We analyzed the characteristics of activities conducted by seniors in gardens. For example, some activities are less active and some are more active. Based on the analysis results, we decided to design the large public area into a garden for the seniors to conduct various activities.

结构设计 || Structural design

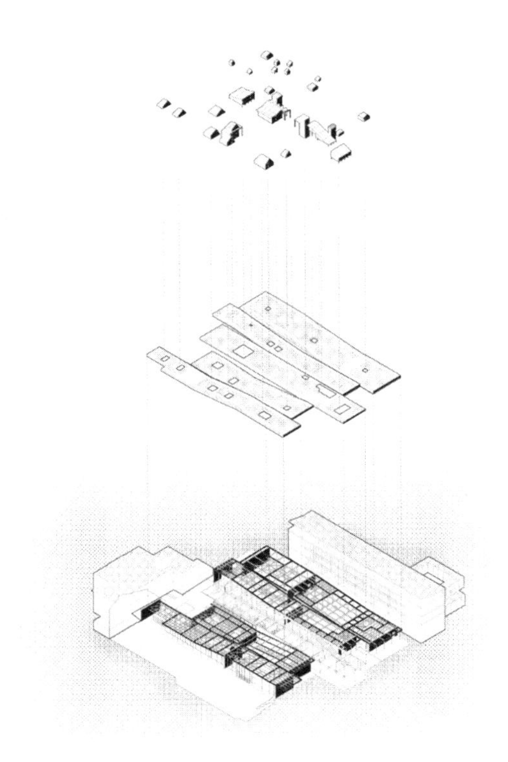

改造说明 || Renovation description

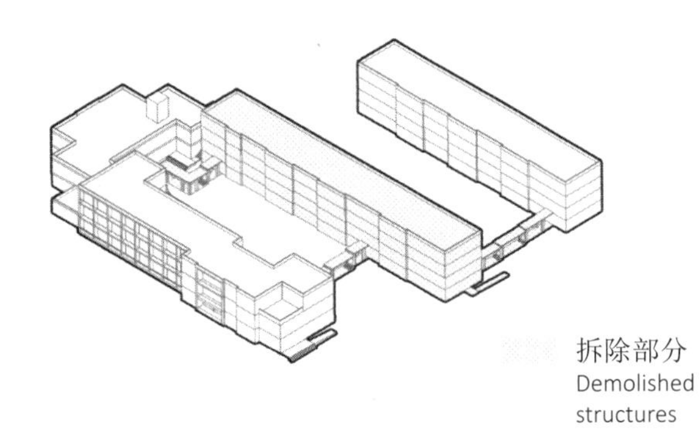

拆除部分
Demolished structures

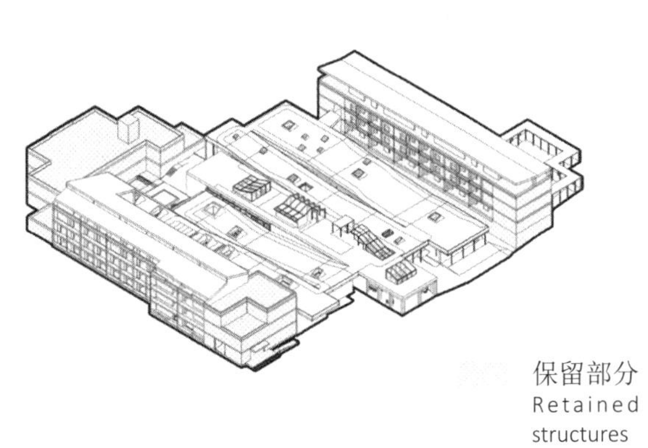

保留部分
Retained structures

平面图 || Floor plan

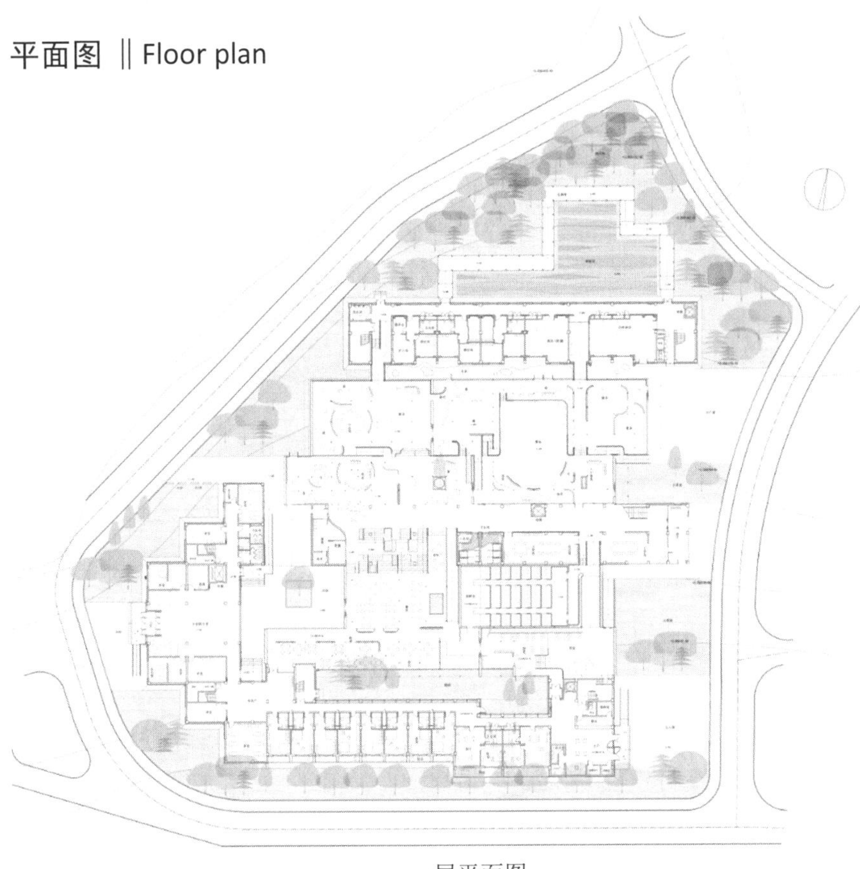

一层平面图
Plan of the 1st floor

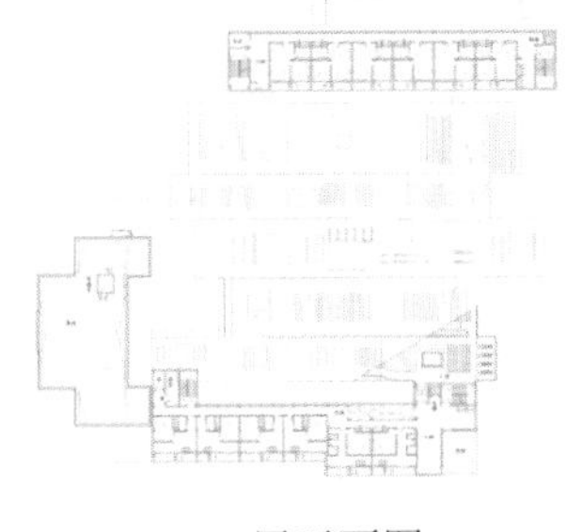

二层平面图
Plan of the 2nd floor

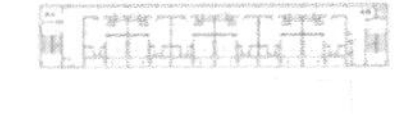

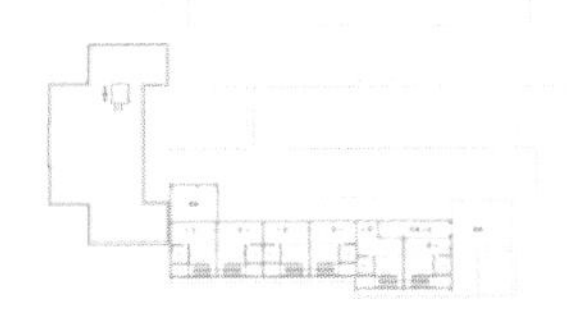

三层平面图
Plan of the 3rd floor

功能分析 || Functional analysis

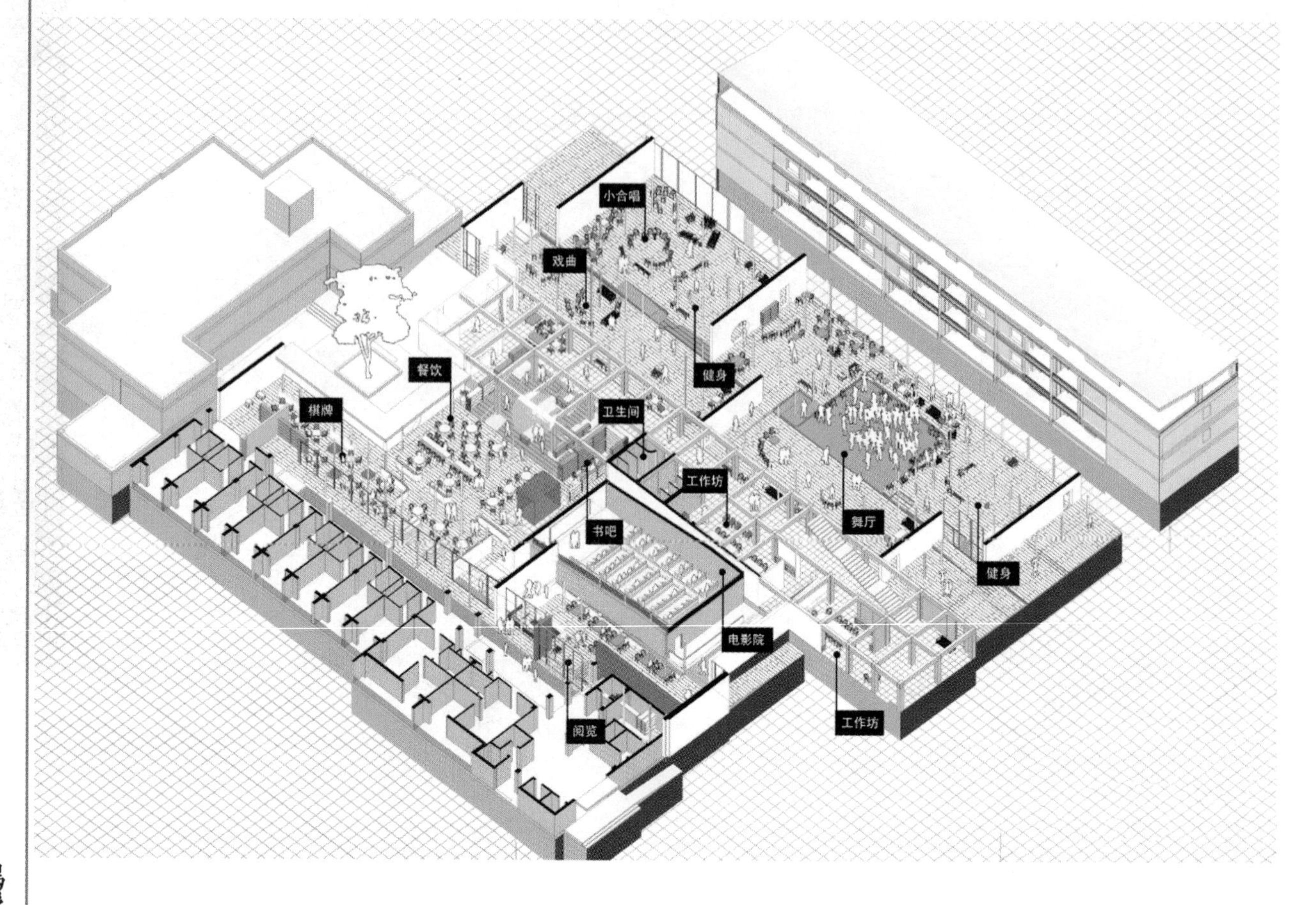

剖面图 || Section

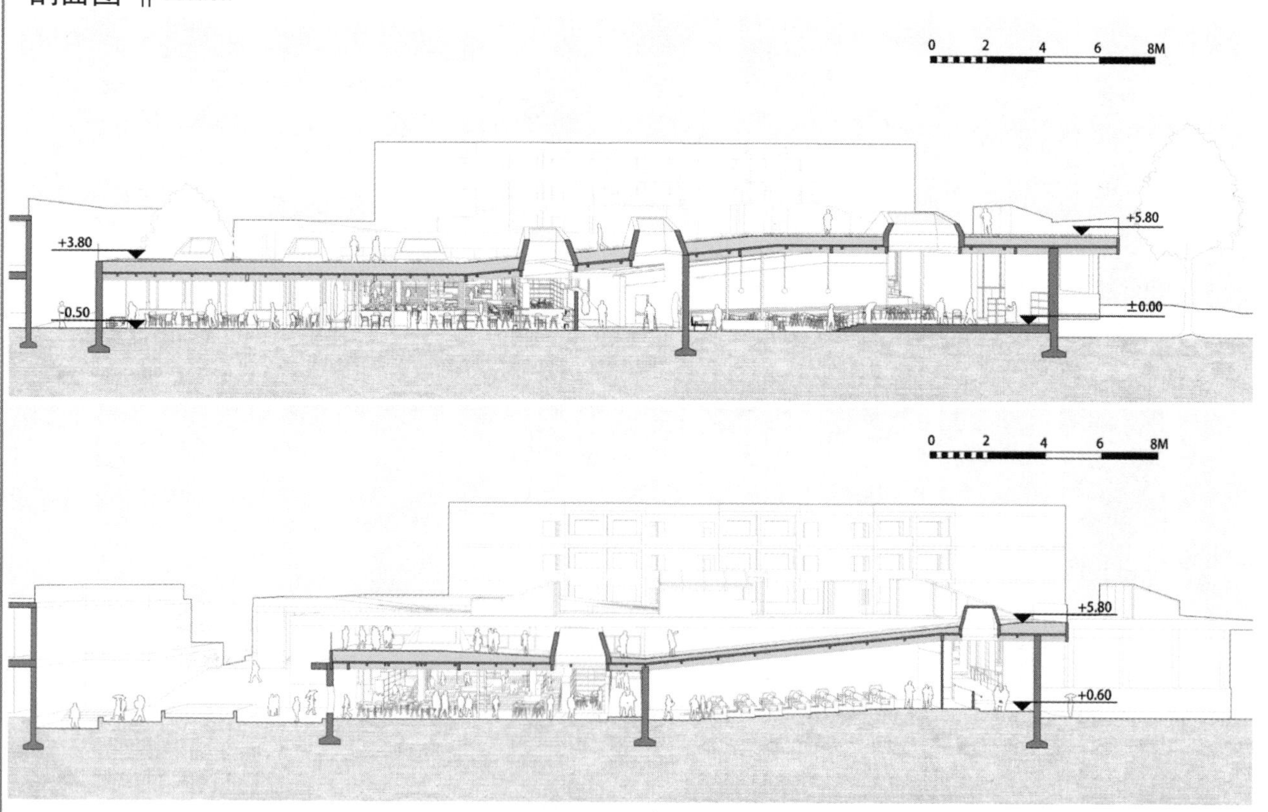

导师点评：

“耦耕苑”设计方案，是在对现场调研和分析的基础上，拆除 2 号楼的二、三层，扩充了首层平面，使 1 号楼、2 号楼、3 号楼连成一个整体，在 1 号楼、3 号楼之间创造了一个为老年人交流活动的大型公共休闲空间，功能丰富，空间有趣。同时改造后，建筑间距加大，采光效果也随之更好。方案利用改造后形成的大面积首层屋面，开辟了蔬菜种植场地，让老年人在耕种中既获得身体锻炼，又成就了老有所用的心理体验，体现了一种人文关怀。

由于首层公共区域过于集中，在消防疏散、通风组织等方面会有一定困难，而室内装饰氛围表达与主题的呼应上略显不足。

周立军

Advisor's Comments:

According to “Lotus Garden”, based on site survey and analysis, the 2nd and 3rd floors of Building 2 are removed, and the 1st floor is expanded, so that Building 1, 2 and 3 are integrated. Besides, a large public recreation space with diverse functions is arranged between Building 1 and 3. The distance between neighboring buildings is increased, creating better lighting. The expanded 1st floor is used to grow vegetables, so that the seniors can do exercises and realize their value during the process of planting. It's a good way to meet the seniors' physical and psychological needs.

One problem is that various functional zones of the public area on the 1st floor are arranged too intensively, creating difficulties in fire escape and ventilation. Another problem is that interior decoration fails to well reflect the design theme.

Zhou Lijun

专家点评：

同学对建筑的空间理解较深刻，通过对原有建筑空间进行合理改造，使空间形式灵活多变，通过合理加建，将原有庭院空间变得更加灵动与实用，增加了老年人更多户外体验的可能性。

在重组户型空间中，充分体现了建筑学专业对空间的把控动力，将户型分为多种组合单元，布局合理，空间利用率得到了有效提高。整套方案从立意到方案生成，再到方案深化都看得出设计团队的同学们具有较深厚的建筑理论修养与建筑设计能力。

在单元户型的内部空间设计中，垂直步梯的设置上，还欠缺将无障碍设计导入室内居室空间，忽视了老年人普遍行动受制的现实问题。

马本和

Expert's Comments:

Based on an in-depth understanding of buildings' spaces, the scheme rationally redesigns the original architectural spaces by reducing, transforming or adding spaces, realizing a more flexible spatial division. The yards become more ethereal and practical, attracting the seniors to walk out of the buildings.

The redesign of house types is solid evidence of architecture students' ability to use spaces. Various house types are rationally combined, greatly improving the utilization rate of spaces. From the conceptual scheme to preliminary scheme to detailed scheme, all the schemes reflect the extensive theoretical knowledge and high design ability of the students in architecture.

For the interior design of apartments in units, stairs are not accompanied by wheelchair ramps leading to apartments. The fact that many seniors need the help of wheelchairs is ignored.

Ma Benhe

Opera and Clouds

CIID“室内设计 6+1”2016（第四届）校企联合毕业设计

CIID"Interior Design 6+1"2016(Fourth Session)University and Enterprise Joint in Graduation Design

高　　校：　南京艺术学院
College：　Nanjing University of the Arts
学　　生：　华彬彬　魏飞　冼佩茹
Students：　Hua Binbin　Wei Fei　Xian Peiru
指导教师：　朱飞
Instructors：　Zhu Fei
参赛成绩：　医养建筑改造与室内设计组三等奖
Achievement：　Third Prize for Aged care Building Renovation and Interior Redesign

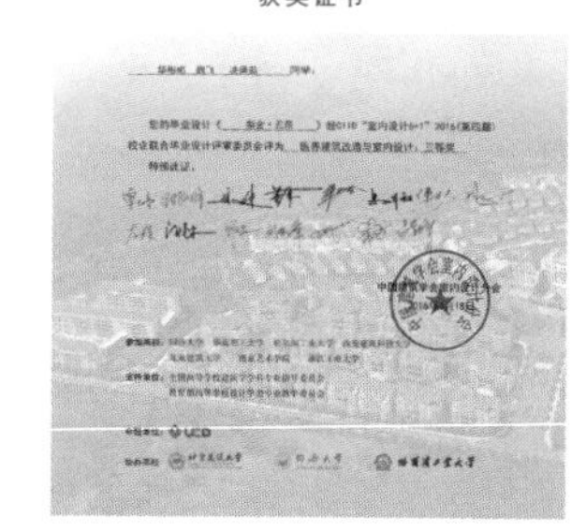

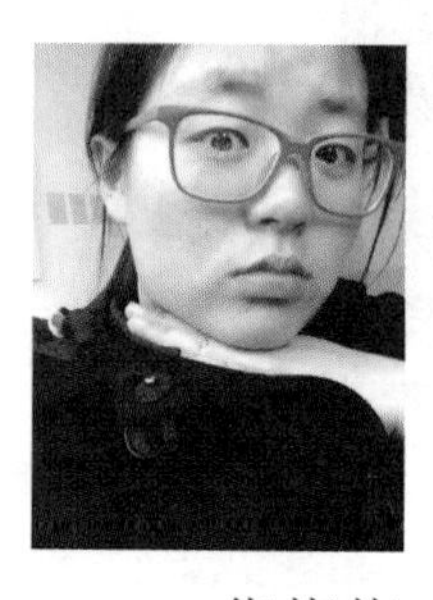

华彬彬
Hua Binbin

魏飞
Wei Fei

冼佩茹
Xian Peiru

学生感悟：

我们经过前期对本次课题的调研和资料收集，将对象定位在有自理能力的久居老人和失智老人身上。整个老年公寓的定位为以疗养为主的医养结合的老年公寓。通过参加这个联合毕业设计，尝试了一次自己从未涉及过的领域，使我接触到了建筑设计、室内设计、景观设计，并且从中学到了很多知识，学会用建筑的思维去思考室内设计，并从室内设计的角度考虑信息的传达，在设计过程中了解了许多京剧文化、养老机构的规范等多方面的知识。

Students' Thoughts:

Based on early stage survey results and other information that we collected, we chose able-bodied seniors and seniors with dementia as the target group, and gave the functions of medical care and recuperation to the buildings. The competition gave me an opportunity to touch a new design field, and in this process, I learned a lot of new knowledge about architectural design, interior design, landscape design, as well as Peking opera and regulations concerning retirement homes. I also learned how to understand interior design from the perspective of an architect, and how to deliver information from the perspective of an interior designer.

空间分析 || Spatial analysis

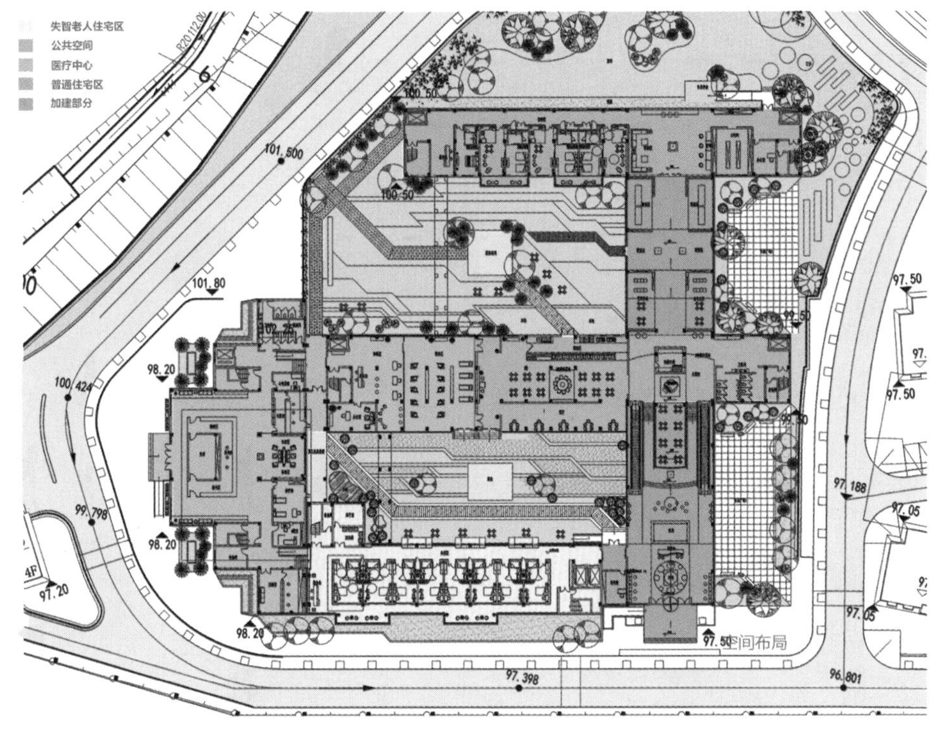

整个空间中主要有五个区域，灰色代表的是加建的门厅公共部分，黄色是失智老人住宅区，红色为主要的公共活动空间，紫色为自理老人住宅区，蓝色为医疗中心。

The project site is divided into five zones. Grey zone is a newly built lobby as a public area, yellow zone is the living area of seniors with dementia, red zone is the main pubic area, purple zone is the living area of able-bodied seniors, and blue zone is a medical center.

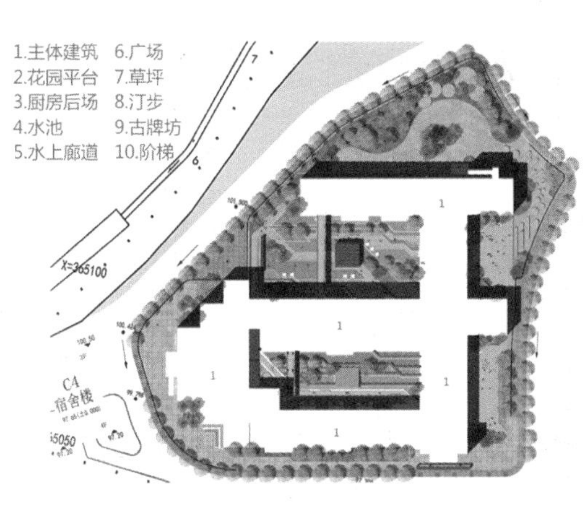

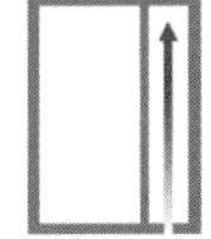
呈长方形走廊
过于单一狭长

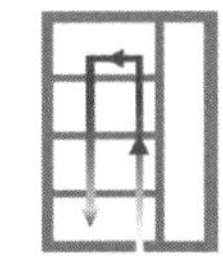
多为相同设计
空间重复率高

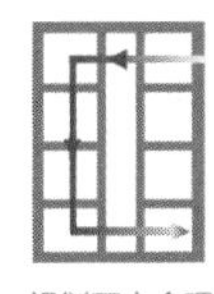
规划不太合理
雷同医院病房

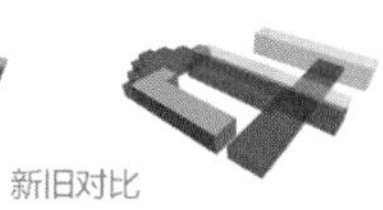
新旧对比

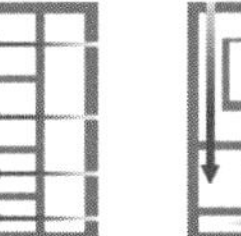
老人行动不佳
增设休息座椅

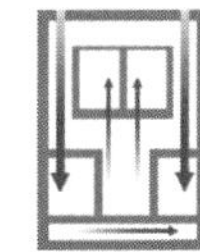
布局并非相似
满足不同需求

重新规划空间
划分更加合理

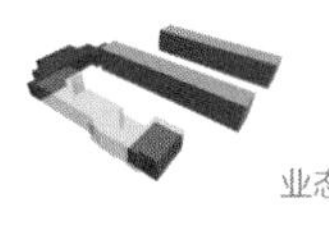

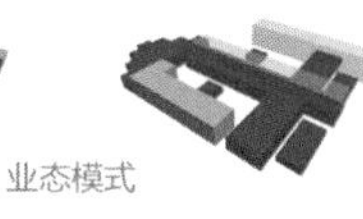
业态模式

人群分布

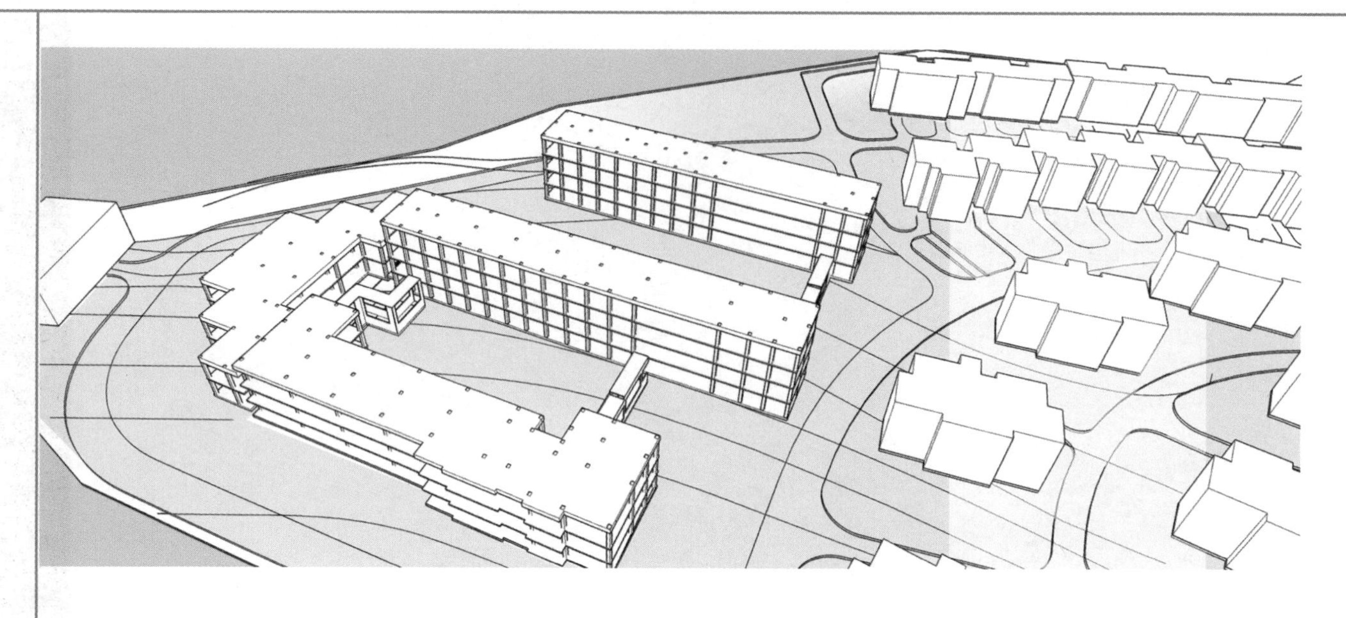

室内设计 || Interior design

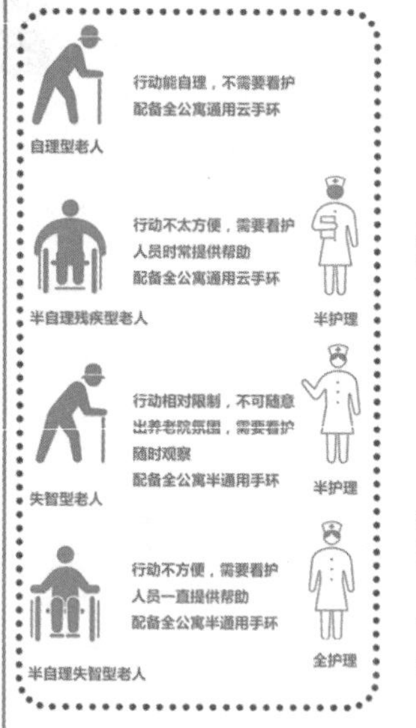

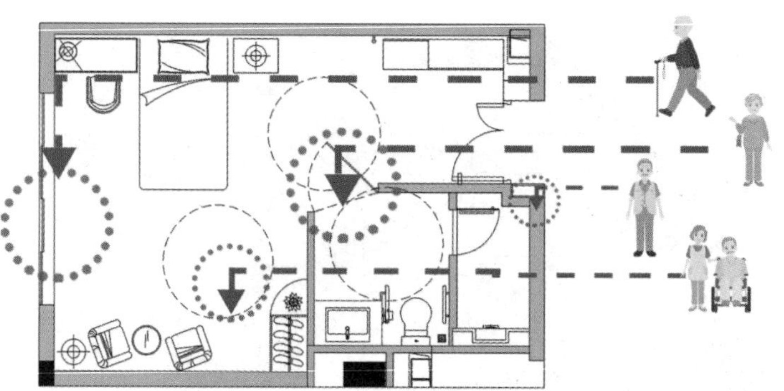

根据老人的日常行为记录，我们将设计群体定位为四种老人。

The target group is divided into four types of seniors according to their daily behavior.

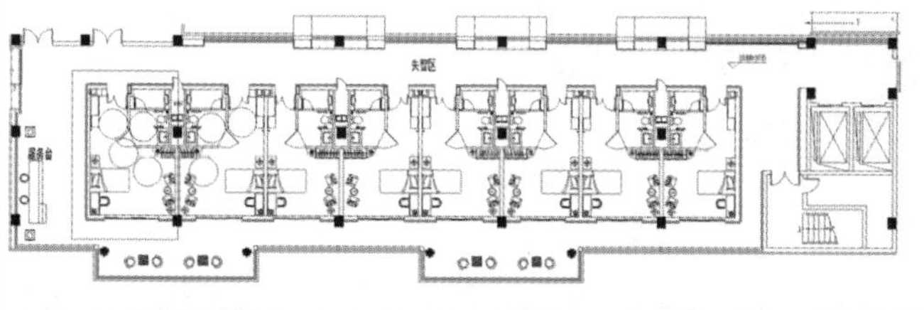

室内效果图
Interior design rendering

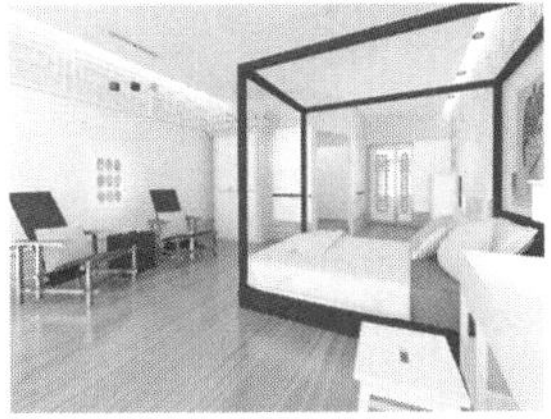

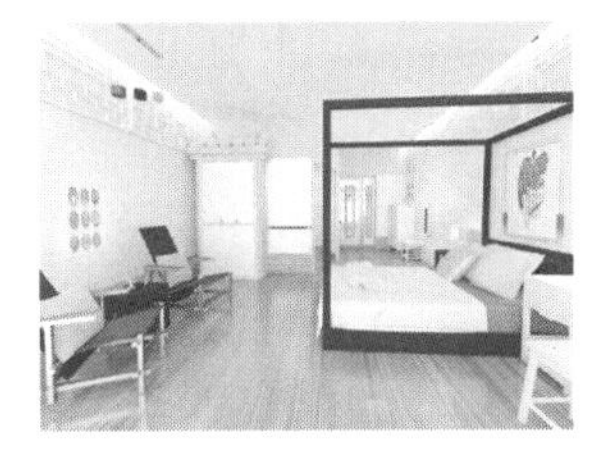

流线分析
Traffic flow analysis

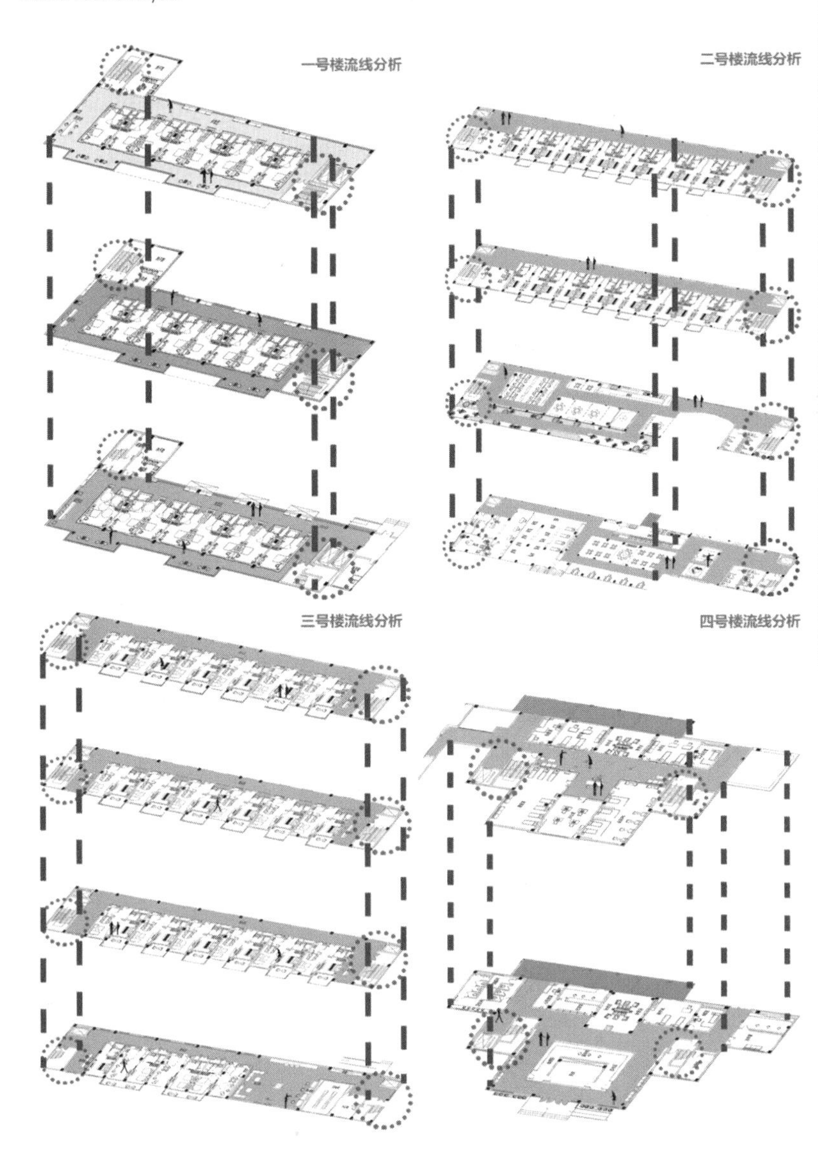

在人流动线上，我们每一栋楼的流线都经过了仔细的考虑，失智老年空间设计回字型走廊，在左右两边都设置了消防通道，应对特殊情况的发生，普通住宅区中增设了一部电梯方便老人行动，人流通道以北面的灰色通道为主，医疗中心设置了两种通道，分别是灰色的主要人流通道与红色的工作人员应急通道。

We've carefully studied the traffic flow of the seniors in each building. In the living area of seniors with dementia, a hollow rectangular corridor is designed with fire escapes on the left and right sides in case of a fire. In the living area of able-bodied seniors, an elevator is arranged for convenient transport. The grey walkway in the north is the main walkway in the project area. Two walkways are arranged in the medical center: grey walkway as the main walkway and red walkway as the emergency access.

门厅效果图
Lobby rendering

在室内的设计中我们考虑到了一些细节的设计，例如门在空间中的使用，在主入口的前方设置了牌坊，增添京味，同时迎合主题营造戏剧感。

Design of details is included in interior design. For example, a memorial arch is arranged in front of the main entrance, presenting a typical characteristic of Beijing, and responding to the theme dramatically.

立面分析图 || Facade analysis graphic

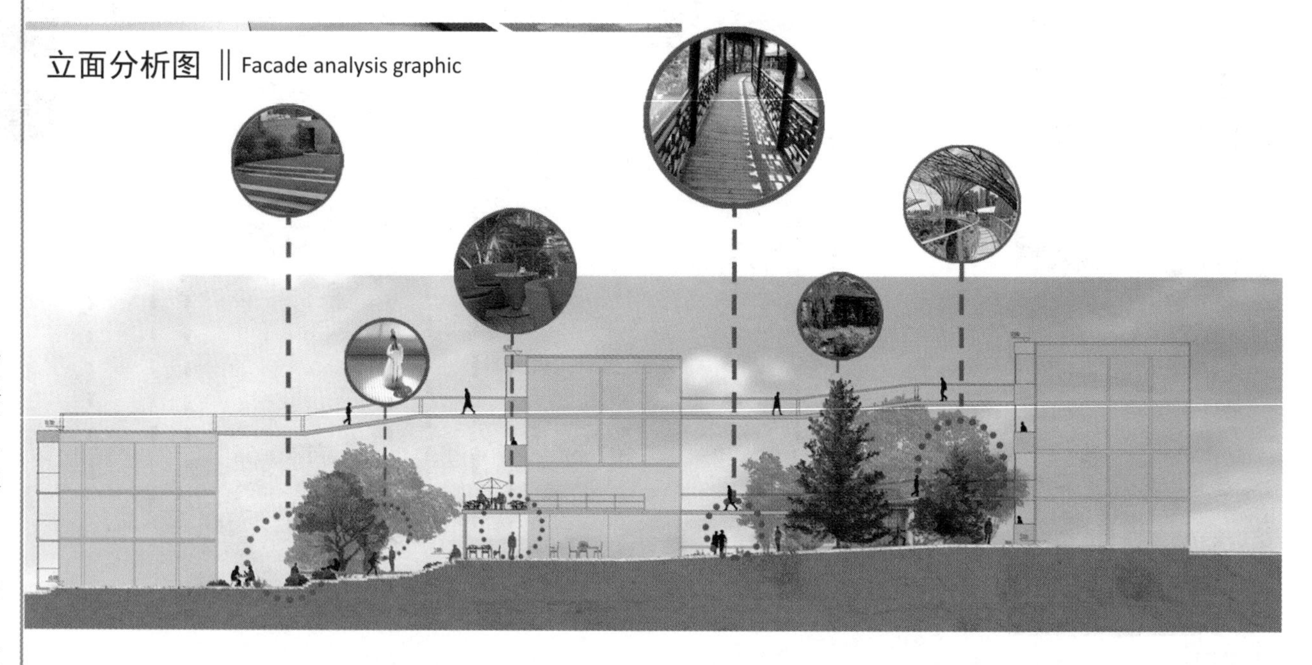

导师点评：

“梨舍·云居”以中国传统京剧元素展示老年公寓改造方案的设计主题，设计者从改变空间水平及垂直交通的角度去改善空间流线的通达性，试图改变建筑室内空间形态的联结关系。设计方案的特点在于将京剧元素符号化运用到空间标识系统及色彩方案中，以点、线、面的构成语言呈现在室内空间。在老年居室空间的设施设计中引用了蒙特里安的色彩方案，并加以变化和延伸，旨在突出居室室内与公共空间中西合璧的视觉感受。整体方案在概念主题的逻辑关系、整体格调把握、受众群体——老年的生理和心理适宜性分析、空间设施的舒适性方面待进一步调整和深入。

任彝

Advisor's Comments:

"Opera and Clouds" renovates the retirement community through exhibition design themed on Peking opera elements. The scheme tries to create smoother traffic flows horizontally and vertically, thus changing the connection relations between interior spaces. The highlight of the scheme is incorporating Peking opera elements into the design of the sign system, and into interior color design in the form of points, lines and surfaces. Mondrian's color scheme is applied in interior facility design. Through the change of color and spatial extension, Chinese and western cultural elements are harmoniously mixed, creating visual impacts. Further adjustments and deeper analysis are required in terms of the logical relationship between the theme and the content, overall style control, the physiological and psychological suitability of the target group (the seniors), and comfort of facilities.

Ren Yi

专家点评：

“梨舍·云居”的设计者以中国的国粹——京剧元素为主线展开设计概念，将京剧的内涵文化思想意识同老年公寓环境进行了巧妙融合，主题明确清晰，方案结构完整，并且从多方面考虑到老年人生理、心理的需求，进而延续到空间布局及装饰，突出了医养结合的特色，同时也考虑到经营方的投入与产出的关系，是有一定可实施性的设计作品。

设计者“精心”选定主题，创造了设计的亮点也局限了设计本身，在局部空间设计及色彩使用上，还是较多的“想当然”占据了主导，养老设施的特殊性促使设计者时刻将自己作为老年人去思考和进行设计。方案最终的呈现还具有一定的可操作空间，还需要对老年人的生活规律及习惯进行进一步的调研。

陈天力

Expert's Comments:

Adopting Peking opera, the quintessence of Chinese culture, as primary design element, "Opera and Clouds" skillfully incorporates Peking opera culture into the environmental redesign of the retirement community. The theme is clear, and the scheme structure is complete. Besides, the physiological and psychological needs of the seniors are considered in the design of various aspects, including spatial layout and decoration. The scheme realizes the integration of medical and aged care resources, and analyzes the input-output relationship of the project. It's feasible to some extent.

The carefully selected theme is the highlight of the design, but also a restriction. Regional spatial design and color selection are kind of divorced from reality. After all, the design of retirement homes requires designers to think and design from the perspective of seniors. Generally speaking, the final scheme is feasible to some extent, but a further survey on the lifestyles of seniors is still required.

Chen Tianli

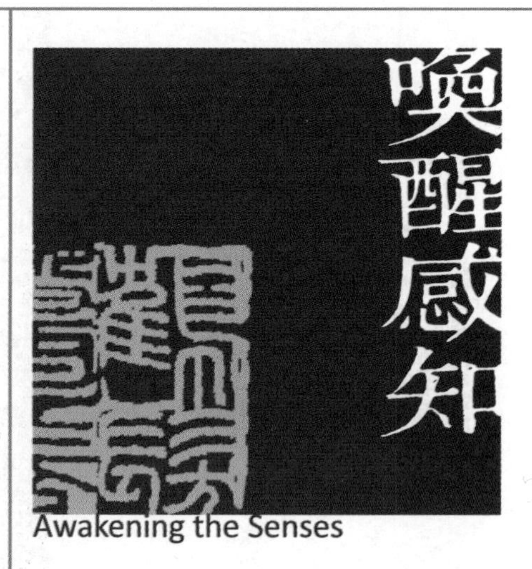

CIID“室内设计 6+1”2016（第四届）校企联合毕业设计
CIID“Interior Design 6+1”2016(Fourth Session)University and Enterprise Joint in Graduation Design

高　　校：西安建筑科技大学
College: Xi'an University of Architecture and Technology
学　　生：李坷欣　刘雨鑫　曾嘉
Students: Li Kexin　Liu Yuxin　Zeng Jia
指导教师：刘晓军　何方瑶
Instructors: Liu Xiaojun　He Fangyao
参赛成绩：医养建筑改造与室内设计组三等奖
Achievement: Third Prize for Aged care Building Renovation and Interior Redesign

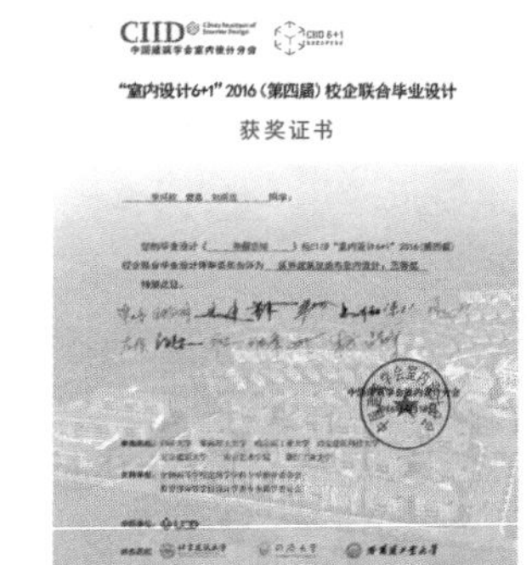

李坷欣
Li Kexin

刘雨鑫
Liu Yuxin

曾嘉
Zeng Jia

学生感悟：

很荣幸能够参加这次的“6+1”室内设计比赛作为我最终的毕业设计，在此次的毕业设计中我们是初次接触老年医养空间的设计，我们从第五维空间即感知出发，探讨了适合老年人医养的生活方式，从中学习到了许多、感受到了许多。首先感谢刘晓军老师和何方瑶老师的指导，让我们明白了很多专业知识，同时为我们感知主题指出了很好的研究方向。通过这次比赛，我们更深刻理解了国内室内设计研究和发展的动向。

Students' Thoughts:

It's an honor to participate in the CIID "Interior Design 6+1" University-Business Graduation Projects Competition. Retirement home spatial design is a brand new topic for us. Themed on the fifth dimension, or the senses of human beings, we discussed lifestyles good for the health of seniors, and have learned a lot during the process. I'd like to give my thanks to Mr. Liu Xiaojun and Ms. He Fangyao for their guidance. They imparted a lot of professional knowledge to us and gave very usefully directions to our theme presentation. Through the competition, I had a deeper understanding of the trends of China's interior design.

组团生成 || Group formation

“向心性”的住宅聚落

感知，是为了让老年人的身体对自然重新恢复“敏感”，从而达到医养的目的。我们将感知划分为邻里社团感知和老年人的自身身体感知。在邻里社团感知中，老年社区组团形成朋友圈效应，使得老年人加强交往，彼此需要，也就能彼此感知。

Awakening the senses of seniors is to help them recover bodies' "sensitivity" to nature, promoting their health. We want to create a sense of neighborhood for the seniors, and improve their physical senses as well. For the former, group living will create circles of friends in which they can communicate, help and understand each other.

基地位于京郊山脚下，三面环山，园内冬暖夏凉，植被丰富，空气新鲜，是京郊的避暑养生胜地。

Located at the foot of a mountain in the suburb of Beijing, the retirement community is surrounded by mountains on three sides which give it comfortable temperatures all year round. Good weather and bush vegetation make it an ideal place to spend hot summers.

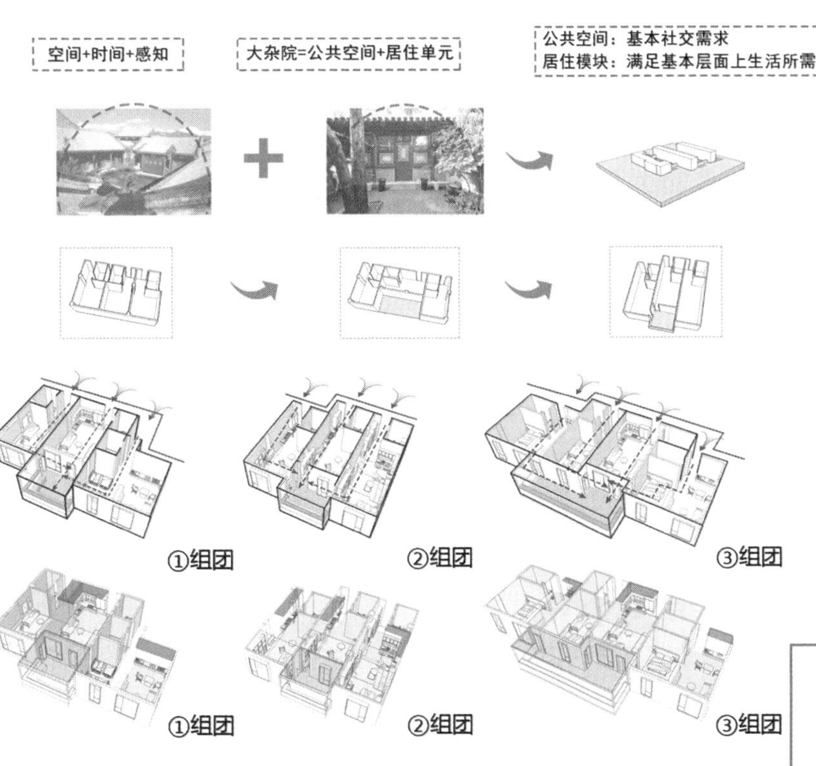

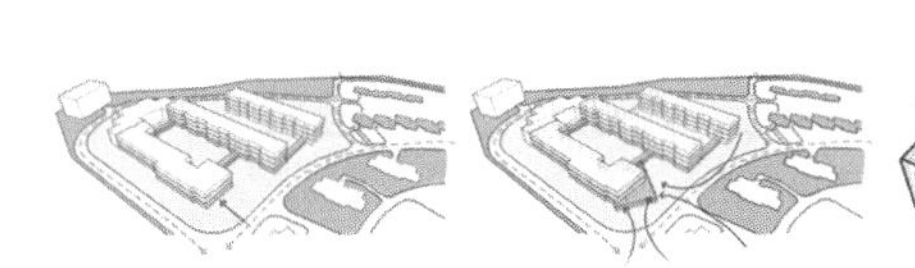

红色箭头表示建筑来往人群的趋势，为顺应这种趋势，建筑入口朝向发生改变。

The red arrows show the traffic flow of the occupants which points to the need to change the entrances of the buildings.

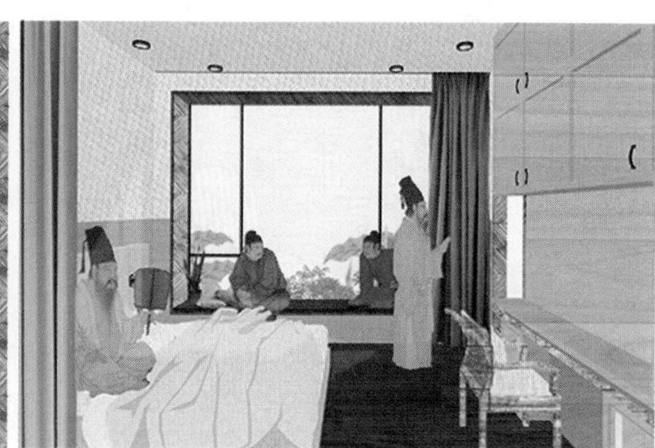

平面图 || Floor plan

1 c3中庭
2 护士站
3 连廊
4 餐厅 / 活动室
5 厨房 / 烹饪教室
6 客房
7 咖啡厅
8 康复室
9 spa / 中医馆
10 瑜伽室
11 接待厅
12 门厅
13 超市

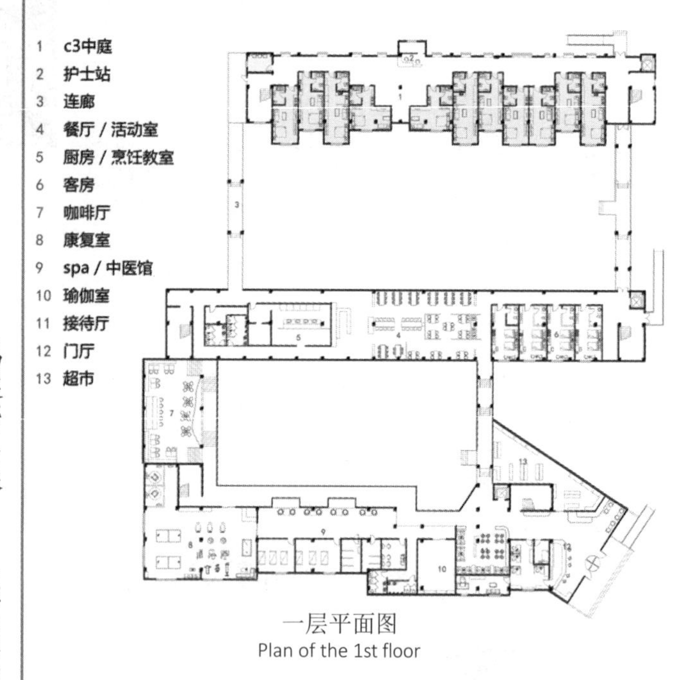

一层平面图
Plan of the 1st floor

1 护士站
2 中庭活动空间
3 组团中庭
4 员工休息室
5 电梯间
6 洗衣房
7 晾晒台

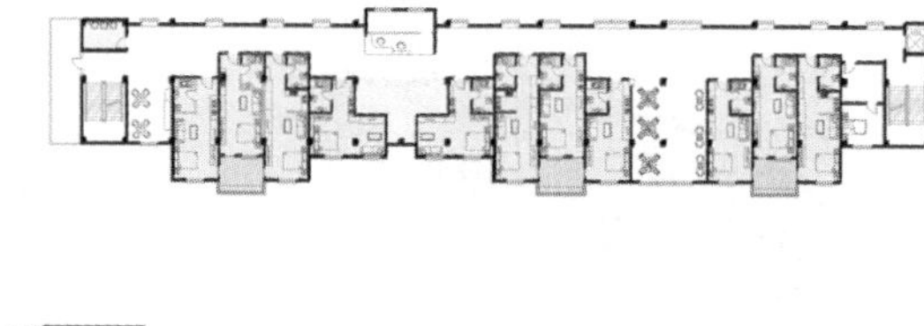

二层平面图
Plan of the 2nd floor

1 护士站
2 中庭活动空间
3 组团中庭
4 员工休息室
5 电梯间
6 洗衣房
7 晾晒台

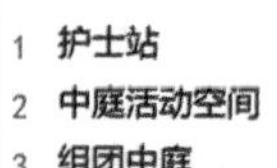

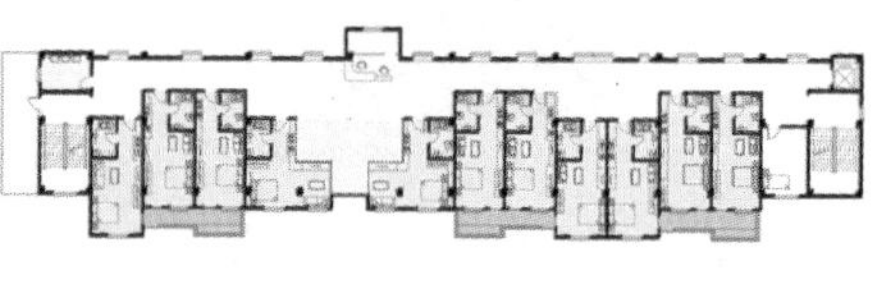

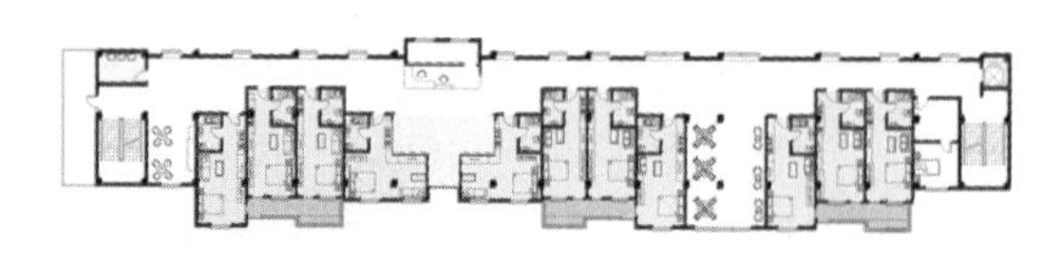

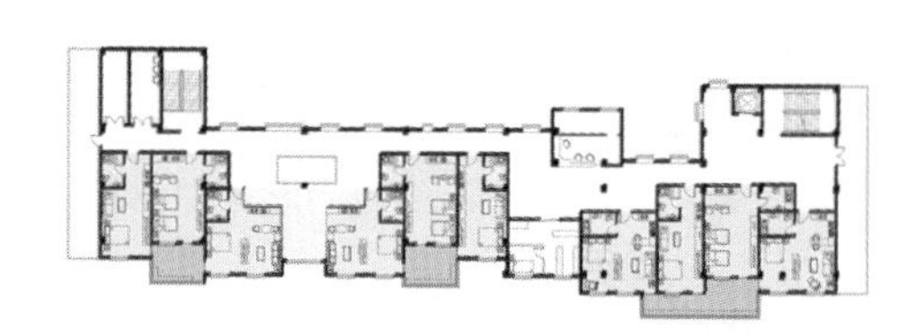

三层平面图
Plan of the 3rd floor

1 护士站
2 中庭活动空间
3 组团中庭
4 员工休息室
5 电梯间
6 洗衣房
7 晾晒台

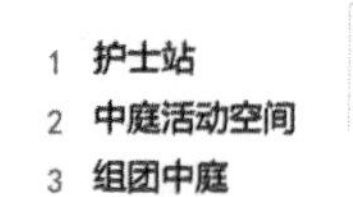

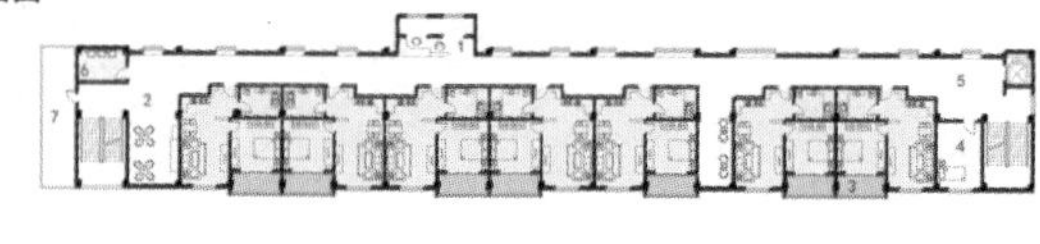

四层平面图
Plan of the 4th floor

空间分析 || Spatial analysis

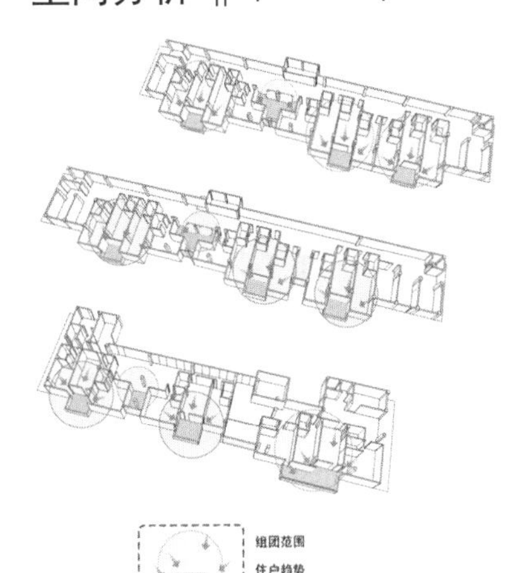

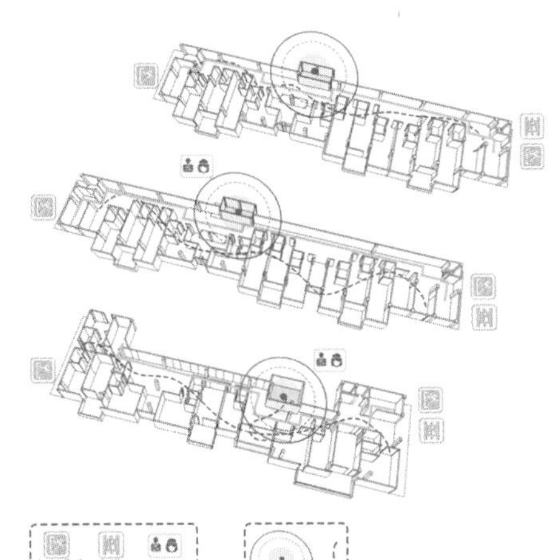

组团范围
住户趋势
公共活动区

安全通道 电梯 护士站
交通流线

辐射范围

走廊拥有更大的公共空间，作为枢纽串联起各个组团，形成一个层级利于管理和服务的空间系统。为了有更便捷的管理方式，将护士站和其他辅助空间设置在北侧。

The corridors form large public spaces leading to all the groups, and thus act as a system of management and service. To ensure more convenient management, a nursing station and other supporting facilities are arranged to the north of the buildings.

不同人的行走模式

Movement modes of different types of occupants

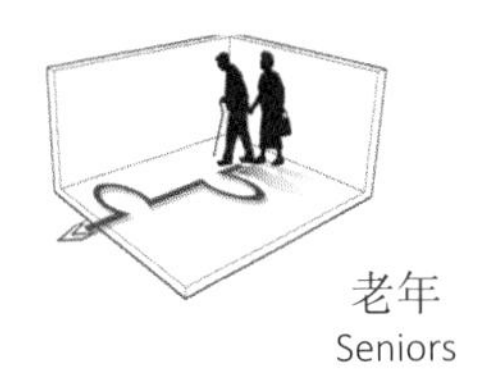

老年
Seniors

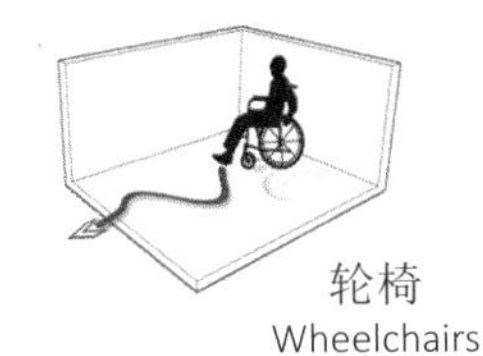

轮椅
Wheelchairs

视觉障碍者
People with visual disabilities

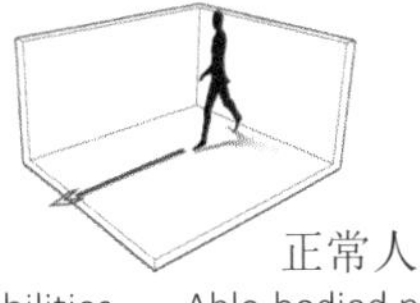

正常人
Able-bodied people

儿童
Children

7 8 9 10 11 12 13 14 15 16 17 18 19 20 21 22

居室
门厅
其他公共空间
户外散步
食堂
老年公寓外

老年活动分析

Analysis of seniors' activities

根据不同的身体状况综合考虑老年人的活动方式，做出动线分析。

We considered the activities of the seniors according to their physical conditions, and then analyzed corresponding traffic flows.

将居住流线和使用流线区分开来，这样的设计便于空间区分，保证了相对的独立性。同时加强老年人的运动量，并促使他们进入更多的空间，感知场所环境。

The traffic flows in the living and public areas are separated to realize better spatial division, and to keep the independence of the areas. Such a division encourages the seniors to walk more, and to feel and perceive different environments.

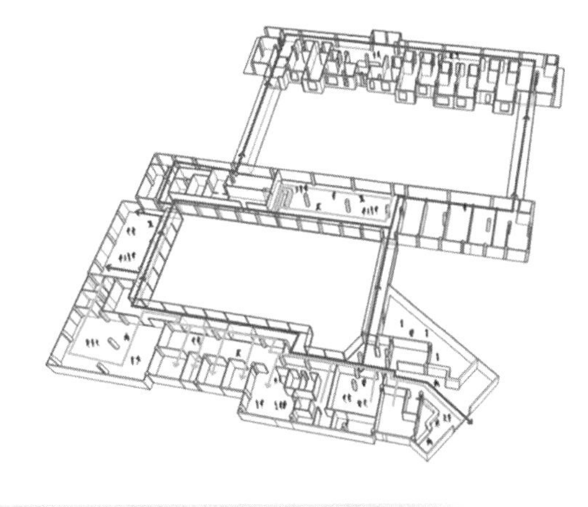

走廊作为交通部分，我们选择让出部分居住空间，在建筑中间形成一个开放的活动区域，可以种植绿植，同时保证中通的环境。

Corridors act as a transport system in the buildings. In the design, rooms in the middle are turned into a landscaped public area as a transport node.

好玩空间 || Entertainment spaces

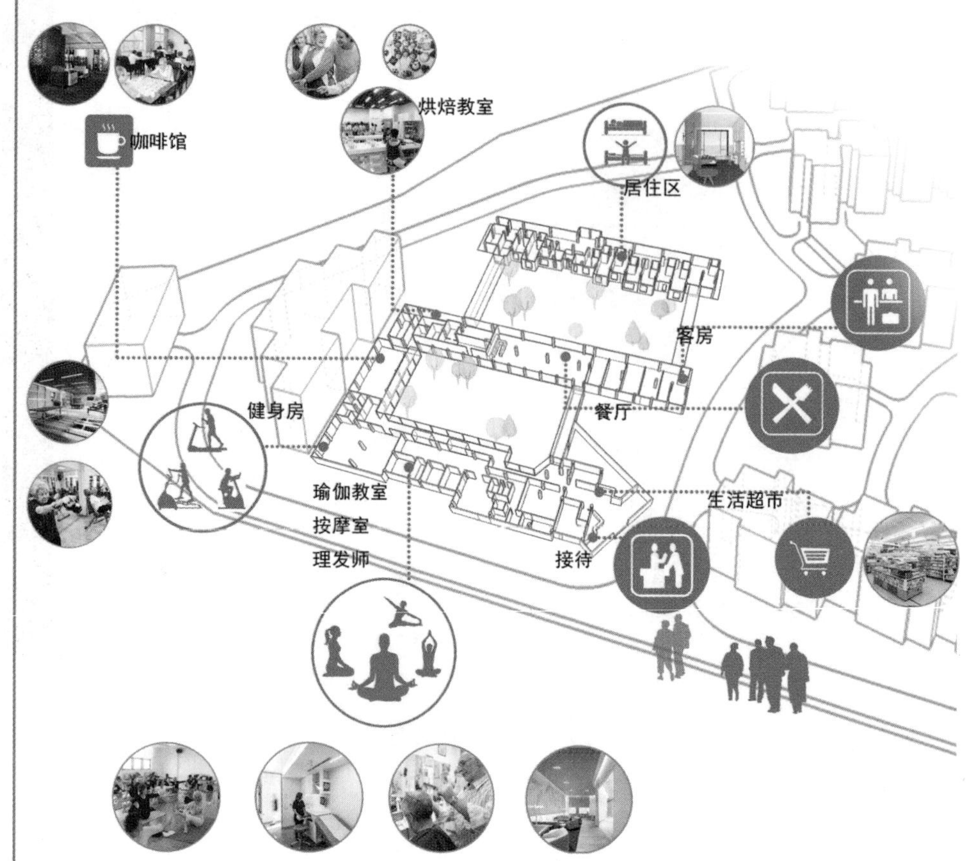

一层空间我们设置了接待空间，提供老年人进行购物的生活超市，模拟实际的社会生活状况。
丰富生活的瑜伽教室和健身房让老年人拥有更多的活动形式，按摩室和理疗结合体现医养的目的。

A reception area is designed on the 1st floor. Besides, a supermarket is designed to meet the occupant's daily needs and to create a social environment.
Yoga rooms and gym are arranged to enrich the life of the seniors, and massage and physical therapy rooms are designed for health care and health maintenance.

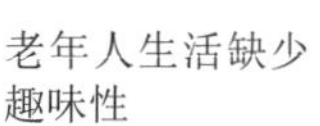

老年人生活缺少趣味性
The seniors' life in the retirement community is relatively monotonous.

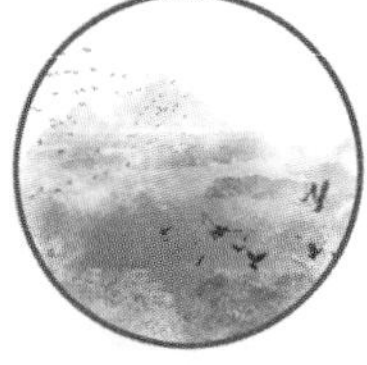

基地生态环境较好，是许多鸟类的栖息地
The retirement community is home to various kinds of birds thanks to its good natural environment.

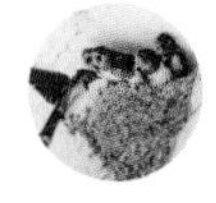

老年人可以跟动物产生一种感知
It's easier for seniors to build an emotional connection with animals.

走廊扶手的研究 || Study of corridor handrails

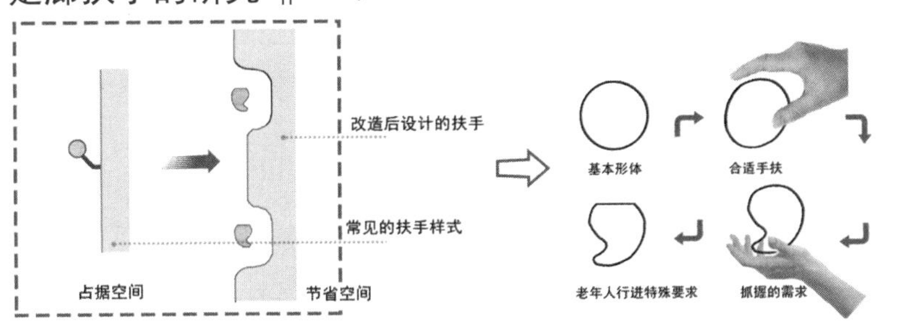

我们发现，人不仅对于形体有着明确的感知，同时，对于材质质感也有着十分明确的感知能力。

Study results show that people are sensitive to the shapes and textures of things through the sense of touch.

柔软、有点扎手　光滑 坚硬　有弹性　柔软 光滑

麻材质　浅色木材质　橡胶材质　棉材质

感知护墙板

墙面
色带
扶手

1 2 3 4

在触觉方面，人对于材质也有着感知能力，因此每层做不同材质的扶手。

Given this, different materials are used to make corridor handrails on different stories.

我们选择大杂院形式，将多户集中布置，用庭院进行组织，从而产生更多的交流，同时，将公共空间实现综合利用。

The group living model realizes concentrated residence around the yard which promotes communication and realizes the comprehensive utilization of public areas.

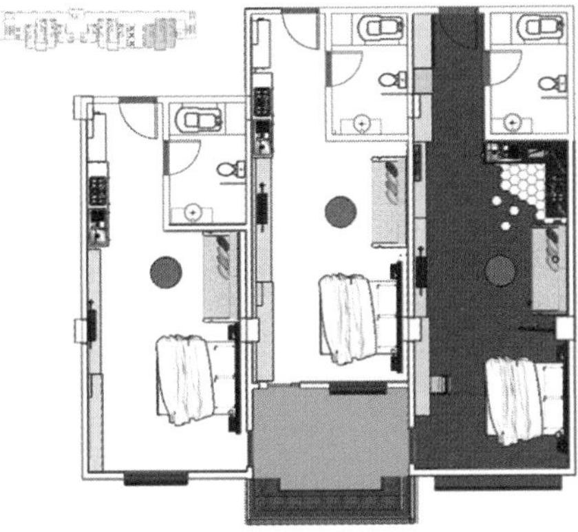

橱柜设置扶手
Cabinet handles

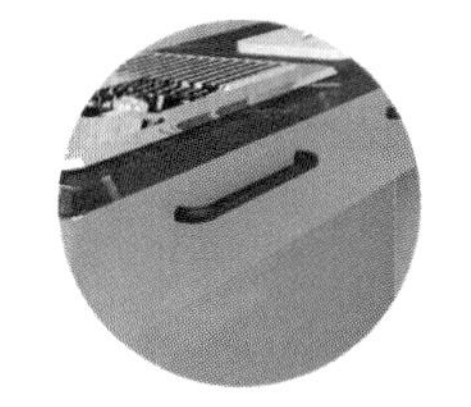

操作台让轮椅方便使用
Wheelchair accessible operations areas

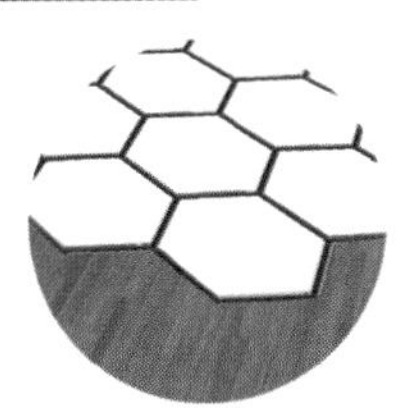

用铺装区分干湿区
Dry and wet areas are divided using different pavement materials.

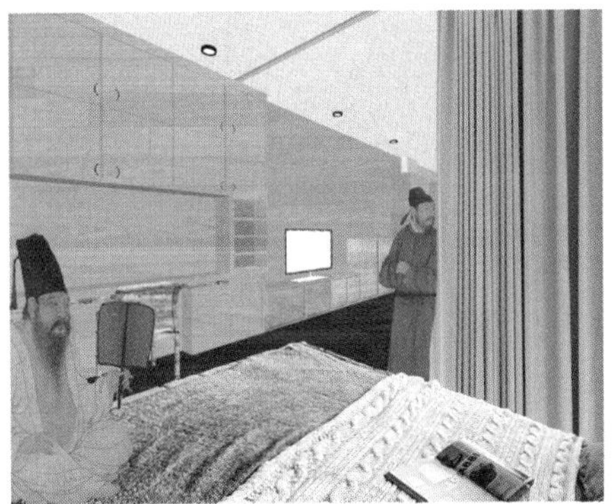

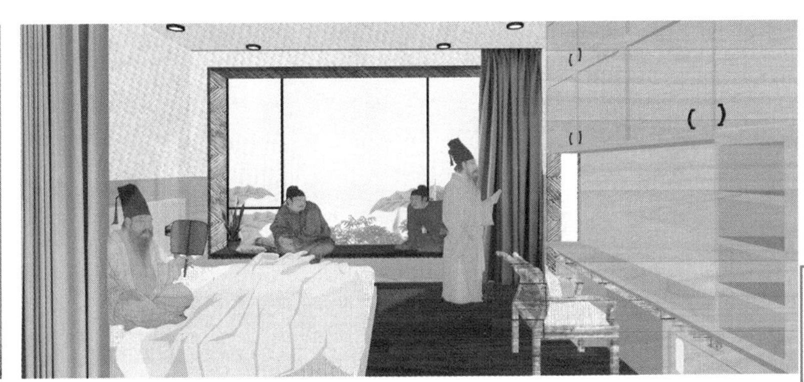

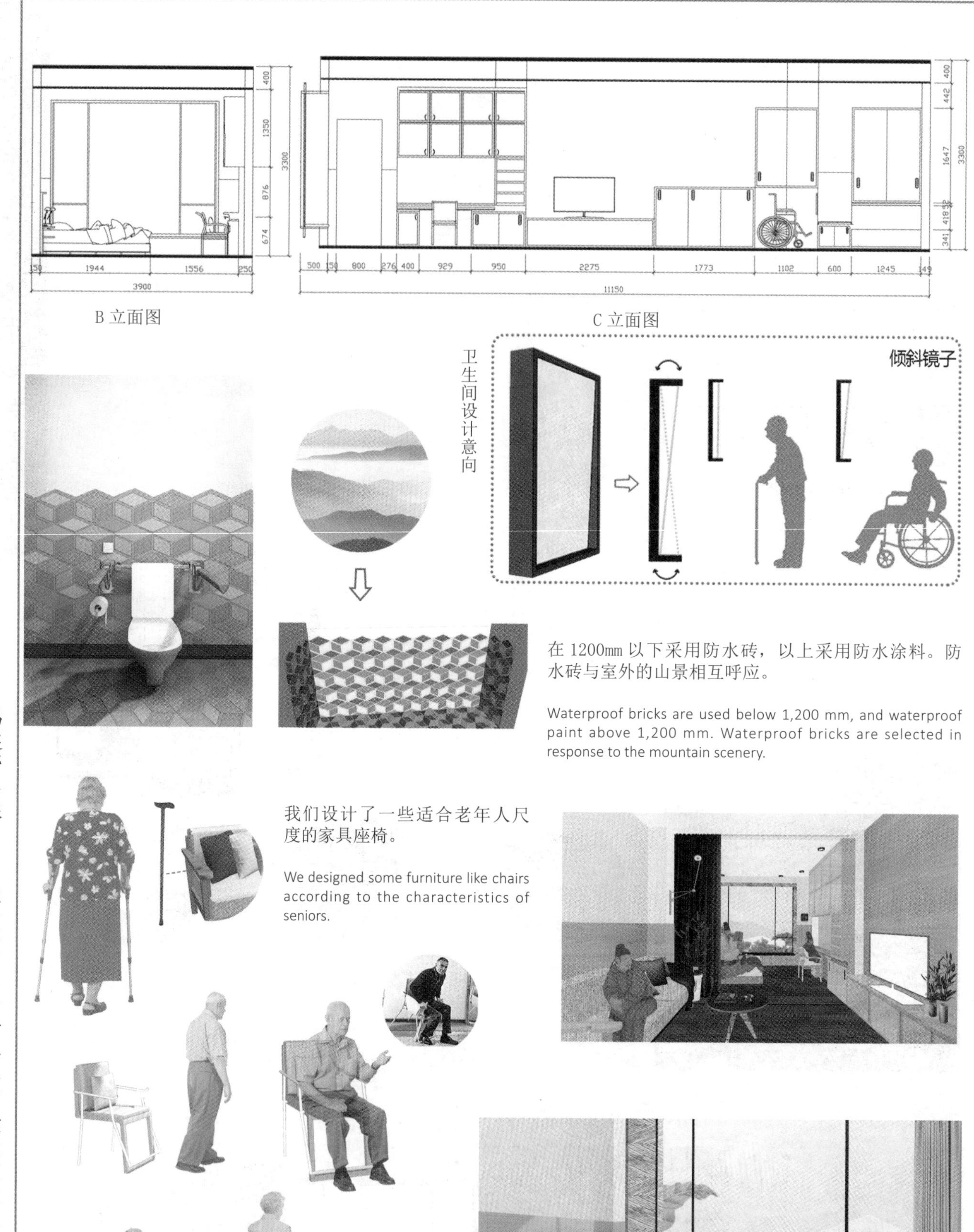

在 1200mm 以下采用防水砖，以上采用防水涂料。防水砖与室外的山景相互呼应。

Waterproof bricks are used below 1,200 mm, and waterproof paint above 1,200 mm. Waterproof bricks are selected in response to the mountain scenery.

我们设计了一些适合老年人尺度的家具座椅。

We designed some furniture like chairs according to the characteristics of seniors.

导师点评：

通过“唤醒感知”的立意，以老年人五感渐退为出发点，并通过空间与设施的设计使其延长感知的消退过程，从而实现健康养老的目的。

空间组织合理，在现有的平面布局中，提取部分空间整合出相对独立的公共空间，为原有封闭式单元住宅增加了共融性与交互性的空间，通过对空间使用功能的改变使老年人的行为方式有所转变，以此增进彼此的关系，从而唤醒老年人渐渐消退的感知。

渲染图表现细腻、精致，通过使用唐代绘画中的古装人物为表现图内的场景人物，体现了同学们在艺术方面的情趣与追求，同时也取得了一定的视觉表现力。

在空间局部规划与设计方面还缺少较严谨的表现，缺少将老年人的需求要素、行为要素始终如一地融入到整个空间设计的每一处。

马辉

Advisor's Comments:

Given the fact that seniors are troubled by declining five senses, “Awakening the Senses” tries to, through spatial and facility design, slow down the speed of the decline, thus promoting the health of the seniors.

The scheme features rational spatial organization. Based on the exiting plan layouts, the scheme integrates some spaces into independent public spaces, promoting the communication and interaction among seniors originally living in independent units. The change of spatial functions will lead to the change of the senior's behavior. In the process of more and more frequent communication, the declining senses of the seniors will be awakened.

The renderings are detailed and exquisite. Painting skills and style of the Tang Dynasty are used to create figures, which indicates the designers' artistic ambition and creates impressive visual effects as well.

Regional spatial planning and design should be more rigorous. The scheme fails to incorporate the needs and behavioral characteristics of seniors into the entire design process.

Ma Hui

专家点评：

根据本有建筑及空间环境分析利弊，结合老北京人的生活习惯，设计空间内部结构，空间合理整合，这种群落式的生活空间，方便老年人相互沟通和交流。丰富的公共活动空间，拉近了邻里之间的相互关系，形成了老年人的朋友圈，解决了老年人孤单寂寞的实际问题。

明确的交通路线、清晰的人流走向使功能空间联系更加紧密。其间的公共交流空间方便了老年人的活动和休息。

通过各感官的具体分析，结合物理环境进行细致的设计，满足老年人日常生活的实际需求。

方案的整体设计从人本角度出发，诠释了人与人、人与建筑、人与自然的相互情感。

张红松

Expert's Comments:

Based on an analysis of the original buildings, including their spaces and environments, as well as the lifestyle of the old timers in Beijing, the scheme redesigns the interior structures and rationally integrates the spaces, creating group living spaces which promote the communication among the seniors. Besides, more public spaces are created to build closer neighborly relations, and circles of friends, pulling the seniors away from loneliness.

Clear transport routes and traffic flows remove the boundaries separating various functional spaces. Public spaces are arranged along such routes and flows, for the rest and communication of the seniors.

The five senses are incorporated into the design in the given physical environment of the project, for the purpose of meeting the actual demands of the seniors in daily life.

From a human-oriented perspective, the scheme tries to interpret the relations and emotions between people, between man and architecture, and between man and nature.

Zhang Hongsong

Communal living

CIID“室内设计 6+1”2016（第四届）校企联合毕业设计

CIID“Interior Design 6+1”2016(Fourth Session)University and Enterprise Joint in Graduation Design

高　　校：哈尔滨工业大学

College：Harbin Institute of Technology

学　　生：张相禹 张玲芝 朱梦影 袁思佳

Students：Zhang Xiangyu Zhang Lingzhi Zhu Mengying Yuan Sijia

指导教师：马辉 兆翚 周立军

Instructors：Ma Hui Zhao Hui Zhou Lijun

参赛成绩：医养建筑改造与室内设计组三等奖

Achievement：Third Prize for Aged care Building Renovation and Interior Redesign

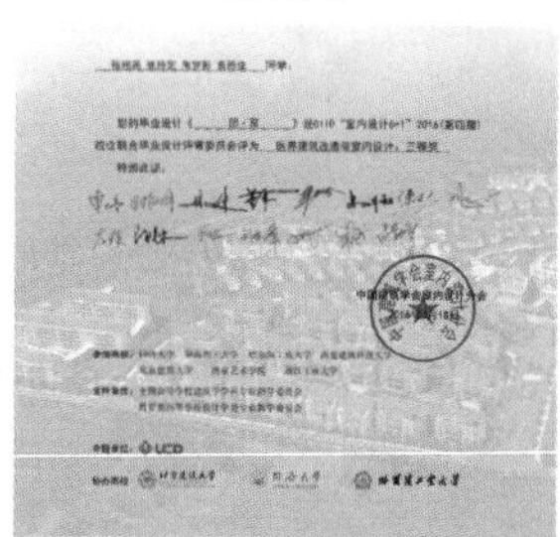

张相禹
Zhang Xiangyu

张玲芝
Zhang Lingzhi

袁思佳
Yuan Sijia

朱梦影
Zhu Mengyin

学生感悟：

经过对原项目定位及策划的分析，我们决定把该项目改造成主要针对半自理老人的、引入休闲度假理念的娱乐养老型老年公寓。北京四合院式的邻里关系非常珍贵，家庭式的入住模式也非常适合老人，基于这两个层面，我们提出了“团聚”主题。“团”代表家庭入住、择邻而居的生活模式，“聚”则代表“有朋自远方来，不亦乐乎”的生活态度。这个地方不再是冰冷的病房，老人们有权力择邻而居、有权力逛街做 SPA，可以有尊严地在这个温馨的地方安度晚年。

Students' Thoughts:

Based on an analysis of original project positioning and scheme, we decided to transform the project into a resort-style retirement community mainly aiming at partially disabled seniors. Given that precious neighborhood in courtyard houses and home-like living environment are what the seniors in Beijing really need, we determined the theme of “Communal Living” – to live with favorite neighbors and enjoy a happy life together with friends. We want to turn lonely and depressing awards into homes where seniors living in have the right to choose their neighbors, to go shopping, to enjoy SPA, and most importantly, to spend their remaining years with dignity and the company of love.

整体鸟瞰 || Bird's-eye view

大面积的公共服务空间
Large public service place
休闲 SPA、养生医疗、拜神祭灵、棋牌娱乐
Areas for SPA, healthcare, medical treatment, and prayer
促进社区范围内的交往
Promote inter-community communication

改造方式 || Method of redesign

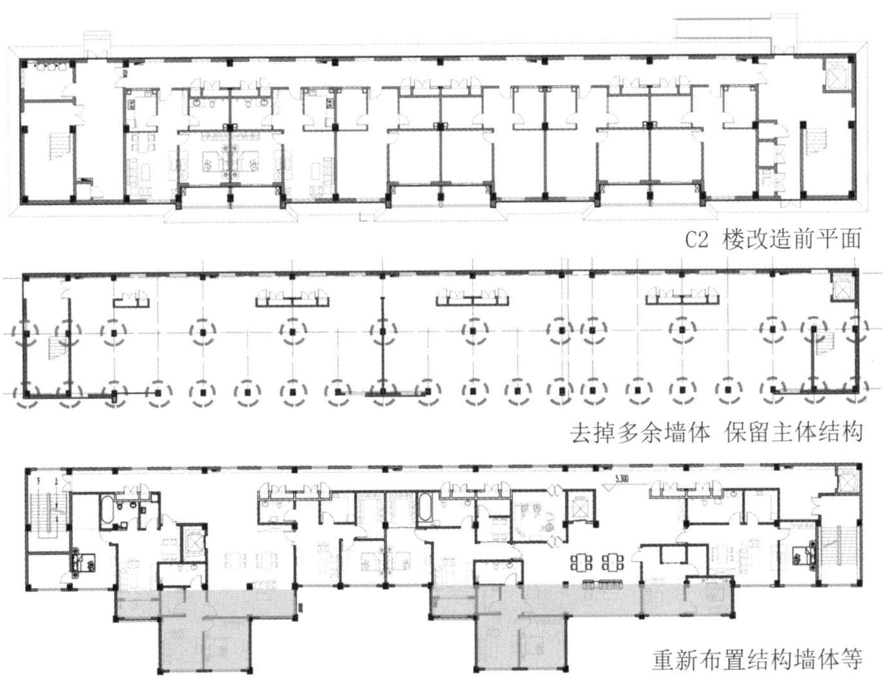

C2 楼改造前平面

去掉多余墙体 保留主体结构

重新布置结构墙体等

生成过程 || Process of redesign

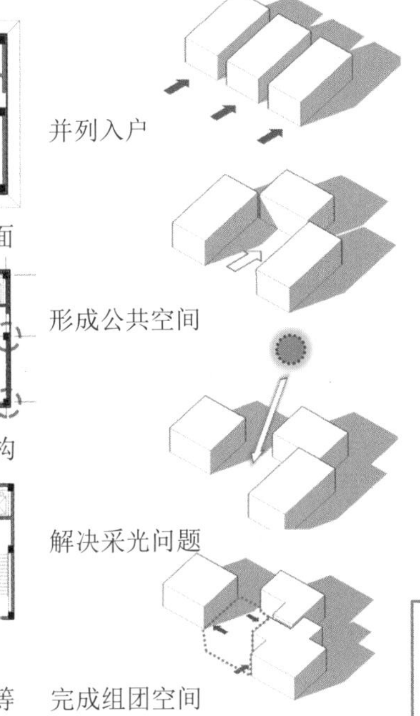

室内透视 || Interior perspective

居住空间阳台透视图

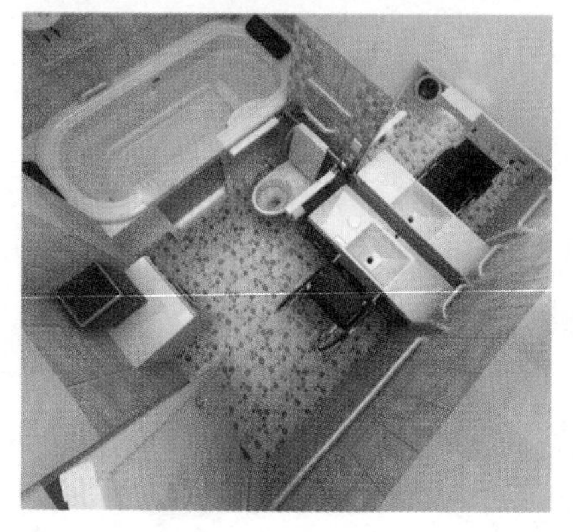

居住空间厕所透视图

居住空间卧室透视图

C2 楼原平面为并列入户的死板空间，作为改造的样板将原有柱网与上下水的功能空间保留，重新布置内部空间。通过空间的左右错动与前后退进。

居住空间：腾出 C2 楼中二层空间为活动空间，居住空间重新打乱重组。

现代居住理念：高私密性＝高档

走廊剖透视

Corridor section and perspective

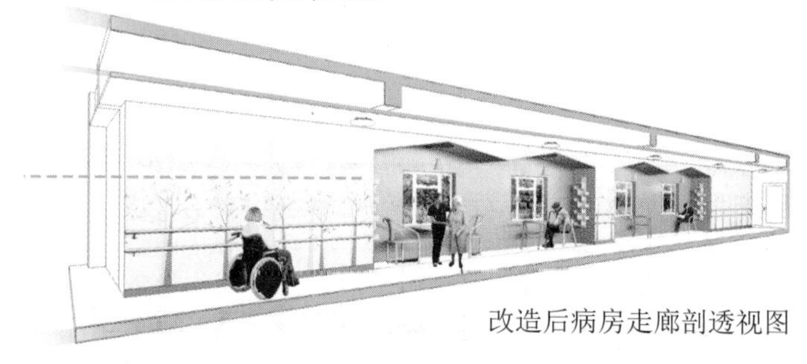

改造后病房走廊剖透视图

70 年代居住理念：
好食物大家分享，
隔壁吵架听得见，
一起洗菜、一起洗衣。

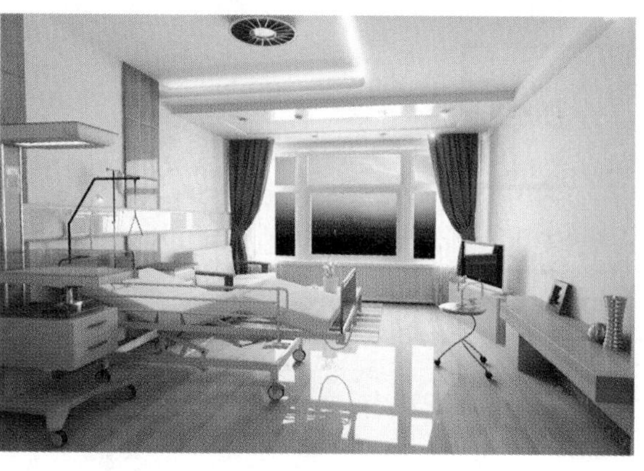

病房透视图
Ward perspective

病房透视图
Fitness center perspective

连廊改造

原园区的三栋楼共用一个在场地东南角的小门厅，门厅的联系作用很弱，从而也阻隔了每幢楼老人之间的交往。改造后的门厅选址在中间栋的端头，联系性加强，同时将门厅扩大，布置丰富的功能，提供一个足够的吸引点，吸引老人走出房门，互相陪伴。

康复医疗区效果图

康复医疗区位置示意图

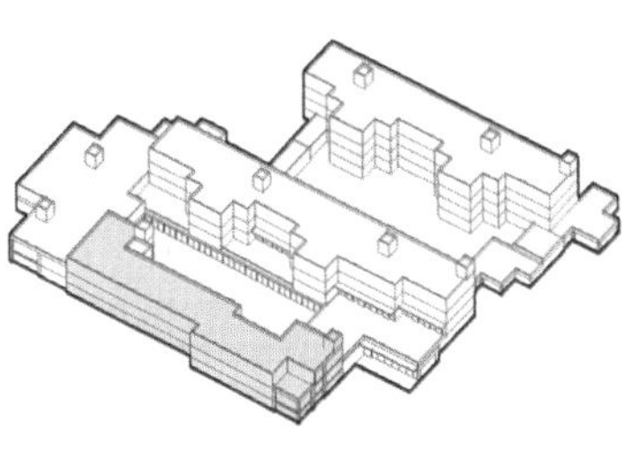

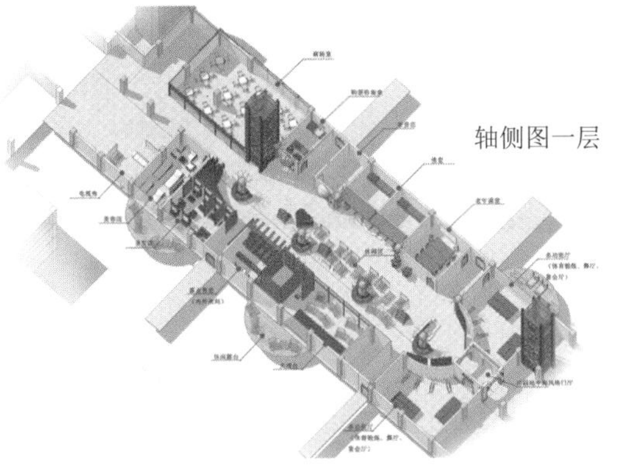

轴侧图一层

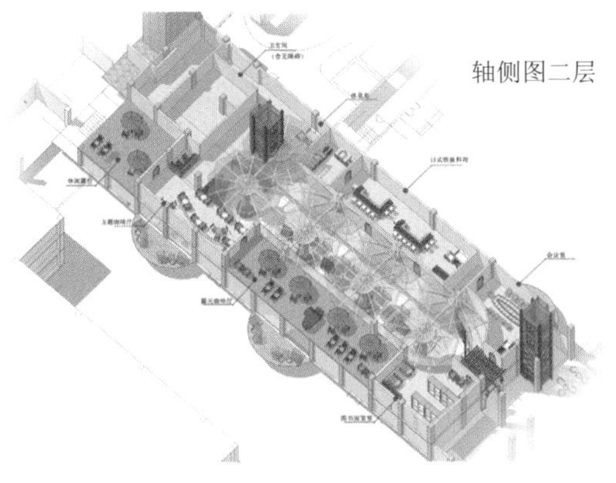

轴侧图二层

用砖墙和隔板墙作为立面材质，围合出了一条八米多宽近五十米长的室内街道。街道与两端的室外庭院空间几乎无缝对接，景色协调一致。

家庭层面——一种养老理念，将密云作为老人的家，每逢假期家人将在这里相聚。

剖面对比

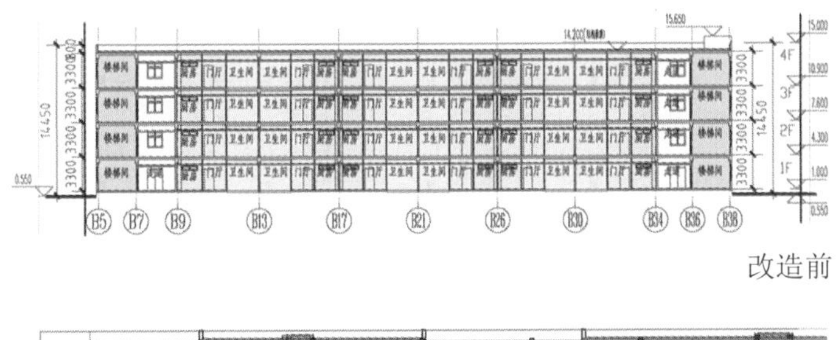

改造前

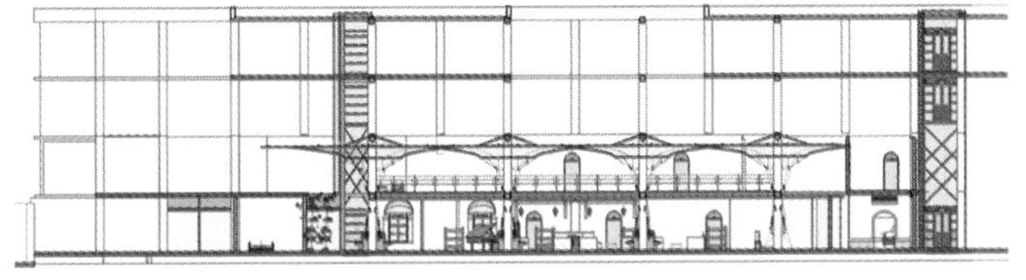

改造后

门厅透视图 A

门厅透视图 B

创造小镇街道化的立体商业娱乐活动空间

Street-based spatial design for business and entertainment activities in a town-sized area

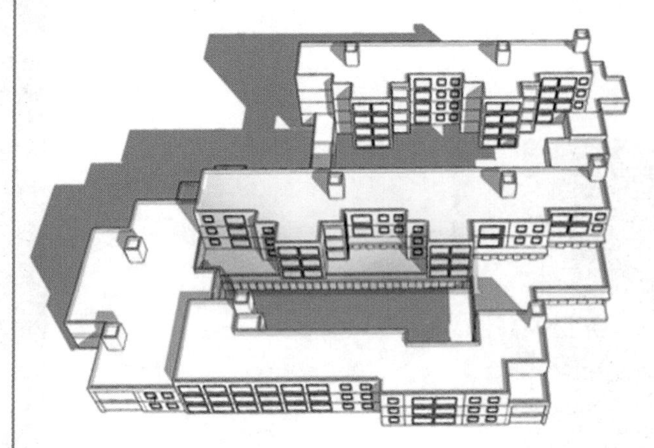

休闲娱乐街所在位置

理念阐释：

“百米长街巷。何处不相见？

一天一小聚，两天一大聚。”

原建筑功能单一，三栋康复楼均为病房，对此，我们呼应最新的娱乐养老理念，增加休闲娱乐空间，丰富老年人的生活。对于老年人来说，公共娱乐空间是十分重要的。本次设计采用“娱乐养老”理念，这里不再是养老院，而是老人与朋友们聚会、与儿孙们共享天伦之乐的场所。专家指出，只有精神上愉悦了，拥有年轻健康的心态、丰富的活动，才能颐养天年。

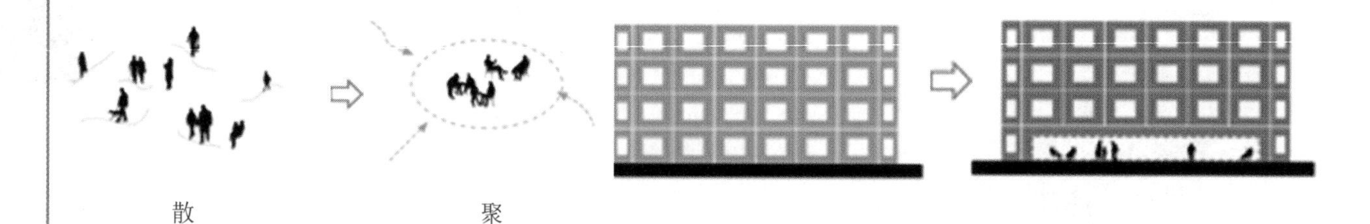

散　　　　聚

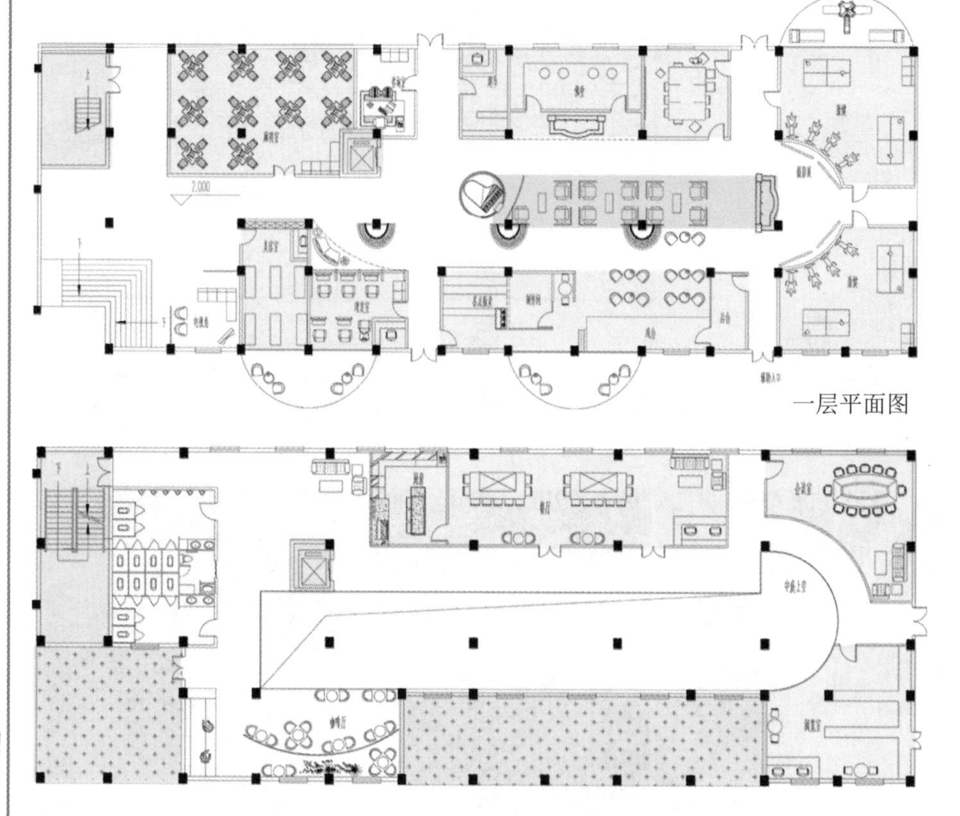

一层平面图

二层平面图

结构体系对比

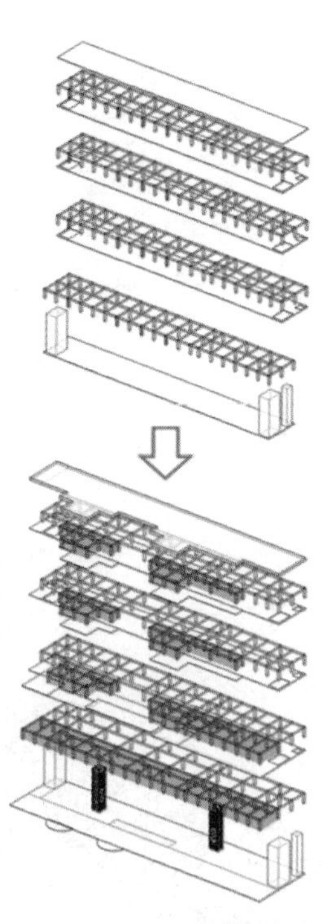

拔掉非承重部分构造柱，打造通高中庭空间，增加空间丰富

导师点评：

设计作品“团·聚”是对原有老年公寓建筑进行的较大幅度的空间改造的设计方案。设计者针对原有单调乏味的建筑空间进行优化与重组，将串联式的空间调整为组团式空间，设计中充分利用阳光、植物、通风等自然因素对空间进行新的塑造，满足老年人在心理和生理上对空间的感受与需求，为老年人创造了富有生命活力的空间环境，体现了设计者较好的空间把控能力和创新设计能力。在设计细节上还能够关注老年人的需求，无论是室内空间还是室外空间都能够结合老年人的特点以需求为导向深化空间环境设计。全套设计成果图式语言表达清晰完整，体现了设计者较扎实的专业设计基础和水平。

吕勤智

Advisor's Comments:

“Communal Living” redesigns the spaces of the buildings to a large extent. The original monotonous architectural spaces are optimized and re-divided. Specifically, spaces originally arranged in a line are divided into groups, and reshaped using natural elements like sunlight, plants and ventilation, for the purpose of meeting the seniors' psychological and physiological needs on an energetic and vibrant environment. The designers' ability to use spaces and innovation can be seen from spatial redesign. Besides, details, no matter in interior design or exterior design, are designed according to the needs of the seniors. The drawings and text clearly and completely deliver the ideas of the designers, indicating the solid basic professional knowledge of the designers.

Lyu Qinzhi

专家点评：

以部分老年人渴望热闹的意愿作为公共空间的设计主线，休闲街空间感强、布局合理、一步一景、层层递进，步行其间让人觉得有很强的叙事性；以另一部分老年人喜爱安静的意愿为起居空间的设计主线，创造了共用起居室的微社区新模式。通过绘制丰富的室内效果图和分析图，在图面上展现出了自己的设计理念和设计工作量。对于原建筑的结构改造也很合理，说明本组师生在设计过程中非常深入地思考和完善了很多设计细节，最终作品的可实施性较强。

卓培

Expert's Comments:

In the scheme, public areas are designed to satisfy some seniors' preference for a vibrant and busy environment. The pedestrian street features a palpable sense of space and a rational layout, with constantly changing landscape. Walking in the street is like reading a story. Living spaces are designed to meet some seniors' preference for a quiet environment. The new model of shared living room is adopted. From a large number of interior design renderings and analysis graphics, we can see the design concepts of the designers, and the great efforts that they made. The structural redesign of the original buildings is also very rational. Generally speaking, the designers have carefully analyzed many details and developed corresponding solutions, and their final scheme is of high feasibility.

Zhuo Pei

遵從老年人居住養老的設計原則，開展將老年人生理、心理與藝術設計相結合的探討。基於老年人的生理、心理和行为特点，借助此次C區更新功能空間環境進行及設施設計、導視系統設計、展示設計、公共藝術等滿足老年人生活、交流及參觀的需求。

零陸

06

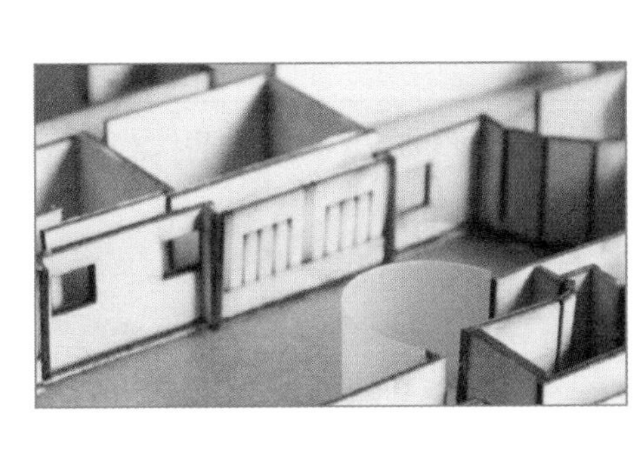

醫養建築設施與展示設計

Medical support building facilities and display design

CIID“室内设计 6+1”2016（第四届）

校企联合毕业设计

CIID"Interior Design 6+1"2016(Fourth Session)University and Enterprise Joint in Graduation Design

鹤发医养卷

——北京曜阳国际老年公寓环境改造设计

White hair volume medical support

—Beijing Yao Yang International Apartments for the elderly environmental reconstruction design

Photosynthesis – Regrowth

CIID“室内设计 6+1”2016（第四届）校企联合毕业设计

CIID“Interior Design 6+1”2016(Fourth Session)University and Enterprise Joint in Graduation Design

高　　校：浙江工业大学
College：Zhejiang University of Technology
学　　生：张怡 罗忆
Students：Zhang Yi　Luo Yi
指导教师：任彝
Instructors：Ren Yi
参赛成绩：医养建筑设施与展示设计组一等奖
Achievement：Frist Prize for Aged Care Facility and Exhibition Design

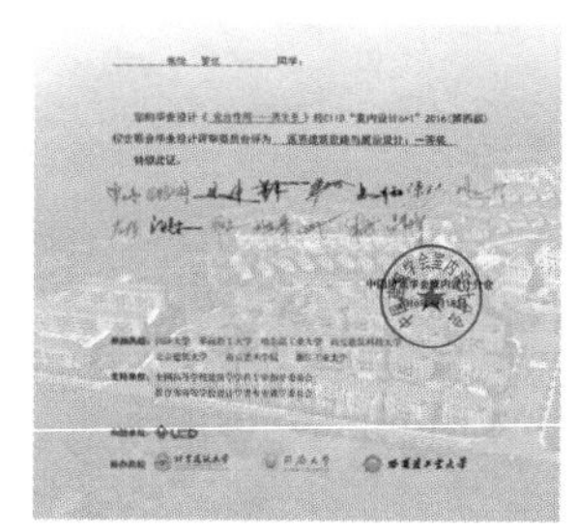

张　怡
Zhang Yi

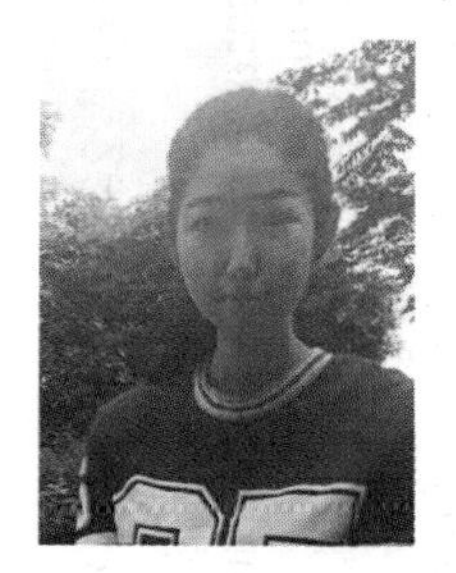

罗　忆
Luo Yi

学生感悟：

很荣幸能够参加CIID“室内设计6+1”2016（第四届）校企联合毕业设计活动，有机会可以和其他六所知名高校同学一起切磋学习，得到专家对我们设计的指导和点评。这是一个非常难得的机会。养老问题是当今的热点问题，对养老建筑的改造设计也是一个很有挑战性的题目。借助北京曜阳国际老年公寓的改造设计，我们不仅了解了一座城市，也了解了一类群体。我们的设计既要保证其合理的使用功能，更要考虑老年人在此环境下生活的心境，重获活力。这次的毕业设计让我们感慨良多，受益匪浅。我们很享受这个过程。

Students' Thoughts:

It's an honor to participate in the 2016 (the 4th) CIID “Interior Design 6+1” University-Business Graduation Projects Competition which gave me a rare opportunity to learn from students from six other universities, and to have guidance and comments from experts. Aged care is a hot-buttoned social issue, so the topic of retirement community renovation actually posed a tough challenge to us. In the process of developing a solution to the Beijing Yaoyang International Retirement Community Renovation project, we walked closer to a city, and also a group of people. We tried to give rational functions to the buildings, creating an environment in which the seniors can live a happy live, and regain their vitality. We learned a lot during the competition and enjoyed the process very much.

头脑风暴 || Brainstorm

概念生成 || Conceptual formation

概念演绎 || Conceptual presentation

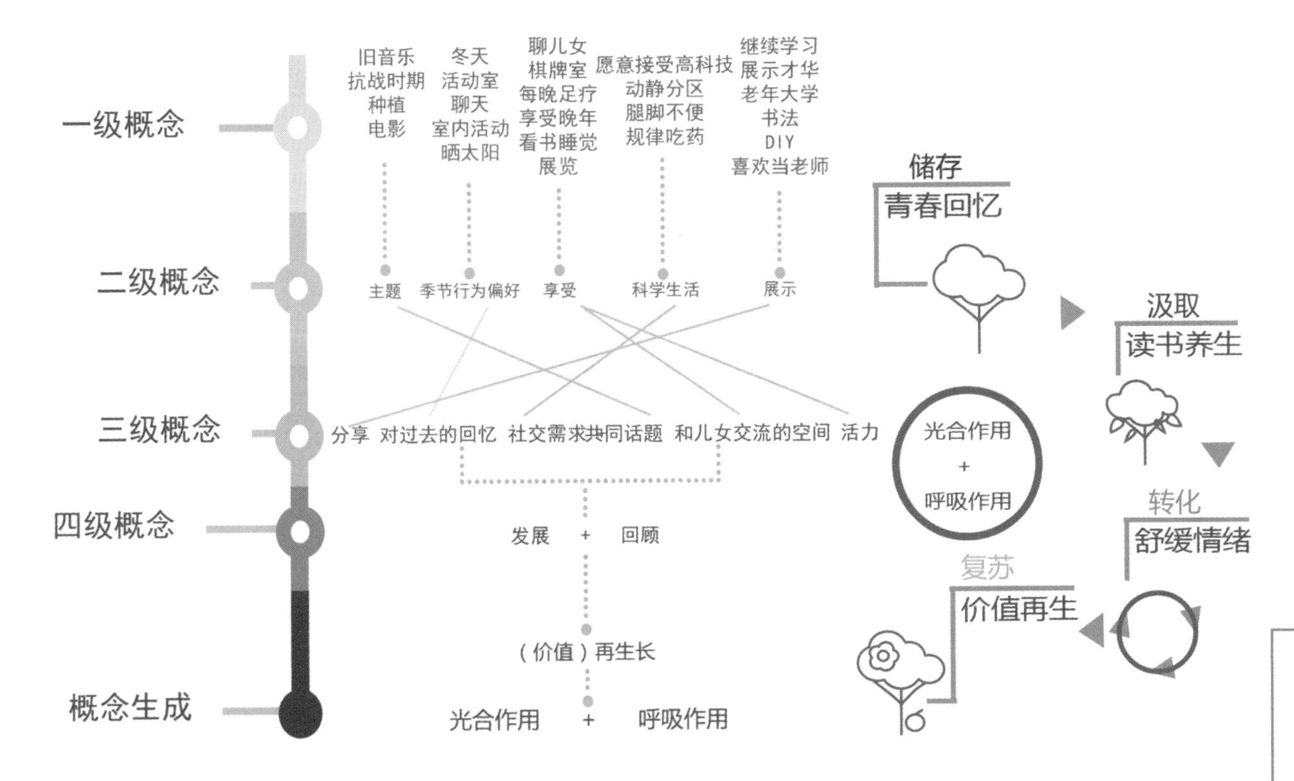

北京曜阳国际老年公寓 C 区地块活动中心改造设计

Beijing Yaoyang International Retirement Community District C Activity Center Renovation

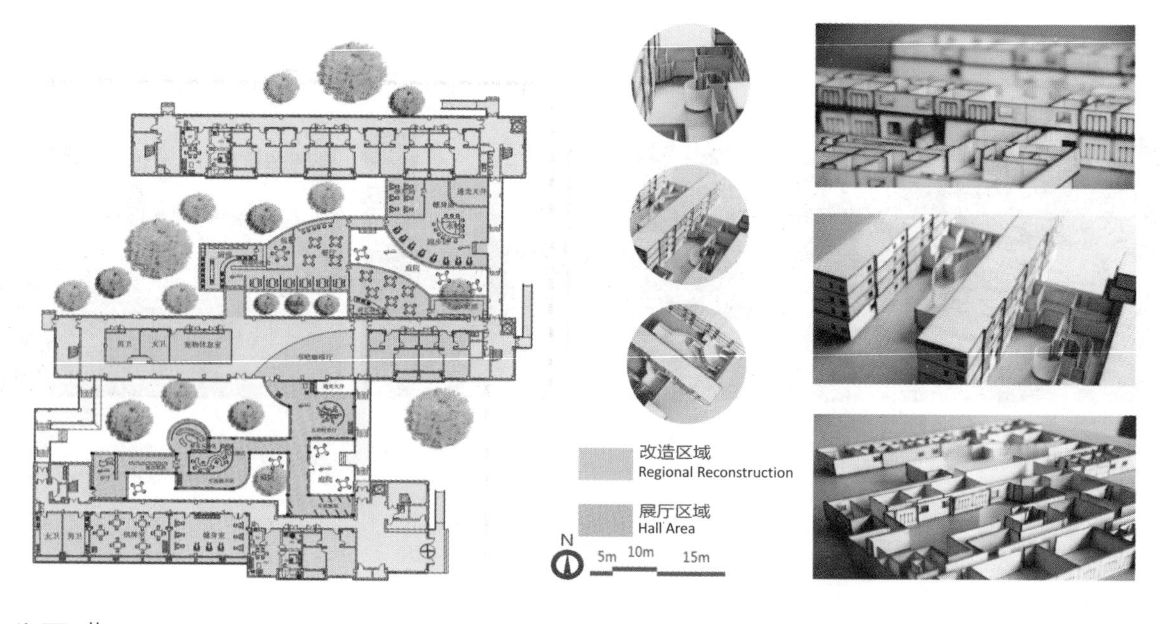

功能分区 || Functional division

交通分区 || Transport division

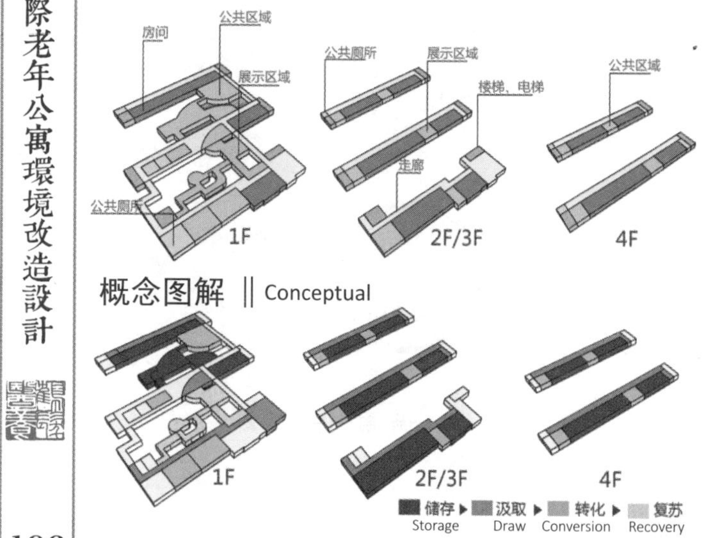

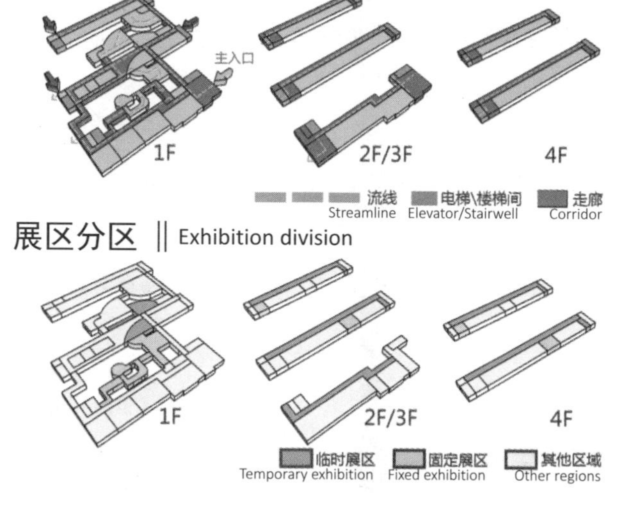

C1-C2 区域一层公共活动中心展厅设计

Design of Public Activity Center exhibition hall in District C1-C2

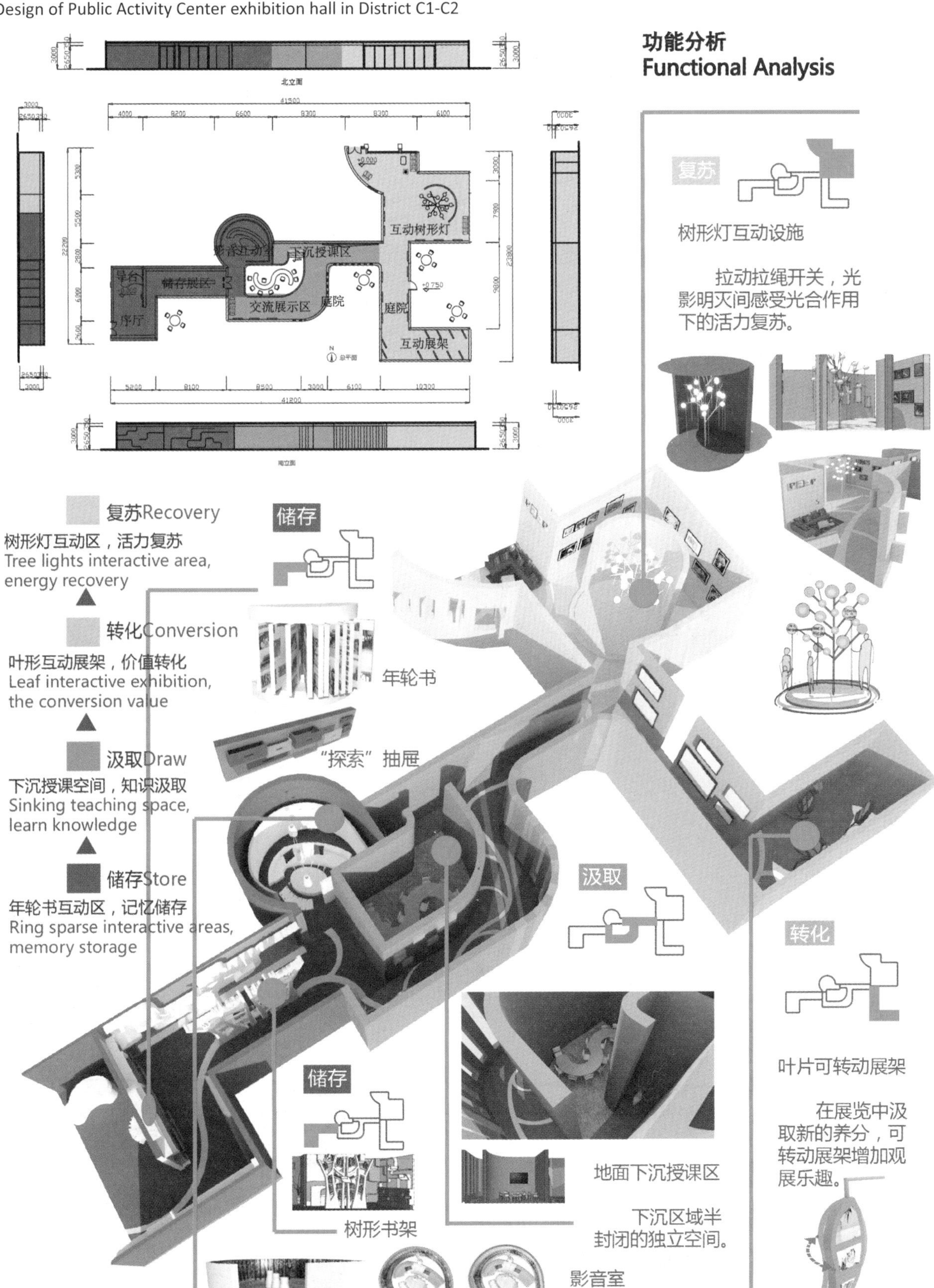

展厅内容分解 || Exhibition hall analysis

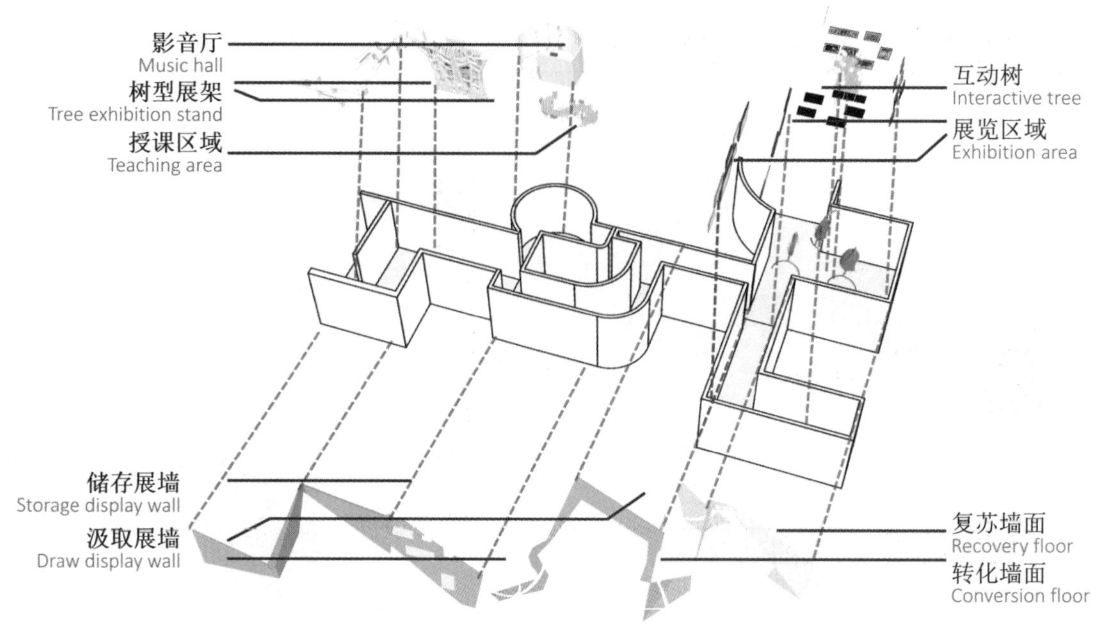

流线分析 Traffic flow analysis

展馆流线
Hall flow lines

出入口
Gateway

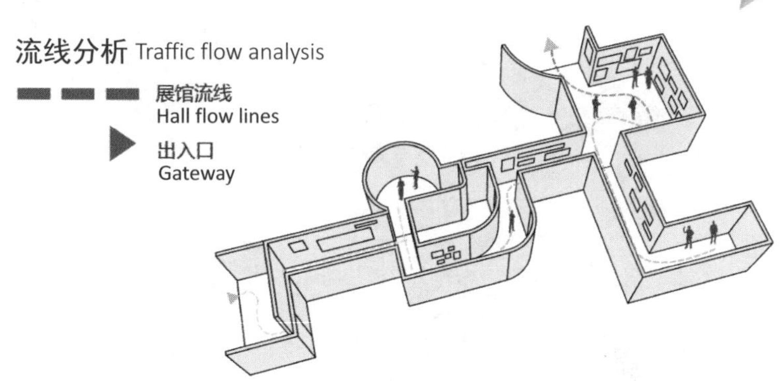

视线分析 View analysis

主要视线范围
Main line of sight

逗留时间长
Long stay

逗留时间短
Short stay

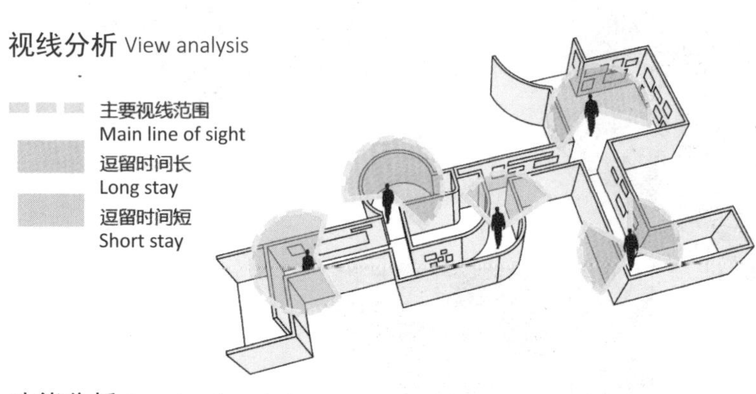

功能分析 Functional analysis

储存
Store

汲取
Draw

转化
Transformation

复苏
Recovery

影音室
Video room

展览区域
Exhibition area

互动区域
Interactive area

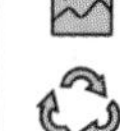

C区域老年公寓设施模块化改造设计 || District C Facility Modular Renovation

实地跟踪调研 Site tracking survey

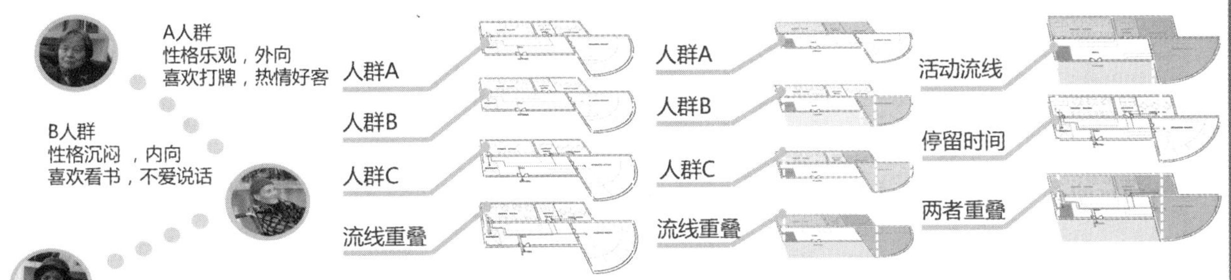

活动流线分析 Activity stream line analysis

停留时间分析 Residence time analysis

总体分析 Overall analysis

模块化概念 Modular concepts

老人根据自己的喜好需求搭配功能空间。无需改变户型结构，用最低的成本让每个户型变成个性化的私人订制。

The design allows the seniors to design spatial functions the way they like without changing house types. In this way, each house type can be personalized at the lowest cost.

个性化模块

厨房模块 Kitchen module

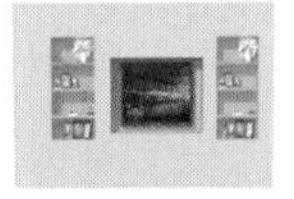

书橱模块 Bookcase module

客厅模块 Living room module

衣橱模块 Wardrobe module

房间模块化制式设计 Modular house type design

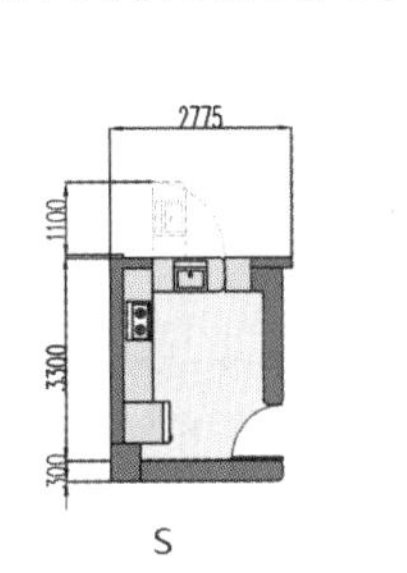

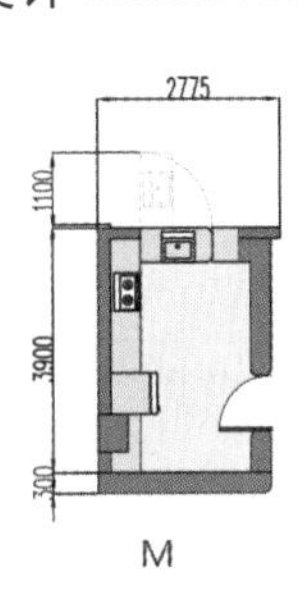

S M L

模块化厨房 Modular kitchen

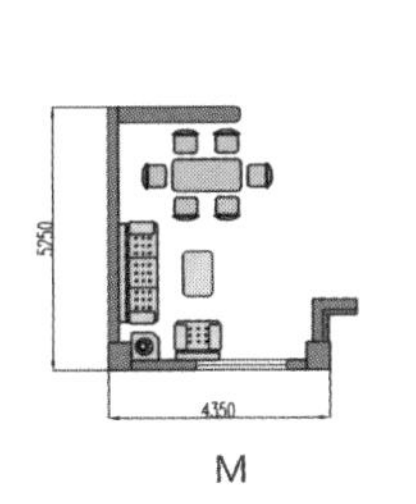

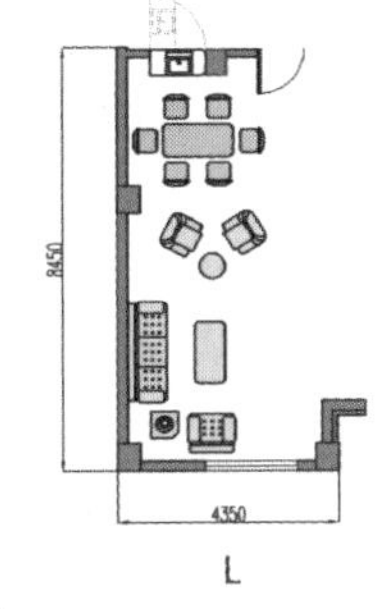

S M L

模块化客厅 Modular living room

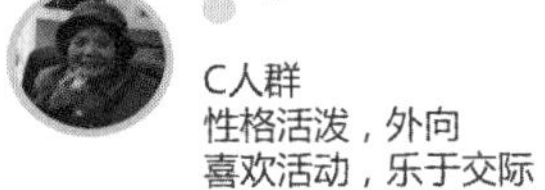

改造前平面图 Floor plan before renovation

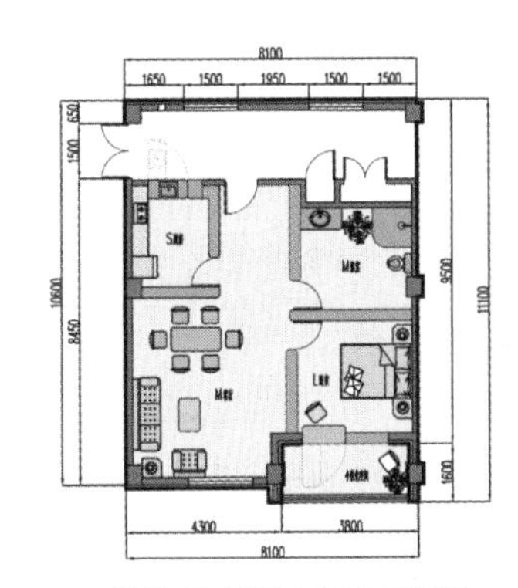

模块化标准间（自理型） Modular standard room (able-bodies seniors)

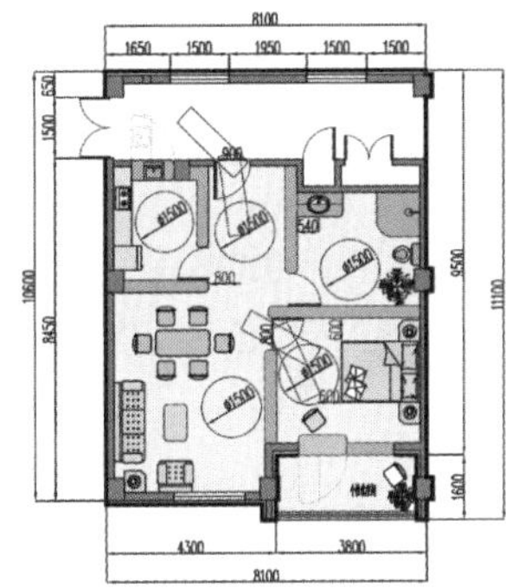

模块化标准间（半自理型） Modular standard room (partially disabled seniors)

中国老年人人体尺度测量图 || Figure Measurement of Chinese Seniors

站姿	(1)身高	(2)正立时眼高	(3)肩峰点高	(4)臂下垂中指尖距地高	(5)胯骨高
	(6)大腿长	(7)小腿长	(8)脚更跟	(9) 肩宽	(10)胯骨宽
	(11)双臂平伸长	(12)上臂长	(13)前臂长	(14)手长	(15)正立时举手高

坐姿	(16)正坐时眼高	(17)正坐膝盖高	(18)正坐大腿面高	(19)正坐时坐凳高	(20)正坐肘高
	(21)正坐时坐凳至肩高	(22)正坐时坐凳至头顶高	(23)正坐时举手高	(24)正坐时前伸手臂长	(25)胸厚
	(26)脚面长	(27)膝弯至臀部水平长	(28)脚长		

老年女性站姿正面图 老年女性坐姿侧面图 老年男性站姿正面图 老年男性坐姿侧面图

环境标识识别系统设计 || Environmental sign recognition system design

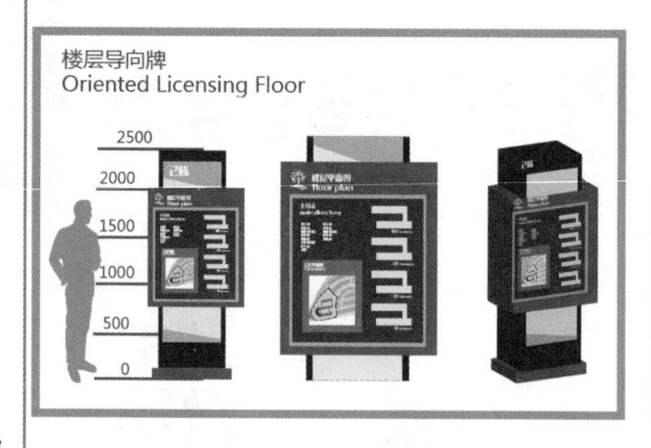

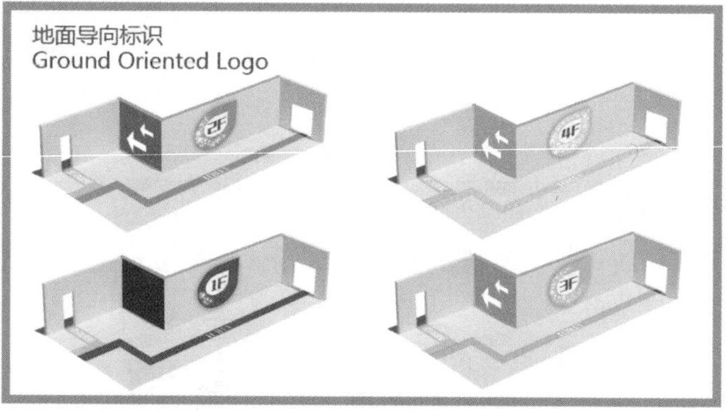

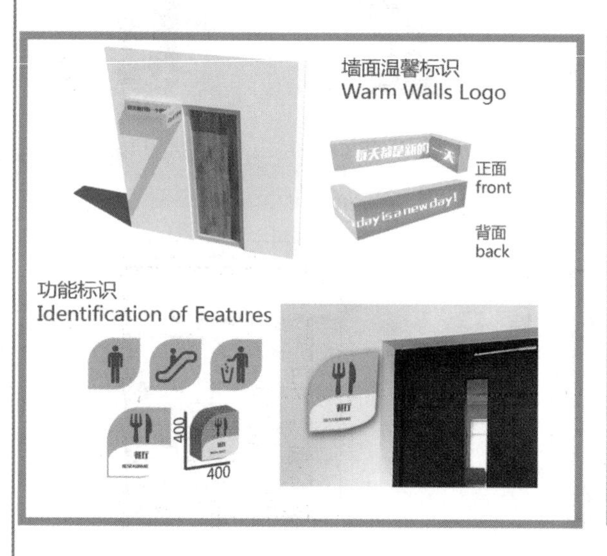

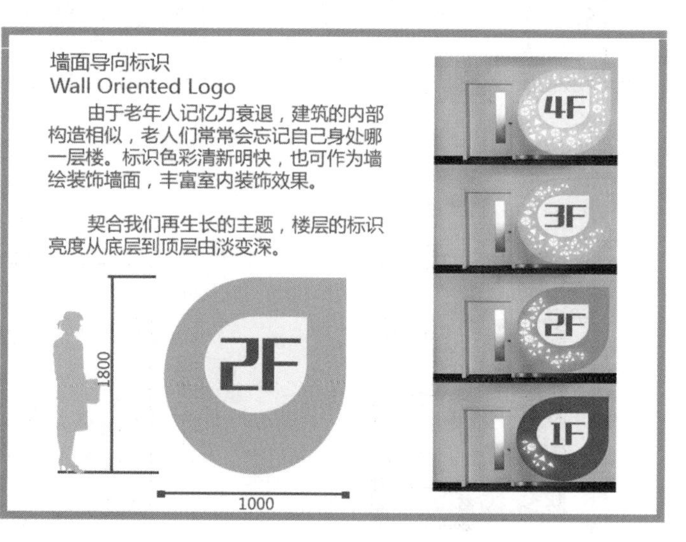

APP 及网页设计 || APP and website design

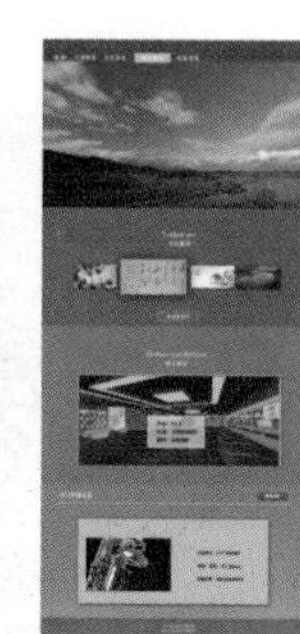

导师点评：

浙工大同学利用原建筑用地之间的绿地，新建了充满阳光和植物的交流展示和健身休闲空间，使原来的三栋建筑，在居住功能之外，拓展了公共交流互动区域，老年人足不出户就可以在四季温暖如春的阳光中健身、娱乐、交流。方案注重发掘老年人的情感需求，以拟人的植物的光合作用寓意，让老人也吸收朋友圈的养分，在相互作用下转化为正能量，积极、乐观、从容地面对老年生活，其乐融融。老年人群需要交流和关爱，由于远离城市，远离亲朋，希望能够在曜阳养老公寓建立起新的朋友圈，让不同兴趣爱好、不同生活阅历的老人，在曜阳养老公寓的温室里，发生新的光合作用，合成新能量。

在方案设计中运用了居住空间模块化功能组合的方法，给出了不同组合的菜单，使老人有了不同空间布置的选择，体现了个性需求的关怀，可以依据自己的爱好，选择居住空间的组合模式，在标准化的同时又体现了个性化。老人居住空间应该尽量便捷，利于人及轮椅活动自如，利于失能照护人员操作，未来还会增加机器人的照护。公寓内家具陈设主要考虑老人自用，来访家属可以利用公共空间及设施。导识设计得很系统全面，尤其是为阿尔茨海默症老人，用熟悉的色彩和物件来辨认自家房间起到了帮助作用，但是略显繁琐，忽略了养老公寓具备的家居属性，相对人群固定和使用重复性高，所以，要简化公共空间性的标识化手法，而是加强分区、分组、分户的识别，便于居住老人使用和提高视觉舒适度。现代老人掌握或学习移动互联网技术，用APP来协助日常康养管理及服务推送，加强与老人们的互动，因此，界面设计应适于老人的视力、手指、操作习惯等，采用适于老人特点的交互设计。希望设计师把好的设计创意用更多的设计方法和实施技术去深化，把细节做好做完美。

杨琳

Advisor's comments:

The students from Zhejiang University of Technology arrange communication, exhibition, exercise and recreational zones on the sunny landscaped areas between neighboring buildings. In this way, the seniors living in the three buildings can have easy access to exercise, entertainment and communication places and facilities, enjoying their life in the warm sun. The scheme focuses on finding and meeting the emotional needs of seniors. Themed on "Photosynthesis", the scheme compares seniors to plants, and love and care to sunlight. "Sunlight" can transform into positive energy which helps seniors live an active, optimistic, relaxed and happy life. Seniors need communication and care especially. Living at Yaoyang retirement community, the seniors are far away from city life, and also their families and friends, so friendship is important for them. The scheme tries to solve this problem by creating an environment which encourages seniors with different interests, professional fields, and life experiences to become friends. In the greenhouse of Yaoyang, new photosynthesis occurs, producing new energy.

The scheme provides modular spaces which can be combined according to the needs and favorites of the seniors, realizing the harmonious coexistence of personalization and standardization. The living spaces of seniors should adopt simple design to ensure a smooth traffic flow of the seniors (including seniors in wheelchairs), medical personnel, and robot nurses which may be used in the future. Furniture and amenities in the apartments should be selected according to the needs of the seniors, and their families can share public furniture and amenities. The sign design is systematic, covering diverse needs of the seniors. The signs for seniors suffering from Alzheimer's are a highlight – they help such seniors find their homes through colors and objects. One problem about the sign system is that it's a little complicated. Retirement homes are not public spaces, which means the signs only aim at a certain group of people, so the sign system should be as simple as possible. Focus should be paid to helping the seniors identify different zones, groups and apartments through eye-pleasant signs in an easy way. In the network era, many seniors can apply or are willing to learn internet technologies. In this context, APP-based health care management and service information pushing will be adopted to promote the communication among the seniors. The interactive interfaces of mobile terminals should be designed according the eyesight, flexibility of fingers, and operating habits of seniors. I hope the designers can expound creative design ideas through more design methods and technologies, and pay more attention to details.

Yang Lin

专家点评：

1. 选择“转化”而非“介入”，以与养老系统同行，从而达成具有相当差异性和价值的设计路径。
2. “转化”的介质是具有生命属性的绿植以及充满感性能量的光。别致且具挑战性。
3. 户型（功能配置）整体不变，而局部（例如依光照条件）界面可变——这是对模块化的活性演绎和有益实验。
4. 以空间为阈限、依流线作建构，顺势而为地牵引绿植的生存姿态及其与生活时光间的关联——这是对“光合”的别样利用和设计转义。
5. 以上的设计逻辑使营造不再被动、建构不再“自成体系”、界面不再“闭环”，功能不再仅限于（狭义的）“使用”。
6. 从此方案中人们能感受到“同呼吸”而非“我教你”的设计，“转化”而非“居高临下”的设计。

赵健

Expert's comments:

1. The scheme chooses to design through "transformation" rather than "intervention" according to the characteristics of aged care. Such a design approach is special and of certain value.
2. Plants which are forms of life and sunlight which provides positive energy work as media for the "transformation". This idea is unique and challengeable.
3. House types (functional configuration) remain unchanged, but regional layouts can be changed according to the seniors' needs (for example, the need for sunlight exposure). This is a good example of and a useful experiment about modular design.
4. Plants are grown in different spaces according to traffic flows and terrains, accompanying the seniors in their life. Photosynthetic occurs to plants, and also the seniors.
5. The above design concepts enable active spatial transformation, integrated architectural structures, interactive spaces, and diverse functions.
6. The designers think and design from the perspective of the users, trying to transforming the living environment of the seniors according to their real needs.

Zhao Jian

Opera and Clouds

CIID“室内设计 6+1”2016（第四届）校企联合毕业设计

CIID“Interior Design 6+1”2016(Fourth Session)University and Enterprise Joint in Graduation Design

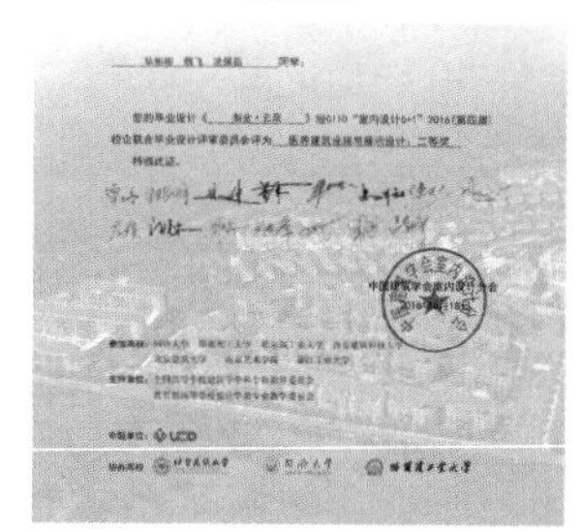

高　　校：南京艺术学院

College：Nanjing University of the Arts

学　　生：华彬彬　魏飞　冼佩茹

Students：Hua Binbin　Wei Fei　Xian Peiru

指导教师：朱飞

Instructors：Zhu Fei

参赛成绩：医养建筑设施与展示设计组二等奖

Achievement：Second Prize for Aged Care Facility and Exhibition Design

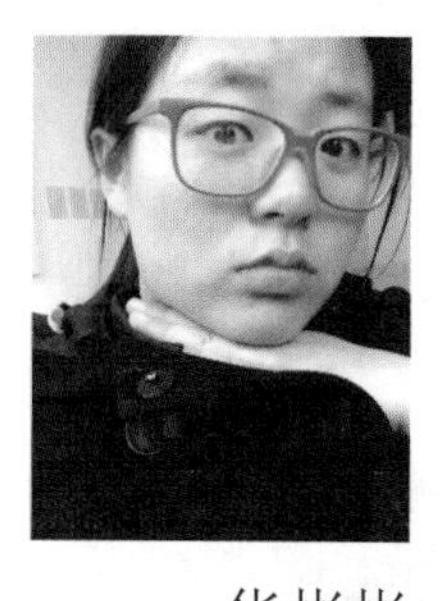

华彬彬
Hua Binbin

魏　飞
Wei Fei

冼佩茹
Xian Peiru

学生感悟：

我们经过前期对本次课题的调研和资料收集，将对象定位在有自理能力的久居老人和失智老人身上。整个老年公寓的定位为以疗养为主的医养结合的老年公寓。通过参加这个联合毕业设计，尝试了一次自己从未涉及过的领域，使我接触到了建筑设计、室内设计、景观设计，并且从中学到了很多知识，学会用建筑的思维去思考室内设计，并从室内设计的角度考虑信息的传达，在设计过程中了解了许多京剧文化、养老机构的规范等多方面的知识。

Students' Thoughts:

Based on early stage survey results and other information that we collected, we chose able-bodied seniors and seniors with dementia as the target group, and gave the functions of medical care and recuperation to the buildings. The competition gave me an opportunity to touch a new design field, and in this process, I learned a lot of new knowledge about architectural design, interior design, landscape design, as well as Peking opera and regulations concerning retirement homes. I also learned how to understand interior design from the perspective of an architect, and how to deliver information from the perspective of an interior designer.

平面分析 || Plan analysis

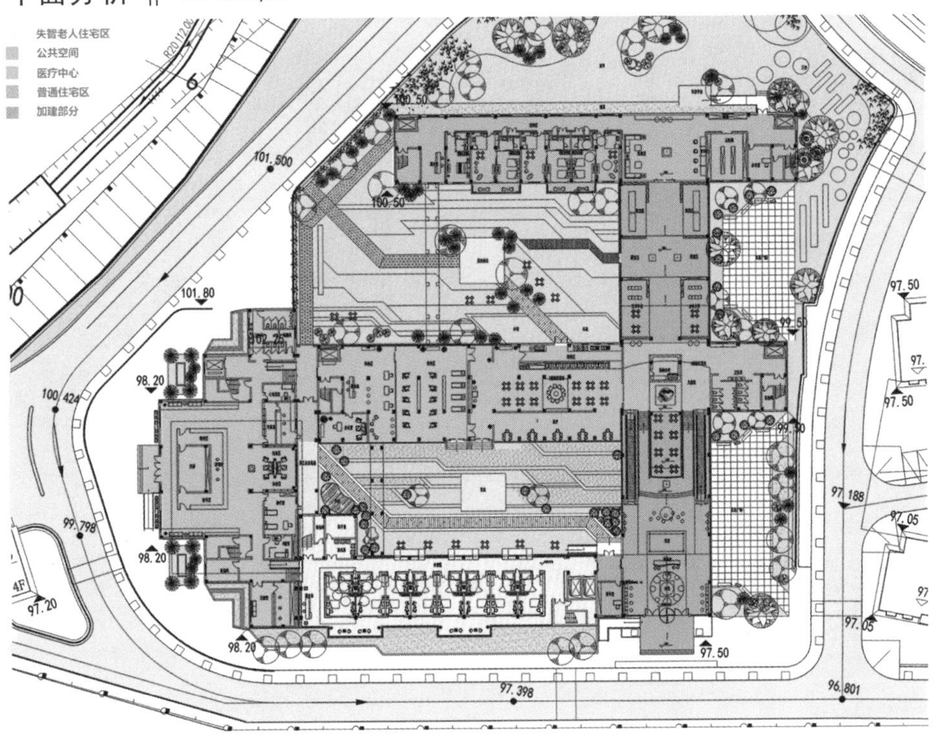

整个空间中主要有五个区域，灰色代表的是加建的门厅公共部分，黄色是失智老人住宅区，红色为主要的公共活动空间，紫色为自理老人住宅区，蓝色为医疗中心。

The project site is divided into five zones. Grey zone is a newly built lobby as a public area, yellow zone is the living area of seniors with dementia, red zone is the main pubic area, purple zone is the living area of able-bodied seniors, and blue zone is a medical center.

文案分析 || Scheme analysis

第一栋楼

湛湛青天不可欺
是非善恶人尽知
血海冤仇终需报
且看来早与来迟

云外的须弥山色空四显
挽翠袖进前来金盆扶定
猛听得旌鼓响画角声震
看大王在帐中和衣睡稳

第二栋楼

第三栋楼

登层台望家乡躬身下拜
鄢坞县在马上心神不定
习天书学兵法犹如反掌
小东人下学归言必有错

扶大宋锦华夷赤心肝胆
听谯楼打罢了初更时分
将酒宴摆至在聚义厅上
我魏绛闻此言如梦方醒

第四栋楼

加建部分

金钟响玉兔催王登九重
为国家哪何曾半日闲空
叹杨家投宋主心血用尽
自那日朝罢归身染重病

门的分析图 || Lobby analysis graphic

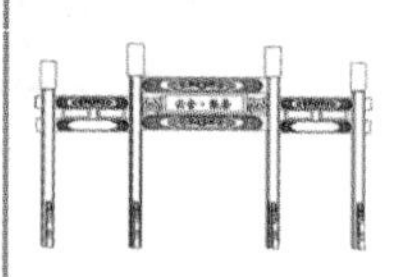

入口前方设置了仿古的迎客牌坊，牌匾上是梨舍·云居的LOGO。

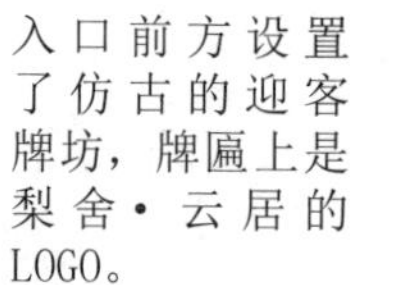

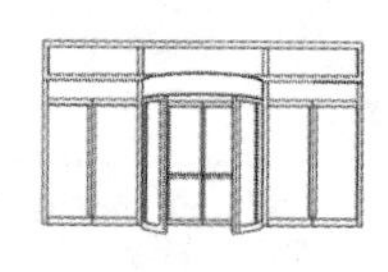

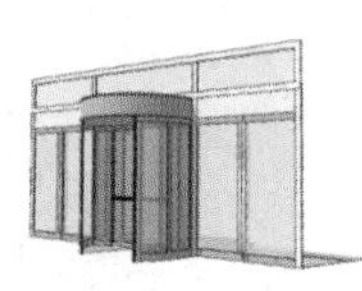

考虑到北方冬季风比较大，使用了旋转门，但两侧使用了双开门。

为营造空间的节奏感与层次感，空间内部增加了很多仿古门。

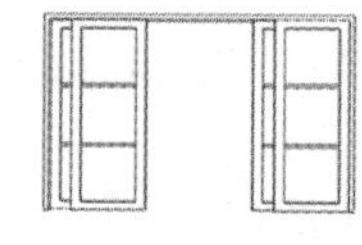

记忆障碍老人区域，考虑到老人走失等安全问题，使用的是刷卡移门。

为了保障住宅区域的私密性，住宅区使用的是刷卡式的移门。

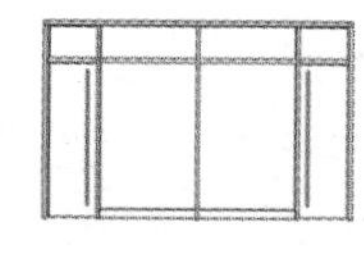

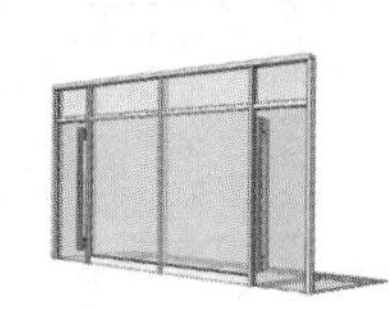

医疗中心考虑到老人行动问题，使用的是智能移门加两侧开门。

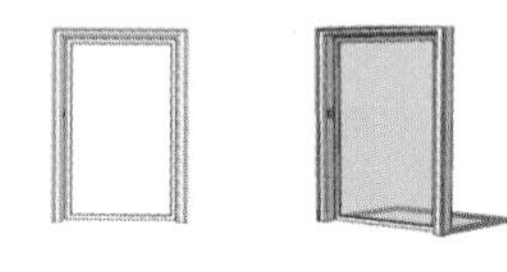

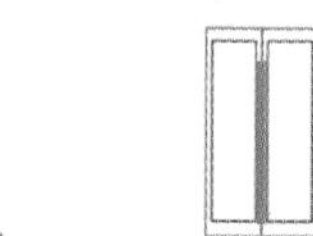

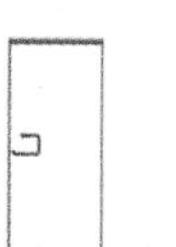

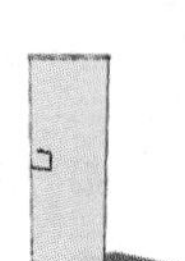

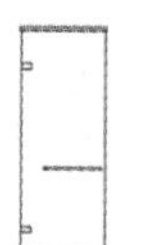

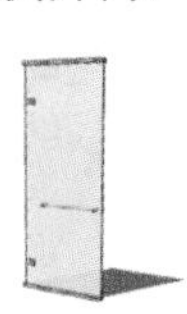

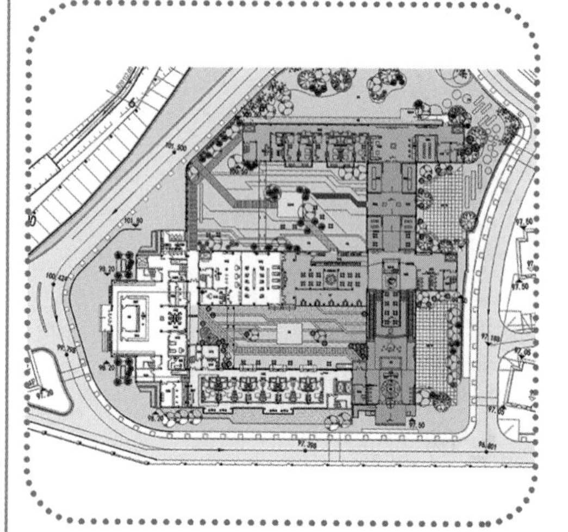

旦角&动

丑角&暖

生角&静

净角&医

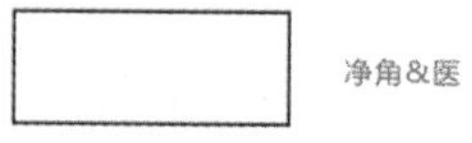

末角&孝

在色彩利用上，我们将京剧中最常见的红黄蓝白黑运用到设计当中，材质的选择上也从戏曲的角度出发，主要以木材为主，凸显适老性，营造家的温馨感。

For the application of colors, red, yellow, blue, white and black, the five colors commonly used in Peking Opera are applied in the design. As for materials, wood, the material used to build traditional opera stages, is widely used in the design to create a warm and home-like atmosphere.

信息分析 || Information analysis

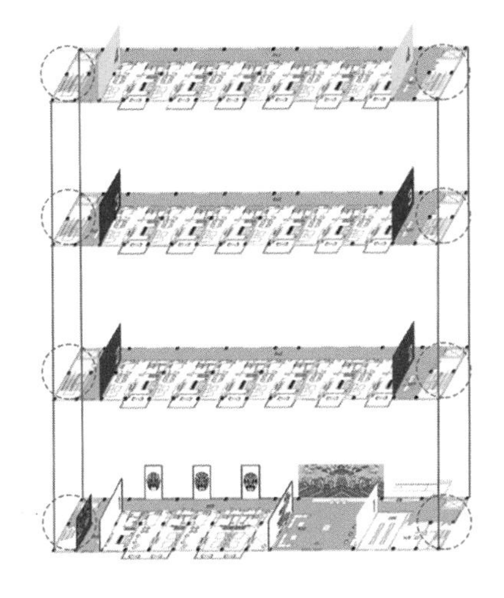

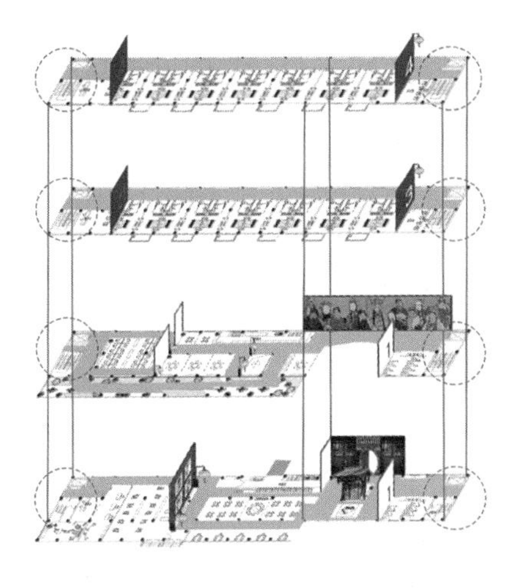

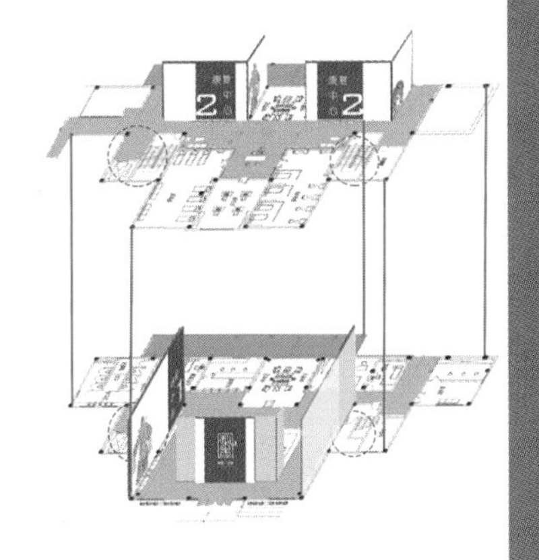

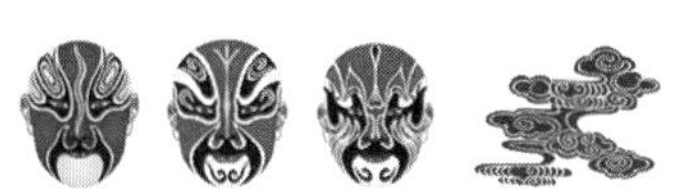

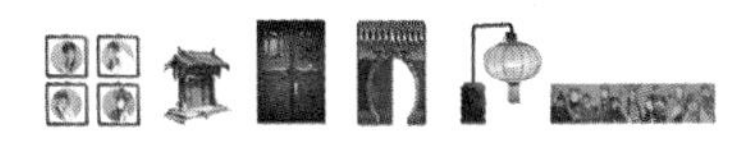

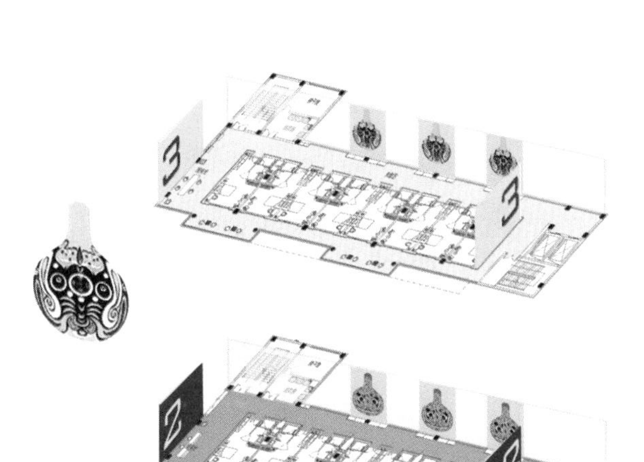

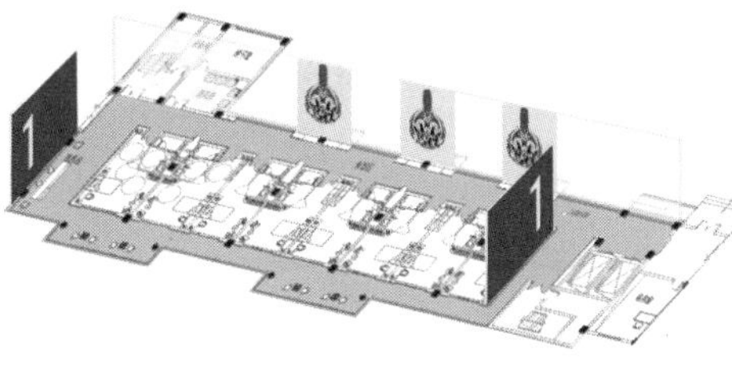

生角的扮演者多为青壮年男子，三号楼中主要为自理型老人的住宅区域，生角符合其中的空间氛围。旦角色彩丰富，表现能力强，其生动柔美的形象与公共空间的氛围相符合。

In Peking opera, Sheng characters are always played by young men; therefore, this element is applied in the design of Building 3 which is occupied by able-bodied seniors. The element of Dan characters, which are related to vibrant colors, expression and femininity, is applied in the design of public areas.

LOGO 设计 ‖ Logo design

这是我们根据颜色与材质 LOGO 设计的导视系统，大量的运用在每个空间，尤其公共空间运用的最多。取梨舍二字，代表戏如人生。取自于戏曲《斩经堂》。人生如戏，何不戏说人生。再取云居中的云纹作为 LOGO 的设计元素。

The sign system is designed according to the colors and material of the logo. A large number of signs are arranged in the project area, especially public areas. “Opera” indicates “life is like a play, why not enjoy your life”, a line from Peking opera “Murder in the Temple”. The cloud pattern on the logo stands for “clouds” in the name of the project.

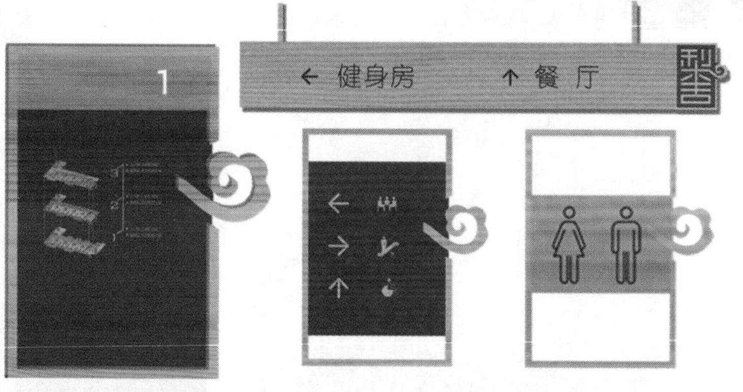

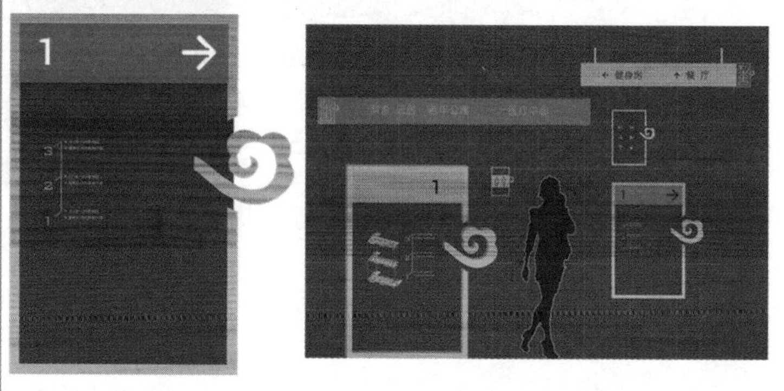

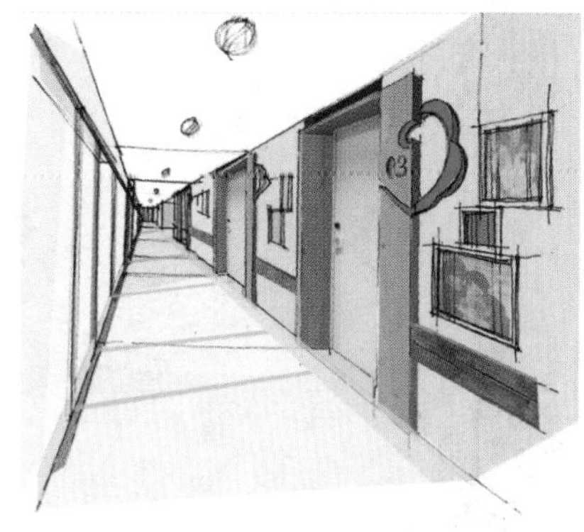

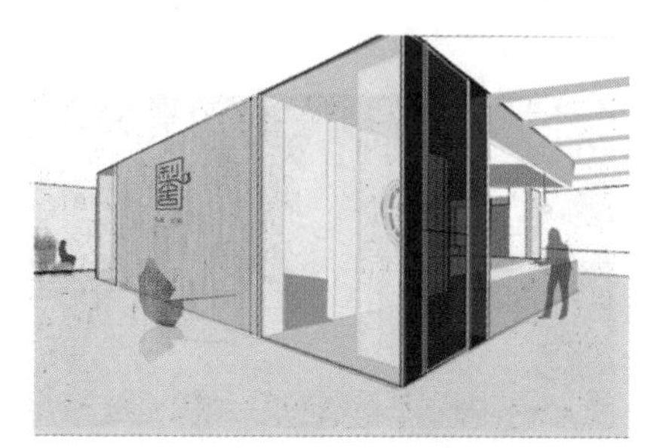

导师点评：

“梨舍·云居”：中国传统京剧元素展示老年公寓改造方案的设计主题，设计者从改变空间水平及垂直交通的角度去改善空间流线的通达性，试图改变建筑室内空间形态的联结关系。设计方案的特点在于将京剧元素符号化运用到空间标识系统及色彩方案中，以点、线、面的构成语言呈现在室内空间。在老年居室空间的设施设计中引用了蒙特里安的色彩方案，并加以变化和延伸，旨在突出居室室内与公共空间中西合璧的视觉感受。整体方案在概念主题的逻辑关系、整体格调把握、受众群体——老年人的生理和心理适宜性分析、空间设施的舒适性方面待进一步调整和深入。

任彝

Advisor's comments:

"Opera and Clouds": renovates the retirement community through exhibition design themed on Peking opera elements. The scheme tries to create smoother traffic flows horizontally and vertically, thus changing the connection relations between interior spaces. The highlight of the scheme is incorporating Peking opera elements into the design of the sign system, and into interior color design in the form of points, lines and surfaces. Mondrian's color scheme is applied in interior facility design. Through the change of color and spatial extension, Chinese and western cultural elements are harmoniously mixed, creating visual impacts. Further adjustments and deeper analysis are required in terms of the logical relationship between the theme and the content, overall style control, the physiological and psychological suitability of the target group (the seniors), and comfort of facilities.

Ren Yi

专家点评：

“梨舍·云居”的设计者以中国的国粹——京剧元素为主线展开设计概念，将京剧的内涵文化思想意识同老年公寓环境进行了巧妙融合，主题明确清晰，方案结构完整，从多方面考虑到老年人生理、心理的需求，进而延续到空间布局及装饰，突出了医养结合的特色，同时也考虑到经营方的投入与产出的关系，是有一定实施性的设计作品。

设计者“精心”的主题选定，创造了设计的亮点也局限了设计本身，在局部空间设计及色彩使用上，还是较多的“想当然”占据了主导，养老设施的特殊性促使设计者时刻将自己作为老年人去思考和进行设计。方案最终的呈现还具有一定的可操作空间，还需要对老年人的生活规律及习惯进行进一步的调研。

陈天力

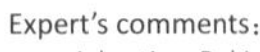

Expert's comments:

Adopting Peking opera, the quintessence of Chinese culture, as primary design element, "Opera and Clouds" skillfully incorporates Peking opera culture into the environmental redesign of the retirement community. The theme is clear, and the scheme structure is complete. Besides, the physiological and psychological needs of the seniors are considered in the design of various aspects, including spatial layout and decoration. The scheme realizes the integration of medical and aged care resources, and analyzes the input-output relationship of the project. It's feasible to some extent.

The carefully selected theme is the highlight of the design, but also a restriction. Regional spatial design and color selection are kind of divorced from reality. After all, the design of retirement homes requires designers to think and design from the perspective of seniors. Generally speaking, the final scheme is feasible to some extent, but a further survey on the lifestyles of seniors is still required.

Chen Tianli

Six Harmonies

CIID“室内设计 6+1”2016（第四届）校企联合毕业设计

CIID“Interior Design 6+1”2016(Fourth Session)University and Enterprise Joint in Graduation Design

高　　校：南京艺术学院
College：Nanjing University of the Arts
学　　生：张梦迪　龚月茜　戚晓文
Students：Zhang Mengdi　Gong Yuexi　Qi Xiaowen
指导教师：朱飞
Instructors：Zhu Fei
参赛成绩：医养建筑设施与展示设计组三等奖
Achievement：Third Prize for Aged Care Facility and Exhibition Design

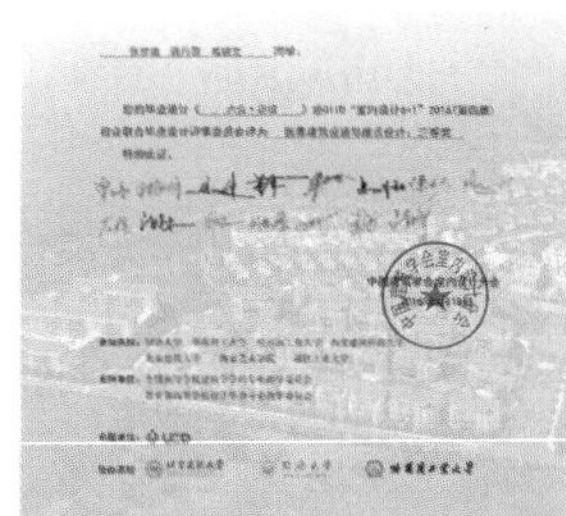

张梦迪
Zhang Mengdi

龚月茜
Gong Yuexi

戚晓文
Qi Xiaowen

学生感悟：

经过大半年的奋战从北京到上海再到哈尔滨，我们经历了很多，也收获了很多。在没有参加“6＋1”之前，我们觉得毕业设计只是对这几年来所学知识的单纯总结，但是通过这次锻炼，我们发现我们的认识过于片面，毕业设计不仅仅对前面所学知识的一种检验，更也是对自己能力的一种提高。通过这次互动，我们明白了自身的知识和经验还是缺乏的。虽然在设计的过程中，我们遇到了很多的困难，方案改了一次又一次，也被批评了很多次，但是在方案完成的那一刻，我们是快乐的。只有坚持，才会有真正的收获。

Students' Thoughts:

From Beijing to Shanghai to Harbin, during the competition which lasted for more than half a year, we experienced a lot and also learned a lot. Before participating in the competition, we thought that a gradation project is just a test of the knowledge that we'd learned in the university years. Later we realized that our understanding is too partial – a graduation project is not only a test, but also a process of improvement which helps us realize our lack of experience and knowledge. It's true that the process of design was hard. We met a lot of difficulties, changed our scheme again and again, and were criticized for many times. However, at the moment when the scheme was finally completed, we were very happy and excited. Success cannot be made without insistence.

方案说明 ‖ Scheme overview

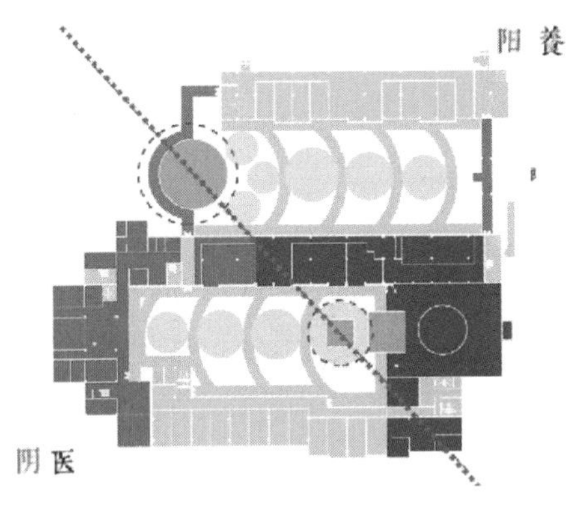

此次设计改造的是一个有关于老年人的医疗养老空间，老年人是一个特殊的群体，因此在设计的过程除了将与展示专业相关的内容运用到设计中，同时还要时刻考虑目标人群。在密云曜阳国际老年公寓这个项目中，不仅是室内的改造，还有庭院的部分，要打破原先几栋楼各自为盈的模式，增加交流的空间。

The project is about the renovation of aged care buildings. Seniors are a special group; therefore, in the process of design, aside from applying knowledge related to exhibition design in the design, we should also consider the actual needs of the seniors. The renovation comprises both interior renovation and yard innovation. For the latter, our focus is linking independent buildings to promote the communication among the seniors.

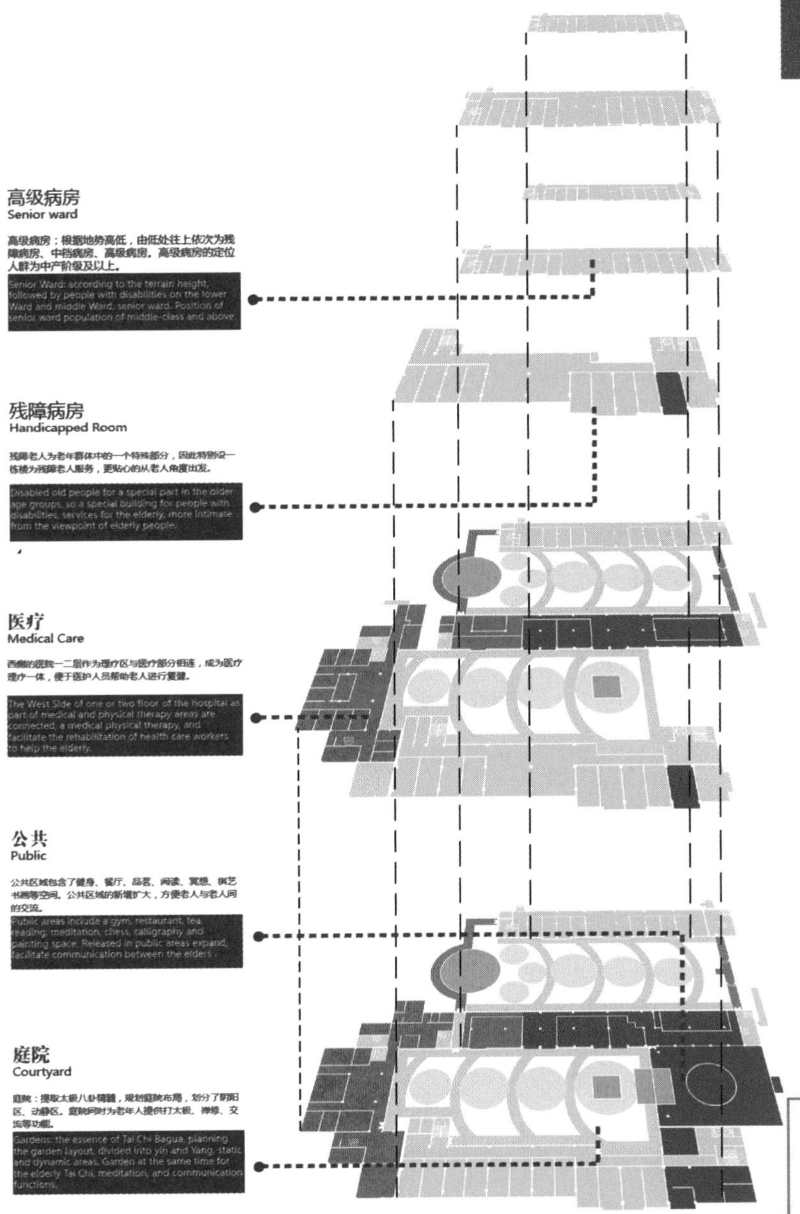

公共区域 ‖ Public areas

老有所养，是中国人自古即寄语的美好愿望。随着我国老龄化人口的不断加剧，老年人的医养空间的建设吸引了人们的眼球，同时重视改善目前老年人的居住环境。“六合·宓馆”的设计是依据现代中老年公寓体系中比较完善的机制和设施，以人为本，倡导中医与道家的传统养生法则，围绕太极主题进行规划的方案。

Being properly cared for after getting old is a traditional Chinese concept. In the context that population aging is increasingly becoming an issue for concern, the construction of aged care facilities has attracted a lot of attention, and consequently, the living environment of seniors has been improved. Themed on Taiji, Six Harmonies is a human-oriented design scheme which adopts modern retirement home operation system and facilities, and advocates TCM and Taoist health maintenance principles.

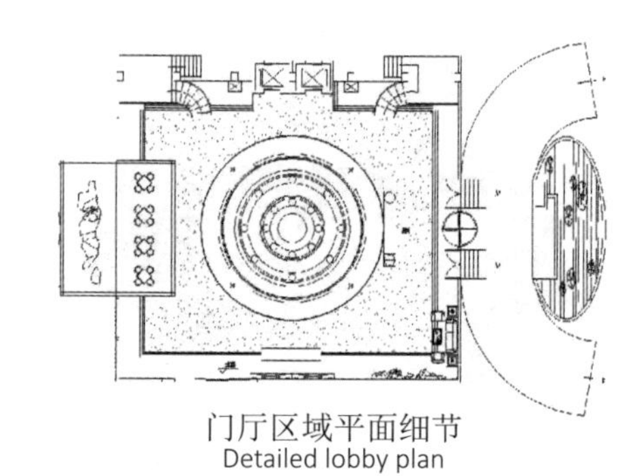

门厅区域平面细节
Detailed lobby plan

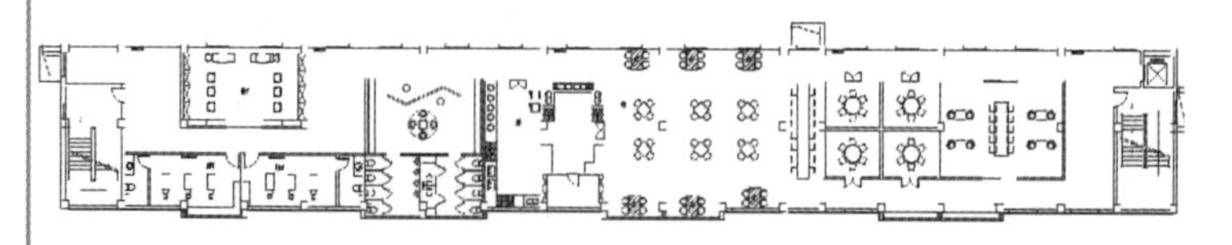

一层公共区域平面细节
Detailed plan of the public area on the 1st floor

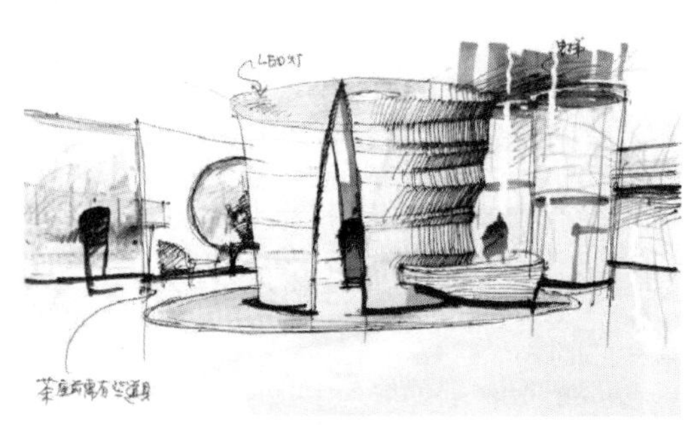

手绘效果图
Freehand rendering

二层公共区域平面细节
Detailed plan of the public area on the 2nd floor

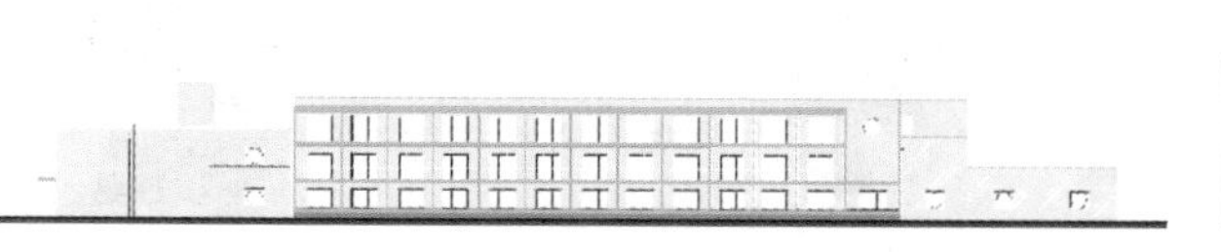

导师点评：

“六合·宓馆”命题新颖，充满东方意味，设计方案从中国传统中医养生切入，“顺气一日分为四时，中医养生、养气，气和而神形、阴阳平衡”，并提出宅物、宅己、宅生、合形、合气、合心的设计思想，方案带入感极强。

方案充分考虑了普通健全老人、残障老人、旅游人群、家属子女、医疗服务人员、物业管理人员六类受众人群，并创新性地将老人分为展示型老人和观赏型老人，注重搭建展示公共平台，方便老年人交流互动，采用竹、木、石、土、水的设计元素，使得空间充满淡雅、传统的中式味道。平面方案规划合理，注重阴阳结合，病房区域分类设置，合理布局，入口门厅重点设计，效果尤为突出。庭院设计有机结合太极八卦的意向，设置两处凉亭，体现中式建筑的特征。

方案设计新颖，紧扣中式建筑、园林、室内的特征，有一定创新性。但方案图纸部分有所欠缺，制图规范性不足，需进一步提升。

朱宁克

Advisor's comments:

"Six Harmonies" is a typical oriental theme. Adopting TCM as the highlight, the scheme incorporates such TCM concepts as "variation of qi in a day", "health maintenance", "cultivation of qi", "qi-spirit harmony", and "yin-yang balance" into the design. It also discusses the relationship between homes and objects, man, and life, and the coexistence of different forms, types of qi, and ideas in the context of design. The scheme is very attractive.

The needs of six groups including able-bodied seniors, seniors with disability, tourists, occupants' families, medical personnel, and property management are discussed in the scheme. The scheme innovatively divides seniors into seniors preferring to show and seniors preferring to watch, and corresponding spaces for performance and communication are designed. Besides, elements like bamboo, wood, stone, earth and water are employed to create an elegant environment filled with traditional Chinese culture. The plan design features yin-yang integration, ward classification, and rational spatial division. The entrance hall is a highlight of plan design. The design of the yards applies the concepts of Taiji and Eight Diagrams, with two pavilions presenting yin and yang in each yard.

The scheme is creative, incorporating traditional Chinese culture elements into architectural, landscape and interior design. The drawings should be more normative, and relevant improvements are needed.

Zhu Ningke

专家点评：

该方案通过对原有建筑的门厅、公共活动区域、各层病房和室外庭院进行改造，试图改善目前的居住环境。门厅的改造，引入视觉冲击力较强的体块，其实质的功能、空间感受尚待商榷；公共活动区域的引入能增加使用人群的交流，但应做到动静分区，设置更符合老年人生理及心理需求的内容；病房的分区规划有一定的合理性，建议根据各种设定人群的需求推导病房护理单元可提供的服务，从而进行更为细致的空间调整和划分；室外庭院的形式及材料的选择等，可紧扣提出的“太极”“中医”“道家”这几个主题进行深入设计。方案思路明确，各图纸在规范表达及突出表现方面欠佳，版面略松散。

李莉

Expert's comments:

In the scheme, the halls and wards of the buildings, public areas, and yards are redesigned, for the purpose of creating a better living environment. The hall design, including functional and spatial design, is impressed, but its actual effects still need further discussion. More public areas are created to promote the communication among the seniors, but areas for active and quiet activities should be separated to meeting the physiological and psychological needs of different occupants. The zoning of wards is kind of rational. My suggestion is that the ward area can be divided in a more specific way that seniors needing different medical services are arranged in different zones correspondingly. As for the design of the yards, forms and the selection of materials should be themed on Taiji, TCM and Taoism. The scheme expresses design ideas clearly and logically, but the drawings are less normative, impressive, and compact.

Li Li

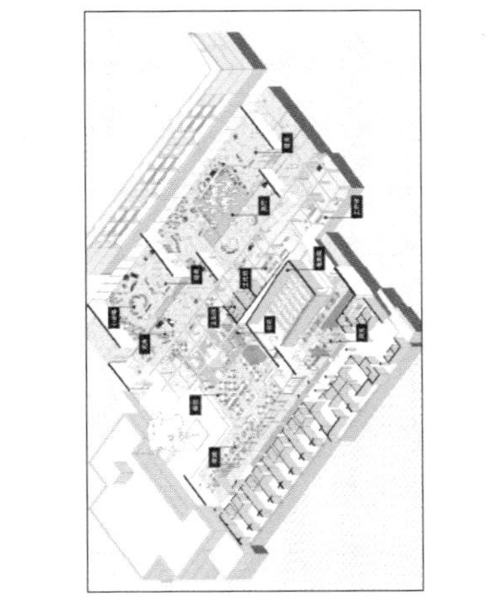

演講人發言

The speaker spoke

CIID“室内设计 6+1”2016（第四届）校企联合毕业设计

CIID"Interior Design 6+1"2016(Fourth Session)University and Enterprise Joint in Graduation Design

鹤发医养卷
——北京曜阳国际老年公寓环境改造设计

White hair volume medical support
—Beijing Yao Yang International Apartments for the elderly environmental reconstruction design

Ying Zhi Operation Mode with the Combination of Rehabilitation Treatment and Aged Care

Under the current situation that the vacancy rate of national aged care institutions is more than 50%, the government hopes to inject new vitality into aged care by combining medical treatment with aged care. The deep fusion of rehabilitation treatment and aged care may be one way to be chosen. As the pioneer of this choice, Ying Zhi may be the only state-controlled light assets operation management institution capable of output management and large-scale chain operation, which deeply fuses rehabilitation treatment with aged care. Its 10 years operating experience can bring some enlightenment to this industry.

The conventional aged care institution fees include bed fee, nursing fee and meals fee, but generally it's unlikely to get profits from bed fee and meals fee. The operation mode combining rehabilitation treatment and aged care guides architectural design and planning in the early stage, and so reduces construction costs of aged care institution to some extent. Meanwhile it also brings 10-15% additional medical income, which contributes to the income of aged nursing institution. But the expenditure of aged care institution is not increasing with it. Medical service income is provided to aged care institution by medical rehabilitation team, so the income of institution increases while the whole operation costs still retain the original level.

In rehabilitation hospital, the rehabilitation process is developed by therapists; while in the aged care institution, the rehabilitation plan is also made by therapists. This has not only realized the concept of combining rehabilitation treatment and aged care, but also effectively improved the accuracy of service, and reduced fixed costs at the same time. The reasons why the aged people have chosen Ying Zhi for their later life include the decreasing of fixed costs proportion, the high efficiency of professional talents, as well as the accuracy of its market positioning.

The ability of independent operation of medical institution supports and drives the operation of aged care institution, and thus stably and quickly raises the occupation rate of apartment for the aged, bringing stable cash flow to the institution. Meanwhile, through professional services provided by medical personnel, the elderly can enjoy secure and happy later life, which provides necessary support to prevent the happening of emergency. Thus the medical treatment cost of the aged is reduced. At the same time, due to the unified management of rehabilitation bed and nursing bed, the smooth referral and mutual adjustment operation mode has been formed.

The operation mode with the deep fusion of rehabilitation treatment and aged care has both increased the output value per bed in rehabilitation hospital, and tremendously improved the potential value of aged nursing institution.

Speech from Professor Hao Luoxi at Tongji University

In the context that China's population aging is increasingly becoming an issue for concern, the architectural design industry has been thinking how to enrich the life of seniors and improve the quality of their life with the power of design. Senior-oriented design covers various aspects of seniors' life, and in the case of architectural design, senior-oriented design, including its connotation and denotation, has been changing with the development of relevant theories. However, its core has remained unchanged, that is, to solve practical problems facing seniors in their daily life through design, improving the quality of their life.

Architectural lighting environment can affect the sense of visual comfort, moods and circadian rhythms of seniors to some extent. Therefore, how to design a healthy lighting environment suitable for seniors is the problem that every designer should think about. For the sense of visual comfort, since seniors respond to light and color differently compared with young people due to the degeneration of their visual systems, their living environment should be lighter with a higher lighting uniformity, so as to reduce stimulation to their visual systems. As for moods, researches have proved that a healthy and comfortable lighting environment can reduce the negative moods of seniors, and appropriately arranged colored lighting and dynamic lighting can improve the moods of seniors. Besides, light also has the function of treating some common diseases among seniors. In terms of circadian rhythms, high-intensity light stimulation in the daytime can lengthen seniors' circadian rhythms, improving the quality of their sleep, and preventing depression at the same time. High-intensity light can also relieve the symptoms of Alzheimer's disease and cardiovascular diseases.

It can be seen from the above that senior-oriented healthy architectural lighting design alone is a huge system. So, for the sectors of senior-oriented architectural design and interior design, there is an even longer road of exploration ahead.

医康养结合的英智运营模式

在目前全国养老机构空置率超过 50% 的背景下，政府希望通过医养结合为养老注入新的活力，将康复医疗和养老深度融合或许是一条可以被选择的道路，英智作为这一选择的先行者，可能是目前国内唯一一家将康复医疗与养老深度融合、并且有能力输出管理、规模连锁化运营的国有控股的轻资产运营管理机构，其十年的运营经验可以为行业带来一些启示。

常规的养老机构收费一般是：床位费、护理费和餐费这三项，但是一般不太可能通过床位费和餐费获取利润。而“康养结合”运营模式的服务理念先期引导建筑设计规划，使得养老机构的建设成本在一定程度上降低，同时又有 10% 到 15% 的额外得医疗收入，加大了护理型养老机构的收入。但是从支出来讲，并没有增加养老机构的支出。医疗服务收入是医院的医疗康复团队提供给养老机构，所以收入有所增加，但是没有增加机构的整体运营成本。

在康复医院中，康复治疗流程由康复师制定，在养老机构中，老人的康复计划也是由康复师制定。这不仅使得康复和养老结合的服务理念得以实现，还有效地提高了服务的准确性，同时起到降低固定成本的作用。固定成本所占比重减小，专业人才的高效化，机构市场定位准确，是吸引老人来到“英智”安度晚年的原因。

医疗机构独立运营的能力，支持带动养老机构的运营，从而稳定快速地提高了老人公寓的入住率，为机构带来稳定的现金流。同时通过医疗人员的专业服务，给老年人带来了安全愉快的晚年生活，为预防急症的发生提供必要的支撑，减少了老人的就医成本。同时由于康复医疗和护理型床位的统一管理，形成顺畅转诊和相互调节的流畅运营模式。

康复医疗与养老深度融合的运营模式，既增加了康复医院的床均产值，也大大提高了护理型养老机构的潜在价值。

关晓立
英智康复健康管理集团总裁
中国医促会健康养老专委会副主任
瑞典吕勒欧大学人体工程学硕士专业

同济大学郝洛西教授：

面对中国严重的老龄化问题，建筑行业一直在思考如何通过适老性的设计丰富老年人的生活，提高老年人晚年的生命质量。适老性设计涵盖了老年人生活的方方面面，就建筑设计而言，它的内涵和外延随着理论的不断更新而不断地延展，但是其核心却一直没有改变——通过设计，解决老年人生活中面临的现实问题，从而提高他们的生活品质。

建筑光环镜对老年人的视觉舒适、心理情绪及生理节律均有一定的影响作用，如何设计一个适合老年人群的健康光环境是我们每一个设计者始终思考的问题。视觉舒适度方面，由于老年人视觉系统的退行性变化，导致其对光与色彩的感受和年轻人有所差异，因此，老年居住空间的光环境应该适当提高照度，同时提高均匀度，降低眩光对老年人视觉系统的刺激；心理情绪方面，研究表明优质的照明环境能改善老年人的负面情绪，适当的彩色光和动态光也能给老年人的生活增添更多的乐趣，同时光照对一些老年人高发的病症也有一定的治疗作用；生理节律方面，白天高强度的光照刺激可以改善老年人的生理节律，提高老年人的睡眠质量，减少老年人抑郁情绪的产生，对老年人阿尔茨海默病及心血管疾病也有缓解作用。

由此可见，适老性设计单就健康光环境而言，就是一个庞大的体系，对于适老性建筑及室内设计更是一条漫长的探索之路。

郝洛西
中国照明学会副理事长
国际交流工作委员会主任
同济大学教授

Speech from Wang Chuanshun

The 4th CIID "Interior Design 6+1" University-Business Graduation Projects Competition has come to an end successfully. This year, the topic points to the improvement of the aged care environment, a social concern of China, and also of the world. It raises a question that needs to be urgently answered by us. Shanghai Xian Dai Architectural Decoration &. Landscape Design Research Institute Co., Ltd. has been committed to the exploration and research of how to improve the aged care and health environment, a topic widely and warmly discussed currently. Over the recent years, we have carried out many major projects, and in the process of which, we have made several bold explorations and innovations related to various aspects such as the living environment and habits of seniors, the improvement of seniors' physical and psychological health, the allocation public social environmental sources, and the relationship between seniors' quality of life and advanced technologies and Internet Plus.

In the process of improving the aged care environment, design thinking plays a decisive and leading role. An excellent design scheme will not only bring happiness and satisfaction to seniors, but also create a new commercial aged care model. More importantly, it embodies social responsibility and human emotions. Featuring business-university cooperation in carrying out graduation projects, the competition provides an opportunity for today's students to express their concern and give innovative solutions to a hot-button social issue. And it is therefore of great social and teaching significance. I'd like to, on behalf of the CIID Second (Shanghai) Specialized Committee, and ARCPLUS Shanghai Xian Dai Architectural Decoration &. Landscape Design Research Institute Co., Ltd., extend my graduations to the success of the event, and express my sincere thanks to all the universities, colleges, advisors, students, and other parties that participated in the event for their active supports.

王传顺老师寄语：

由 CIID 主办的第四届“室内设计 6+1”校企联合毕业设计活动已圆满落幕，今年的设计主题聚焦老年养老环境，这既是中国社会所关注的问题，也是全球社会所关心的问题，是紧贴时代之需的命题。现代环境院一直致力于养老与医疗环境这一前沿课题的探索与研究，近年来先后完成了多项重大工程项目，在老年人的居住环境和生活习惯、老年人起居行动安全和身心健康的保障、社会公共环境资源配套、老年人与科技进步和互联网 + 等，进行了大胆的探索与创新。

在改善养老环境的过程中，设计思维是最重要的一环，具有引导性的责任。一份优秀的设计，不仅能为老年人带来幸福感、满足感，而且能引领新型的养老商业模式，更是社会责任与人类情感的共同担当！本次活动通过校企联合开展毕业设计的形式，展现出当代大学生对这一社会性问题的热切关注和创意思维，具有良好的社会意义和教学价值。我谨代表“CIID 第二（上海）专业委员会”、华建集团上海现代建筑装饰环境设计研究院有限公司对活动的圆满举办表示热烈祝贺，对各大院校、指导老师、参赛学生、社会各界的积极参与表示衷心感谢！

王传顺

中国建筑学会室内设计分会副理事长

上海现代建筑装饰环境设计研究院有限公司总工程师

聯合畢業設計的教學探討

Teaching Graduation Design

CIID“室内设计 6+1”2016（第四届）

校企联合毕业设计

CIID"Interior Design 6+1"2016(Fourth Session)University and Enterprise Joint in Graduation Design

鹤发医养卷

——北京曜阳国际老年公寓环境改造设计

White hair volume medical support

—Beijing Yao Yang International Apartments for the elderly environmental reconstruction design

Tutorial Process of the Retirement Community Environmental Redesign Project: Reflections and Explorations

Foreword

The 2016 CIID "Interior Design 6+1" University-Business Graduation Projects Competition adopts the topic of "Healthy Aging – Beijing Yaoyang International Retirement Community Environmental Redesign" given by Beijing Urban Construction Design & Development Group Co., Ltd. Architecture Institute. In the context that population aging is increasingly becoming an issue for concern, the topic is undoubtedly something needing discussion. Covering three rehabilitation nursing buildings in a senior living & health care real estate project located near the Miyun Reservoir in the outskirts of Beijing, the topic aims at finding out how to meet the real needs of seniors through redesigning the functions and interior & exterior environments of the three buildings.

This year's topic, which is about redesigning the functions and interior & environments of three rehabilitation nursing buildings in a 10-year-old retirement community in the suburb of Beijing, is of profound social significance. It will start a hot conversation over the growing problem of retirement living in cities. To solve the problem, reasonable strategies and approaches regarding interior and exterior environmental redesign are required to build safer, more comfortable and more home-like living environments where seniors can be well taken care of, and enjoy their life.

A city is considered to be ageing when 7% of its population is aged 65 or over, or 10% of its population is aged 60 or over. In China, super cities like Shanghai, Beijing and Tianjin have started to age. And according to UN's demographic data, China will become an aged society around 2024-2026. China is aging at an alarming rate which is as fast as the rate of its urbanization. The fact is that China's aging population will keep growing at an increasingly faster rate which will unavoidably impose far-reaching influence on China's economy, society, families and other aspects. As the population of seniors living alone and the very elderly grows larger and larger, the traditional mode of seniors being cared for by younger generations will be more and more unpractical. Consequently, the relation between generations will change in ways to support, house, take care of and communicate with seniors, and communities and the society will unavoidably play a bigger role in aged care which will lead to growing needs in this aspect. Survey data issued by Beijing Statistics Bureau in 2011 show that 58.6% respondents thought that retirement communities can barely or cannot meet their needs; and 69.4% respondents were not very satisfied with community medical and health services. Respondents with the needs for seniors' restaurants offering delivery service, community medical and health care facilities for seniors, nursing homes, and community senior activity centers accounted for 78.1%, 69.4%, 59% and 52.2% of total respondents respectively. 31.6% respondents expressed the desire to live in retirement homes. These figures indicate that Beijing was not ready to provide aged care to a growing number of seniors, which, actually, provides an opportunity and also the possibility to carry out the above-mentioned graduation project.

According to the topic and site characteristics of the project, two principles are given as instructional guidance:

1.Incorporate regional factors into building function determination

Since the project site is surrounded by mountains near the Miyun Reservoir in Beijing, the natural environmental factors like climate, geographical conditions, wind directions and sunshine duration of the project area need to be taken into account when determining the functions of the three buildings. Besides, the living habits and behavior characteristics of seniors living in northern China, primarily those living in Beijing should also be considered. Put simply, highlighting the regional characteristics of the project is a key to the solution of the topic. Although currently most seniors still rely on younger generations for care, more and more old people have chosen to live in retirement homes or travel between homes and other places seasonally like migratory birds. During the survey stage of the project, the students looked up relevant foreign and domestic materials, and conducted online and offline (e.g. at public parks, community health centers, and retirement homes nearby) questionnaire surveys, trying to find functions fit for the buildings. At last, given the climate characteristics of Beijing and the common chronic diseases among seniors, two schemes were developed. One is turning the buildings into a nursing home aiming at seniors with some special diseases and the need for seasonal aged care. The other is transforming the buildings into a high-end rehabilitation hotel that provides outdoor rehabilitation services to seniors suffering from diseases like pulmonary diseases and orthopedic diseases using the large area and geographical advantages of the project site. Specifically, unique outdoor elevated walkways and garden trails will be built; more interior and exterior public spaces will be added; and the styles, functions, furniture, lighting etc. of rooms will be changed in a human-centered way, for the purpose of creating a safer and more comfortable environment for the seniors. The high-end rehabilitation hotel, named Lotus Garden, is a multifunctional complex composed of gardens, farm, nursing home, community center, hotel and other facilities, providing TCM-WM experiential health care with the assistance of necessary medical equipment.

2. Revitalize filial piety culture with the power of classics

As the saying goes, filial piety is the most important of all virtues. Chinese people have the tradition of respecting and caring the elderly. This year's topic, compared with similar previous ones, focuses more on social ethics and the culture of filial piety – it's given the mission to promote a traditional Chinese culture and a social moral. For graduates of the '90s, growing up with the love and care from their parents and elders, they have gotten used to receiving their love and care, taking it for granted without feeling grateful or shameful. Some time ago, there was a hot nationwide debate over "should we help seniors up from falls?", a proposition that reflects a current problem the China is facing a lack of social ethics and morals. Given the above, as the project started, I told my students that to fundamentally solve the problem of aged care at social level, the most important thing is bringing awareness to the virtue of filial piety and its practice, and in this project, what they should do is put their filial piety and gratitude into the design, helping the senior improve their living environment in a practical way, and

适老环境改造设计教学中的反思与探索

左琰
中国建筑学会室内设计分会
理事
同济大学建筑系教授

引言

2016年中国建筑学会室内设计分会（CIID）“室内设计6+1”2016（第四届）校企联合毕业设计选题由北京城建设计发展集团股份有限公司建筑院提出，为“鹤发医养——北京曜阳国际老年公寓环境改造设计”。该课题切中当下老龄化社会的热点，对位于北京城郊密云水库附近以老年人居住和养生为主题的大型开发楼盘中的三栋康复护理大楼重新进行功能定位和室内外环境改造，以求能真正满足以老年群体为主的市场需求。

今年的毕业设计选题有着深刻的社会现实意义，通过对北京郊外一个已建成十年的老年公寓社区中的三栋康复楼重新进行功能定位和环境改造，来引发对老龄化社会日益突出的城市老人居住问题的热烈探讨，寻找合理的室内外环境适老改造策略和方法，以期使老人的居住环境品质在安全、舒适和归属感等方面有所提升，从而能实现老有所养、老有所乐的设计目标。

老龄化城市的定义是65岁以上人口占城市人口总数7%以上或者60岁以上人口占城市人口10%以上的城市，上海、北京、天津等国内超级城市已率先进入老龄化社会。根据联合国的人口统计数据，中国将在2024——2026年前后进入老龄社会。可以看到，目前人口老龄化速度之快令人担忧，其与城市化的快速发展是同步的。随着老年人口规模的不断扩大以及老年人口高龄化速度的持续加快，势必对经济、社会、家庭等各方面产生深远影响。随着“空巢”老人、独居老人和高龄老人不断增加，传统的家庭养老功能逐步弱化，将引起代际关系在供养方式、居住方式、照料方式、交往和沟通方式等方面的变化，家庭养老功能部分向社区和社会养老转移已成为必然，导致社会化养老服务需求不断加大。据2011年北京市统计局调查，58.6%的老年人认为社区养老条件一般或不能满足要求，69.4%的老年人认为社区医疗卫生服务还不能令人完全满意；希望开设老年 餐桌并提供上门送餐服务的占78.1%，希望建立社区老年医疗保健机构的占69.4%，希望社区办托老所的占59%，希望建立社区老年活动中心的占 52.2%；有31.6%的老年人希望将来入住养老机构。以上数据显示北京在几年前还未做好接受日益庞大的老人群体养老服务的准备，这为接下来的课题设计定位提供了适老环境改造的契机和余地。

根据此次的设计命题和场地特点，在教学指导上遵循两条原则：

一、功能定位突出在地性

此次的设计命题位于北京密云水库旁的群山里，在对设计对象重新进行功能定位时需考虑地域的气候、地理、风向、日照等自然环境因素，也要兼顾以北京居民为主体的北方老人生活习惯和行为特征，功能重新定位时需突出它的地域性，这是设计解题的一把关键钥匙。应对老龄化问题，大部分老人还是选择居家养老的传统模式，而机构养老和候鸟式养老也在逐渐增多。在对大量的国内外文献资料成果的整理、网上发放老人行为活动问卷调查及地区开放性公园、附近街道卫生服务中心以及养老院等多处实地调研后，学生渐渐厘清思路，从当前的养老健康服务产业中找寻出适合此命题的功能定位，最后根据北京气候特点和老人群体中高发的慢性疾病类型，两组方案最终锁定新的建筑功能为针对特殊疾病疗养为目的候鸟型养老院和为患肺部疾病老人提供户外康复运动的疗养型高端酒店，充分挖掘场地宽阔和临山的优势，为骨科疾病和肺部疾病的老人创造了独具特色的户外高架健身步道和种植园庭院步道，增加室内外公共交往空间，居住区域在房型布置、功能配置及家具灯光等方面则考虑了许多人性化的改造措施，大大增加了安全性和舒适性。其中“藕耕苑”方案推出的高端老年养生酒店是将疗养基于体验式养生和必要的医疗服务配备，在中西医结合基础上的一个集城市公园、农庄、疗养院、社区中心、酒店等多功能于一体的复合业态综合体。

二、传统经典阅读复兴孝文化

俗话说“百善孝为先”，孝顺父母、尊老爱幼是中华民族的传统美德。今年适老话题与以往课题相比更偏重于社会伦理和孝道文化，承载着弘扬传统文化和社会风尚的使命。目前的大学毕业生都是90后的年轻人，从小到大都是受父母和祖辈精心呵护和照料，久而久之对父母家人的付出会觉得理所当然，缺乏应有的感恩心和惭愧心，而前阵子热议的关于扶不扶老人这个问题也折射出国内社会伦理和道德的严重缺失。因此毕业设计伊始就和学生达成共识，让学生知道社会养老问题的根本改善是要激发社会

spreading positive social energy in a larger sense.

To help the students have a real understanding of filial piety, during the early survey stage, I recommended them two classics – The Classic of Filial Piety, and Huangdi Neijing. Filial piety, which is the roots of Chinese culture, is also the foundation of personal cultivation and development, and the most basic principle that a man should follow. The fact is that the existence of all social morals and instructions rely on and originate from filial piety. The Classic of Filial Piety, written by Confucius's disciple Zengzi, introduces how filial piety as a basic virtue affects the formation of Chinese culture in the form of dialogues between Confucius and Zengzi. Ms. Yi Jing says in her book Of Filial Piety that "Filial piety as a basic principle of human behavior and life, should not only be memorized, but more importantly, be put into practice. Only when knowledge is turned into action, can it really be learned." When it comes to the project, only when the students carefully observe the physical and behavior characteristics of the seniors based on what they have learned from the classics, can they understand the real physiological and psychological needs of the seniors, thus finding out the entry point and footing of the design.

Huangdi Neijing, which comprehensively discusses the basic theories of TCM from a holistic perspective, is considered as the most important among the four great TCM classics. It explains medicine by using such ancient philosophical concepts as qi, yin & yang, and the five elements, following the unique TCM principle of syndrome differentiation. The formation of this book, the most influential medical text in China, is a long and gradual process from long-term observation of the human body's biological phenomena, and constant verification and summarizing of practical experiences, to turning such objective facts into theories. Seniors are functionally declining, and in their cases, TCM theories can play a guiding role in scientific health care. A factor crucial for health is emotional control, and a comfortable environment is very helpful in changing or adjusting the negative emotions of seniors by promoting their communication and expression. The project site located in the suburb of Beijing boasts geographical and climatic advantages for aged care. In this premise, when TCM culture and the yin-yang theory are combined and incorporated into project design, which is practically feasible, a better living environment can be created for the seniors. And to realize it, it's necessary for the students to read TCM classics, which is also important for the inheritance of traditional Chinese culture.

To the exclusion of the two principles mentioned above, environmental protection and energy efficiency should also be considered in environmental redesign, that is, to make the biggest achievement at the lowest cost. The topic requires the students to, from the perspective of the seniors and their families, think about how to promote the health and protect the rights of the seniors in the entire course of design, out of their true filial piety and love. The process of carrying out the graduation project will, of course, comprehensively improve the professional capacity of the students. However, more importantly, it actually acts as a mirror in which they can see and examine behaviors and thoughts. From caring our own parents and elders, to caring the parents and elders of others, we can use our professional skills to create a better living environment for seniors. When we start to love others and the whole society with action, we are promoting the spread and development of the culture of filial piety.

对孝文化的重视和践行，而此设计需要学生满怀一份虔诚的孝心和感恩心，真正帮助老人改善和提升其居住生活环境的品质，传达社会正能量。

为了使学生能够明理，在早期调研期间推荐他们阅读两本古圣先贤经典——《孝经》及《黄帝内经》。孝道是中国文化之根，孝是修身之根本，是立身之基石，是人最基本的行为，一切社会道德和教化的确立都是以孝为本、以孝为生发点的。《孝经》由孔子学生曾子所作，以师生之间的对白展开，揭示了至德要道贯穿了整个中华文明。易菁在《孝之经纬》一书里讲到："孝作为一种行为规范，作为人生提示，不仅仅在学的阶段要好好学习，更应该在日常生活之中努力去实行。只有把学的内化成自己的行为才叫学习。"[易菁著 . 孝之经纬 . 中央编译出版社 . 2012.115] 因此学生从经典出发，用爱心和耐心对老人日常生活中的身体和行为特征进行细致调查，了解老人生理和心理的真正需求，才能找出设计的切入点和立足点。

《黄帝内经》是中国传统医学四大经典之首，是一部全面论述中医学基础理论的经典著作，从整体上来论述医学，以气、阴阳、五行等古代哲学思想为指导，将辨证论治作为它的诊疗特点，通过长期观察人体生命现象以及反复验证总结医疗实践的过程，并将其上升至理论高度，是中国影响最大的一部医学著作。老人各方面身体机能都在走下坡路，如何使老人健康养生、科学养生，就要以传统中医理论指导实践。无形的情绪控制对健康有着的重要作用，而适宜的环境设计可以改变和调节老人的不良情绪，促进交往和倾诉意识。针对北京郊外这样一个山清水秀的地方，改善和提高老人居住环境品质的一个可行方式就是结合传统中医文化和阴阳学说来综合考虑，而传统经典阅读是学生了解经典、继承传统文化的开始。

除了上述的两个原则，适老环境改造还需要考虑生态环保和节能措施，用最小的改动换来最大的收益。此课题要求我们学生从老人身心健康和权益出发，从父母和家人出发，用一份虔诚的孝心和爱心来贯穿设计过程。这次设计不仅仅被看作是毕业阶段一次专业水平的综合提高，还要将它看作生活中的一面镜子，从中照见和审视自己做人做事的行为和思想。我们师生从孝敬父母和祖辈开始，从小孝到大孝，用所学的专业技能设计出符合老人的宜居生活环境，用实际行动去行孝社会、关爱他人，从而推动孝道文化在社会中的弘扬和发展。

Interdisciplinary Teaching of Basic Architectural Design Knowledge under the Environmental Design Major: An Exploration

Courses introducing basic architectural design knowledge from the environmental design major are a group of architecture-related courses mainly discussing the basic principles and methods included in introduction to architecture, architectural construction and materials, and architectural design. These courses help students to have a correct understanding of design, architecture and space, and to have a command of the most basic architectural design principles. They also teach students how to look up various design documents and specifications, how to understand and develop engineering drawings, and how to develop the ability to transform architectural spaces and design architectural landscape.

Our environmental design major, which is a design discipline with the background of science and engineering, has been offered based on the software (basic educational resources and faculty) and hardware of the architecture major from the very beginning. Therefore, the major presents a combination of arts and sciences required for the development of landscape design and interior design talent. Since artistic skills are necessary in environmental design, students are required to participate in the artistic entrance examination before the college entrance examination. This extra examination ensures the basic artistic skills and knowledge of the students, but their knowledge in science and engineering, especially in mathematical calculation and three-dimensional space, is relative poor. Given this situation, courses introducing basic architectural knowledge have been adjusted to meet actual demands and highlight our instructional characteristics, rather than being borrowed from the architecture major directly.

1. Rational curriculum design which highlights the role of basic knowledge as a platform

Under the environmental design major, the curriculum for the interdisciplinary teaching of basic architectural design knowledge is designed to enhance the logical relations among courses, and to guide the students in terms of design thinking, design creativity, etc. The curriculum of the first semester focuses on the basic knowledge of design, including plane composition, color composition, three dimensional composition, and the principles and skills of perspective, for the purpose of improving students' artistic skills and theoretical level, and laying a foundation for further design education. The curriculum of the second semester starts to include the interdisciplinary teaching of basic architectural design knowledge as described in the table below.

Table 1 Curriculum Schedule for the Interdisciplinary Teaching of Basic Architectural Design Knowledge

Semester	Theoretical courses	Practical courses	
Semester 2	Architectural Graphing; CAD	Living Space Design	Objective: Systematically study the materials in the semester. Task: Complete a living environment design scheme. Requirements: Prepare freehand sketches according to the occupant's requirements, and previously designed spatial functions, utilization streamlines and other elements, and then specify in detail the dimensions and materials using CAD.
Semester 3	Model Design & Making; Elementary Architecture; Architectural Technology & Landscape Engineering; Fundamentals of Materials Science	Exterior Space Design	Objective: Have a command of the methods of spatial confinement. Task: Correctly determine an object to be designed upon an analysis of the exterior environment, and relevant people and functions. Requirements: Complete a standard design scheme (including design analysis, technical drawings, models, and renderings)
		Structural modeling	Objective: Design from the perspective of architectural structure. Task: Model a monumental structure. Requirements: Design a temporary landscape structure, and model it after determining its structure style. Sketches, model, and detailed drawings reflecting regional details are required to be provided.
Semester 4	Fundamentals of Architecture I	Landscape architecture	Objective: Have a command of the process of architectural design. Task: Develop a scenic spot landscape architectural design scheme. Requirements: Develop a complete architecture scheme following the steps of site surveys, site design, early-stage discussion, conceptual design, preparation of sketches and rough model, preparation of final drawings, and presentation of final drawings.
Semester 5	Fundamentals of Architecture II; SU	Workshop practice	Objective: Analyze problems according to real situations. Task: Mapping a rural/urban area with characteristics. Requirements: Collect site data, design survey questions, arrange collected data and pictures, and then recreate the real scenario based on collected data in the form of showcasing or reporting.

环境设计专业建筑设计基础知识课程群集教学探索

李莉
华南理工大学环境艺术设计教研室主任

环境设计专业下开设的建筑设计基础知识课程，是若干门与建筑学知识相关的课程组合，是一个群集，主要讲授建筑概论、建筑构造与材料和建筑设计的基本原理和方法。通过该课程群集的学习，学生能树立正确的设计观，能认识建筑，理解空间的含义，掌握最基本的建筑设计原理，学会查阅各种设计资料集和设计规范，能读懂工程图纸并绘制，同时具有一定的建筑空间改造和景观建筑设计的能力。

本院的环境设计专业是一个在理工科背景下建立起来的设计类学科，确立之初已明确以本校建筑学专业的学科基础、师资力量、硬件设备为依托，文理并重，培养具备景观设计和室内设计能力的专业人才。由于环境设计专业属于艺术类招生，学生在高考前要求参加艺术类考试，故美术功底较为深厚，有一定的艺术修养和绘画基础，动手能力强，但理工科基础知识较差，尤其对于需要逻辑思维的数学计算、三维空间关系等则更为薄弱。针对这种情况，环境设计专业开设的建筑基础知识课程需要根据实际情况和专业办学特点进行调整，而不是简单地将建筑院校的课程直接“拿来”。

一、合理组织教学内容，发挥基础知识平台作用

建筑设计基础知识课程群集的教学内容安排，旨在增强课程前后之间的逻辑关系，以及设计思维、设计创意等方面的引导。在环境设计专业培养计划中，第一学期课程开展设计基础教育，包括平面构成、色彩构成、立体构成、透视原理及表现技法，藉此既加强学生的艺术素养，也起到从高考前的美术教育到本科设计教育的过渡。从第二学期开始，建筑设计基础知识课程群集开始介入，具体安排如表一所示。

建筑制图和计算机辅助设计（CAD）是两门联动的课程：建筑制图要求先进行建筑测绘再进行规范性图纸的绘制，是最直观的认识和研究建筑的途径。

表一　建筑设计基础知识课程群集安排表

学期	讲授课程	实践课程	
第二学期	建筑制图 计算机辅助设计（CAD）	居住空间设计	目的：本学期课程内容的系统训练。 任务：一套居住环境设计方案。 要求：给定一套居住环境设计方案，根据居住对象需求、空间功能、使用流线等进行分析，用手绘表现设计方案，用计算机制图标注详细尺寸及材料
第三学期	模型设计与制作 建筑初步 建筑技术与园林工程 材料学基础	外部空间限定	目的：掌握空间限定的手法。 任务：针对外部环境、人、功能展开分析，找到正确的设计目标。 要求：完成一套规范的方案（含设计分析、技术图纸、模型、效果图）结构造型设计。
		结构造型设计	目的：用建构的角度做设计。 任务：纪念性构筑物“型”的设计。 要求：设计一个临时性景观构筑物，确定结构形式后，通过制作模型的方式，推敲细部（含设计草图、模型、局部节点大样图）
第四学期	建筑学基础一	园林建筑设计	目的：了解并熟悉建筑设计的过程。 任务：风景区观景建筑设计方案。 要求：按照勘探用地、场地设计、前期讨论、方案构思、草图草模修成、正图制作、出图展览的工作顺序开展一个完整的建筑设计方案。
第五学期	建筑学基础二 计算机辅助设计（SU）	设计工作坊	目的：根据真实情景分析问题。 任务：乡村／城市特点区域的mapping。 要求：进行现场数据采集，设计访谈问题、整理收集到的数据和图片等，用展览及汇报的形式再现真实场景

Architectural graphing and CAD are closely linked courses. The former requires surveying and mapping before standard graphing, providing the most direct approaches for the understanding and research of architecture, while the latter realizes a transformation from traditional freehand drawing to computer-aided graphing. The courses require the students to mapping familiar architectural spaces like dorms, class rooms, and campus landscape, as well as articles like furniture, sockets, lamps and small-sized landscape works, and then create technical drawings like plans, sections and axonometric projections according the mapping results. Drawing creation incudes the preparation of freehand black-and-white drawings, and the preparation and printing (1:1) of CAD drawings. The students are also required to know constituent elements, spatial dimensions and structural details through field mapping, and to understand the relationship between spatial perception and specific moduli, realizing the transformation from perceptual thinking to rational thinking.

After the students grasp the basic skills and knowledge to understand and create drawings, such content like the definition and meaning of spaces, methods of spatial confinement, dimensions of design thinking, and spatial organization approaches are introduced, helping the students have a better understanding of spaces. The course of Model Design & Making takes famous architectural and landscape works as the examples of appreciation and modeling. To the exclusion of learning skills needed to make architectural, landscape and interior models, more importantly, the students are required to develop the ability to look up materials and documents, have a better understanding of various design styles, and understand the design philosophy of designers and the relations among related works. They are also required to realize the relationship between architecture's functional spaces and external forms, and between environment and environment; and to have a better understanding of architectural element, landscape elements, and interior boundaries. Evaluation and appreciation of famous works are conducted through group discussions. The course of Elementary Architecture directs the students to understand the process of schematic design in the course of fulfilling external spatial confinement tasks. Such teaching points as survey analysis & data collection, design conception & scheme selection, scheme adjustment & detailing, and the expression of schematic design process are included in the following senior design courses.

Under the architecture major, architectural technology is a very professional course covering architectural structure, architectural construction, architectural materials, architectural physics, architectural equipment and so on which are very complicated and elusive, especially for students studying arts. The environmental design major not aims at developing professional architectural design talent, so there is no need for the students to dig into the above professional knowledge. Given this situation, the teaching objectives are adjusted to helping the students learn the basic concepts of architectural structural design, and mechanics included in architectural structure; to form architectural structures and understand the modeling language of structures using the rules of formative art; and to create art works based on structural modeling and material reapplication, seeing how sciences and arts jointly play their roles in architectural design. The course of Architectural Technology & Landscape Engineering requires the students to, based on existing structural forms, design a series of campus landscape structures, such as bus shelters, footbridges, rest areas, and memorial statues, following the principle of formal beauty. The design schemes shall be turned into models which can reflect ideas regarding structural modeling, and material reapplication. The focus of rough modeling is directing the students to create harmonious relationships among architectural mechanics, architectural materials and structural forms, and among the rhythms, changes and structures of the works.

The above basic architectural knowledge courses are followed by senior design courses. Interior design, which extends and deepens architectural design, forms a part of architectural design. Likewise, landscape design, which aims at external spaces, is also a part of architectural design. The two kinds of design, focusing on exterior and exterior spaces respectively, jointly form the principal part of environmental design. In the nature of things, senior design courses are introduced at the beginning of the 4th semester, first Fundamentals of Architecture I, and then Fundamentals of Architecture II, Senior Interior Design and Senior Landscape Design.

II. Encourage students to put professional knowledge into practice

According to the teaching program for the interdisciplinary teaching of basic architectural design knowledge under the environmental design major, courses are divided into several units in such a way that they are systematically connected over the semesters. However, since the logistic relationships among the courses are barely taught, it's hard for the students to quickly connect the knowledge of different courses with its applications. Without practices, even the tasked are fulfilled, the students still can hardly turn their knowledge, especially basic knowledge, into ideas. Therefore, practical teaching is included in each semester. A design task, which contains all key learning points of the semester, is given to the students, helping them to use what they know into practice, thus improving their practical ability. The practical courses are listed in Table 1 above. Taking the practical course of "Landscape architecture" in the 4th semester as an example, the course requires the student to complete the project following the standard steps of site surveys, CAD mapping, collection of surrounding survey data, site modeling, early-stage group discussions, design specification preparation, and finally the preparation of both design schemes and conceptual models. To complete the project, the knowledge contained in all courses included in the interdisciplinary teaching of basic architectural design knowledge is needed.

Practical courses under the environmental design major are designed to help the students have a correct understanding of design, that is, design is a process to discover, analyze and solve problems, and to study the relationship between man and nature, and man and the society. To better achieve the aim, workshop practices at given urban and rural locations are arranged, requiring the students to walk out of the campus to conduct site surveys, discover problems and take them as key considerations, collect site data, visit target groups, analyze problems through group discussions, and make final reports at last. Workshop practices are arranged in the 5th semester, with the participation of extramural advisors and the teachers and students from other universities and colleges. No specific topic will be given to the practices which focus on the process of discovering and solving problems. In addition, students are organized to participate in various professional design competitions, and the cooperation with students from other majors is encouraged.

计算机辅助设计（CAD）要求将建筑制图课中的尺规制图转化成CAD制图。课程安排学生测绘熟悉的建筑空间如宿舍、教室、校园景观等，并对室内家具、插座、灯具、景观小品进行测绘，根据测量数据绘制平立剖面图、轴测图等技术性图纸，先黑白墨线尺规作图，后计算机作图，等比例打印。通过实地测绘了解建筑的各种组成元素、空间尺度和构造细部；通过图纸绘制，把握空间感受与具体模数之间的关系，思维从感性转向理性。

在学生具备了最基本的识图和画图能力后，开始引入空间的定义与意义、空间限定的方法、设计的尺度思维以及建筑的组织手法，强调对空间的理解。模型设计与制作课程选取建筑、景观名作作为赏析对象，并制作成模型。除了要求掌握建筑模型，景观模型与室内模型的制作方法外，更深层次的目的让学生具备查阅图书资料的能力，对各种设计风格有深入的了解，能体会设计者的设计理念与设计作品之间内在联系；认识建筑功能空间与外在形式的关系；认识建筑与环境的关系；深化对建筑各要素，景观要素，室内界面等的理解；通过对各名作进行评价与赏析，适应小组合作讨论的学习方式。随后的建筑初步课程，以一个简单的外部空间限定设计任务作课程主线，重点引导学生 认识方案设计的过程。调研分析与资料收集、设计构思与方案优选、调整发展和深入细化、方案设计过程的表达将贯穿之后的所有专题设计课。

在建筑院校，建筑技术是一门专业性极强的课程，涉及内容包括建筑结构、建筑构造、建筑材料、建筑物理、建筑设备等，涵盖内容复杂难懂，这对于艺术类学生来说无疑非常困难，对于不以培养专业建筑设计人才为目的环境设计专业学科架构来说，也没有必要掌握如此深入的专业知识。因此，教学目的调整为使学生掌握建筑结构设计的基本概念、建筑结构的力学知识；能运用造型艺术规律来组织建筑结构，了解结构的造型语言；拓展学生以结构造型、材料再运用为切入点进行艺术创造，令其感受建筑技术的科学性和艺术性。课程安排学生根据现有的结构形式，运用形式美的原则进行结构造型设计，采取真题假做的形式，设计一系列校园景观构筑物，包括候车亭、步行桥、休憩角、带纪念性质的雕塑等，成果要求以模型的形式表达结构造型的推敲、 材料的选择再运用。在草模的构思过程中，引导学生把握好建筑力学、建筑材料与建筑结构形态的关系，注重造型的韵律、变化与结构的相呼应，也加深了建筑结构的“理”与艺术美学的“文”的穿插并重。

至此，建筑基础知识已讲授完，进入专题设计阶段。室内设计是建筑设计的延续和深化，是建筑设计的一部分，同理，景观设计是延续了建筑设计外部空间部分，这种内部环境和外部环境的设计共同构成了环境设计的主体。理所当然地，专题设计课在开设伊始即第四学期，先引入建筑学基础一，随后建筑学基础二与室内设计专题、景观设计专题在高年级课程中穿插安排。

二、以专业应用能力培养为目标，引入实践教学

环境设计专业建筑基础知识课程群集的教学计划以单元课程为构架，按照学期先后顺序设置，在一定程度上考虑了系统性和连贯性，但是每门课程之间的逻辑关系缺乏引导，难以令学生在学习中快速地与专业应用能力联系起来，学习任务完成后，不经实践很难将知识点融会在设计思维中，基础知识课程尤为突出。为此，每一个学期均引入实践教学环节，通过一个设计课题将本学期的课程内容做系统的实践训练，课题内容涵盖本学期课程知识点，以此增强学生的专业应用能力。各个学期的实践教学课题具体安排详见表一。以第四学期的实践教学课程——园林建筑设计为例，要求现场勘探用地、CAD测绘，周边调研资料采集，场地模型、开展小组前期讨论，自行拟定设计任务书，进而按照正规的方案设计程序完成此项目，成果包括方案册和概念性模型，这一过程涵盖了建筑设计基础知识课程群集的所有课程内容。

在环境设计专业教育中，充分利用实践教学课题，引导学生树立正确的设计观，即设计是发现问题、分析问题、解决问题的过程，这一过程研究的是人与自然、与社会的关系。为此，在实践课题中增加了设计工作坊环节，要求学生走出校园，基于对选定的乡村、城市某处的特定观察，学会在真实的户外环境中调研，将发现的问题作为思考的重点，同时采集现场数据、访谈使用人群、进行数次小组内部讨论分析会，最后做工作坊终期汇报。设计工作坊安排在第五学期，邀请校外导师和兄弟院校师生共同参加，不针对具体的设计方向和内容开展，强调发现和分析问题的过程。邀请业界精英在工作坊期间穿插进行学术讲座，以此鼓励和推动学生实践及创新能力的提高。在课外实践环节，组织学生参加专业设计竞赛，并鼓励与其他学科、专业的合作者共同参赛。

III. Reform teaching methods and tactics to arouse students' interest in study

For students majoring in environmental design, tasks designed by the teachers are given to assess their knowledge, and their performance is reflected by scores which act as the final scores of corresponding courses. Given that scores cannot reflect the student's knowledge in a directly and timely manner, reviews and presentations are introduced to courses included in the interdisciplinary teaching of basic architectural design knowledge, so that the student's knowledge can be assessed from different angles and in different stages. Reviews are conducted in the middle and at the end of an academic term, called middle review and final review respectively. Students are required to present their schemes, expressing their ideas and concepts. This process is called "self-evaluation", aiming at training the students' oral expression skills, and encouraging them to express their ideas. Besides, "inner evaluation" among the students is also adopted where students evaluate works, which, in their opinions, have obvious problems or strengths. This process requires the students to study the works of others while focusing on their own works, offering an opportunity for the students to learn from each other. To pass the reviews, the students need to collect information supporting their works, schemes and opinions, and know their speeches by heart. The "final evaluation" is given by corresponding teachers, who will, from the perspective of the students, offer guidance and inspirations, and create a natural and relaxed learning atmosphere which arouses the interest of the students in study. The final designs are showcased at the exhibition area to be evaluated by other students, the teachers from the Teaching and Research Office, and the professors at the school. In the course of exhibition preparation, each student needs to find out how to highlight his work according to the situation of the exhibition site, and the visit route and VI need be designed by all the students of a class. Actually, this is a process of turning the knowledge of spatial design into exhibition preparation. The exhibitions required by courses included in the interdisciplinary teaching of basic architectural design knowledge provide a great opportunity of practice for the students, acting as an important part of practical teaching.

Promoted by the Teaching and Research Office at various schools, the Student Research Proposal (SRP) requires the students to study problems, conduct projects, or create designs under the direction of the advisors. SPR comprises the two parts of monograph reading and after-class practices. Monographs are determined by the advisors, and the students are required to selectively read the books, write reading notes or study notes, and communicate with each other. The notes will be scored by the advisors at last. SRP creates opportunities for the students to participate in research training, and engineering and social practices before walking out of the campus, increasing their practical, reach, and innovation abilities in an effective way.

In conclusion, through rational material arrangement, the involvement of practical teaching, and the teaching method and tactic reform, the interdisciplinary teaching of basic architectural design knowledge under the environmental design major can satisfactorily realize the teaching objectives, help the students establish scientific design principles, and realize the harmonious combination of culture, art and engineering technologies.

三、教学方法及教学手段的改革，调动学习主动性

环境设计专业的课程考核多以学生课程作业成绩体现，教师是主导学生作业构思的主体，对课程作业打分，并以此作为该门课程的最终成绩，但单一的分数令学生不能及时了解自己对知识的掌握程度，即课程结果没有迅速反馈。建筑设计基础知识课程群集的各门课程，设立课程作业的评图和展览制度相结合的机制，建立不同角度、不同环节的评价体系。评图环节选取课程的中段和结束两个时间节点进行，即阶段性成果和最终成果的评图机制。学生需要针对课程训练做方案或作业的陈述，表达自己设计的构思，即“自我评价”，旨在训练学生的口头表达方案能力，鼓励其有想法；要求学生之间“相互评价”，对问题突出或者认为较好的作业谈自己的观点，开展讨论，从而促使学生不仅关心自己的作业，更关注周围同学的作业情况，令学生在评价过程中重视相互之间的学习。学生要完成评图过程，需要为自己的作业、方案和观点等收集资料，熟练讲稿。最后，由任课教师做“总结评价”。整个过程任课教师站在学生的角度，重点引导和启发，建立一个自然轻松的学习氛围，从而令被动学习转变为主动学习。学生的最终成果要求在学院的展览空间进行展示，并邀请教研室其他教师和学院教授观展和评价。每个学生在布展过程中需要考虑展览场地的实际情况和自身作业的展示效果，班级全体需要考虑展览路线的设计和VI设计等，这一过程实质上把展示空间环境设计的知识融入到各门课程的成果展览中。经过建筑设计基础知识课程群集各门课程作业的展览运作，学生用实践行动给自己上了最好的一门课，这也是实践教学的重要组成部分。

本校教务处每学年在各学院中推行学生研究计划（SRP），指导教师以SRP项目为依托，引导本科生开展基于问题、基于项目或基于设计的学习。SRP包括专著阅读和课外实践，指导教师制定项目领域内的专业书目，要求学生结合课题，从中选择进行研读，最后以读书笔记、学习心得等形式进行相互交流，以此进行考察和评定成绩。SRP项目的推行，令学生能尽早接触并参与科研训练、工程实践及社会实践，切实提高了综合实践能力和研究创新能力。

环境设计专业开设的建筑设计基础知识课程，通过合理组织教学内容，引入实践教学环节及教学方法手段的改革，做到有的放矢，在有限的课时内达到较为理想的教学目的，帮助学生建立科学的设计原则，实现文化艺术与工程技术的统一。

How to Become a Professional Designer

——To students participating in the 4th CIID "Interior Design 6+1" University-Business Graduation Projects Competition

How to become a designer? How to become a professional designer? What are the basic abilities that a professional designer must have?

In my view, keen insight, clear expression, and wealthy creativity are the three basic abilities that a professional designer must have.

Then, what is insight? What's the difference between insight and observation?

Observation is the capacity to make a rough judgment or have a simple view on something based on what one has seen. An observation is a superficial and shallow understanding of objective things. While insight is the capacity to gain a comprehensive, thorough and inside-out understanding of objective things using all sensory organs. We are not born with insight, but it can be gradually developed like a habit through professional learning and scientific training. Insight is the most basic professional ability that a designer should have. It builds a keen and decisive "design nerve system" which enables a designer to see and feel the nature and changes of objective existence, and mobilizes all sensory organs to have a comprehensive and unique experience of the outside world. What the experience is depends on the cultural level and knowledge of the designer. When such unique experiences are accumulated, they will turn into the "spiritual wealth" and "sources of inspiration" of the designer.

Taking myself as an example, one day when I passed by Jessica Department Store on Central Street in Harbin, some pillars in the store attracted my attention immediately. They looked like common white square pillars edged by black walnut. Although I was some 10 meters away from them, I could feel that the color and texture of the white pillars are kind of special – they could hardly be created by common paint. When I walked up to a pillar, I found in surprise that it was not covered by paint but a kind of soft wall paper with the texture of lambskin. The wall paper fitted very well with the environment of the store: it embodied the connotation of upscale women's clothing – understated luxury, and at the same time, moderately heightened the feminine atmosphere of the store. It created a wonderful design effect much better than what common white inner wall paint could gain. What I want to say is that if I hadn't learned design, or hadn't developed the habit of having an insight into things, I would have possibly missed the detail, and would hardly have felt how that material improved the environment.

As I said before, a designer is not born with insight – it's a result of scientific and professional training for a long time. To develop insight, you need to, from now on, start to observe with your eyes, feel with your hands, listen with your ears, smell with your nose, think with your brain, and feel with your heart things around you, even the little tiny things, until you can view the environment from a unique perspective or angle.

What is expression? What does it really mean to a professional designer?

Expression is another ability that a professional designer must have. In the scope of design, expression can be understood in two ways.

One is verbal expression or the ability to communicate in general. The process of design involves various parties; therefore, communication is required in the entire process of a design project. Smooth communication between the designer and the customer, and its partners and team members in the early stage is very important for the success of a design. If the designer cannot express his ideas clearly, it's possible that the project cannot progress smoothly, or even fails, even if the project itself is excellent. Therefore, a designer should learn to communicate effectively with others about his design ideas using the power of spoken words. The ability to communicate verbally will directly affect the conveying of design information, and even the fate of a design scheme. That's why I say verbal expression is an ability that a designer must have.

The other is design expression which refers to a designer's ability to express his design ideas through drawings by using his "hands". Each able-bodied person has hands, but a pair of skillful hands is not owned by everyone. A designer can only have a pair of skillful hand through professional training. The design education offered by universities contains such training. Through systematic training, each student can develop the ability to draw and model with bare hands which lays a solid foundation for design expression. Then, why should a designer have such ability? Design is a process of innovation. Put simply, to design is to create something unique, personalized and new, so in this case, drawing and modeling with bare hands is the best way of expression. After the September 11 attacks, an invitation from the US was received by the best architectural designers and design teams around the world to rebuild the World Trade Center. Among the design schemes submitted by the candidates, to the exclusion of a few landscape renderings that adopted CAD technologies, most drawings were freehand ones, including watercolors, colored pencil drawings, and even pencil drawings. Why did those top designers and design teams choose to express their ideas of design through freehand drawings at such an important competition? The answer can be given by the following experiment. When several students are asked to simultaneously draw a 10 cm x 10 cm square filled with the same color respectively using identical computers and design software, we will see squares identical in shape and color if the possible difference in display is ignored. However, when they are asked to simultaneously draw a 10 cm x 10 cm square on paper of the same brand and thickness respectively using identical pencils, rulers and paints, we will see differences between every two squares. Some of the differences are errors caused by freehand painting, but it is these organic differences that make the squares personalized and vivid. This experiment indicates that differences are an attribute of freehand expression, and to some extent, these differences turn to personalized and natural expressions which cannot be obtained by computer-aided drawing. This is why all those top designers and design teams chose freehand drawing as their

如何成为职业设计师
——写给CIID“室内设计6+1”2016（第四届）校企联合毕业设计同学们

马辉
哈尔滨工业大学副教授
设计学系副主任

如何成为设计师？如何成为职业设计师？成为一个职业设计师所必须具备的最基本能力是什么？

我认为敏锐的洞察能力、清晰的表达能力和丰富的创造能力是成为职业设计师所必须具备的三种基本能力。

什么是洞察能力？它与观察能力的区别是什么？

观察能力是仅仅局限用眼睛来对客观事物进行的较粗略的一种判断能力和基本看法，它是对客观事物表面的、肤浅的认识。而洞察能力则是调动人体全部感知器官，对客观事物全方位的、彻底的、由内至外的理解能力。人的洞察能力不是与生俱来的，它是可以通过后天的专业学习、科学训练而逐渐形成的。洞察能力是一名设计师应该具备的最基础的职业素养，它会赋予设计师一套敏锐果敢的“设计神经系统”，使设计师清晰明了地看到、感受到客观世界的本质与变化，并调动身体全部感知器官，形成对外界事物的全面而独特的体验。这种体验以设计师所具备的文化、知识为前提，是一种特殊的经历，是一次次感受、一次次积累的综合累加，是设计师的“精神财富”和“灵感源泉”。

例如一次我路过哈尔滨市中央大街的杰西卡商场，在室内一眼就发现商场的柱子有些特别，看起来是普通的白色方柱，四角用黑胡桃木包饰，但我在十米左右的距离就感觉到白色柱身的色彩与质感有些异样，不像普通的涂料刷饰所能达到的效果。走近仔细观察，果然惊奇地发现这种白色并不是白色涂料刷饰的，而是选用了一种手感柔软、带有天然皮革肌理的仿天然羊羔皮式壁纸。这种材料与环境结合得十分恰当，显示出高档女装“低调设计但绝不平庸”的内涵，柔和地烘托出商场整体的女性环境氛围，其表现效果的准确是普通内墙白色涂料所远不能比的。如果没有学习设计的背景、没有洞察周围事物的习惯，我可能很难注意到这种细节，也很难体验到特殊质感材料对环境的烘托作用了。

所以说设计师洞察周围环境的习惯不是与生俱来的，而是要靠科学的方法、专业的训练来实现：从点滴做起，从现在做起，用眼睛观察，用手触摸，用耳朵聆听，用鼻子体味，用大脑思考，用心灵感悟，日复一日，长期积累，就会具备看待环境的独到角度与观点。

什么是表达能力？它对职业设计师的真正涵义是什么？

表达能力是职业设计师必须具备的另一项基本能力，其包括两方面的的涵义。

一方面是指语言表达能力：概括的讲就是沟通能力。设计是一项综合性很强的工作，在设计工作过程中，“沟通”贯穿整个过程。设计师与客户之间、与合作伙伴之间、与团队之间准确顺畅的交流是设计前期非常重要的环节，如果缺少了良好的表述能力，那么设计方案拓展将可能会面临阻力，甚至再好的方案也会导致流产。所以设计师应该学会运用语言进行有效的设计沟通，语言表达的优劣会直接影响设计项目的信息传达，沟通的好坏甚至直接可以关系到设计方案的成败。因此语言表达能力是设计师必备的一种能力。

另一方面是指设计师的设计表达能力：主要是指设计师思想的图纸再现能力。这一能力主要是依靠设计师的“手”来展现的。手是每个健全人都具备的，但灵巧的手则不是每个人都具备的；拥有一双灵巧的手是每个人的愿望，但拥设计师的手是必须依靠专门的训练方法来获取的。大学的设计教育正提供了这一科学训练过程，通过系统的训练使每一名学生都能掌握徒手造型能力，为日后实现设计表达能力打下良好的基础。设计师为什么要具备徒手造型能力呢？设计本身就是创新的过程，我们简单地理解它是与众不同的、个性化的、全新的，那么徒手表现正是这一过程的最好表达途径。“9.11”之后，美国向全世界建筑大师发出邀请，征集重建“世贸中心”建筑方案，在参加投标的建筑设计大师与设计团体中，除少量景观图采用了计算机辅助设计表现外，在设计思路表达方面都不约而同地选择了徒手表现，有的用水彩表现，有的用彩色铅笔表现，甚至还有的用铅笔直接通过素描画的形式表现。这样顶级的设计大赛，这些顶级的设计大师或设计团体为什么要用徒手的形式来表达各自的设计思想呢？做一个实验就可以说明：首先选几名同学分别发给他们配置完全相同计算机，然后选用相同的设计软件，同时制作一个10cm×10cm的正方形，并且填充同样色值的一种颜色，忽略屏幕显示方面的差别，可以看到这几名同学所绘制的图形与色彩是完全相同的。同样，让这几名同学同时用同样的笔、尺和同品牌的颜料，在相同品牌、相同克数的纸上绘制一个相同色值的10cm×10cm的正方形，其结果是每名同学的绘制作品都有或多或少的差别。这些差别有一部分是手工绘制操作上的误差，但是这些有机差异正是个性化的、鲜活的作品差异。从这

way of expression – they chose the best way to show the core values of their works – innovation, uniqueness and personality. In conclusion, freehand expression is a very important and necessary ability for a professional designer.

What is creativity necessary for a designer? How to awaken creativity?

Unlike insight which is not an inborn but acquired ability, creativity is an ability that we are born with. Young children are curious about everything. They always ask "what", "why" and "how", and they are very sensitive to shapes and colors. Their world is a world of imagination which is actually the most important nutrition of creativity. Although we are born with creativity, in many cases, it cannot find its way to grow; actually, on the country, it becomes more and more inactive and even falls asleep together with our imagination as we grow up due to the exam-oriented education. However, design is a process of innovation, and making innovations is the responsibility of a designer. Therefore, creativity is an ability that a designer must have, and consequently, awakening creativity becomes an important objective of designer training.

Then, how to awaken our creativity which declines as we grow up? The answer is systematic and scientific design education at universities and colleges. Curriculum-based professional training can help us find our lost imagination and creativity, and as the education moves from curriculum learning to design practice for professional designers, they will become more and more active and powerful, growing as the growth of a designer.

Insight, expression and creativity are the necessary abilities that a professional designer requires. Actually, we can understand them as the coordination of the eyes, hands and brain. However, to be an excellent designer, having these abilities alone is far from enough. Senior students learning design should believe in themselves, and work hard day after day until they realize the dream of becoming excellent professional designers.

个实验中我们发现，徒手表达是先天带有非同性与差异性的，在一定层面上这些差异就成为了个性化的自然表达。所以这些设计大师或设计团体不约而同地选择徒手方式来表达各自的设计思想，这也正是每一件设计作品力图表达的核心价值——创新、非同、彰显个性。故此，徒手表达能力对职业设计师而言是至关重要的，也是必须具备的。

为什么要培养设计师的创造力？如何唤醒创造能力？

之前我们谈过洞察力不是天生具备的，而是后天培养的一种习惯，那么创造力却是我们人类与生俱来的能力。人在幼年时，求知欲非常旺盛，说的最多的就是"是什么""为什么""怎么办"，幼儿在丰富的想象力伴随下成长，对形态与色彩的感知能力尤其强烈。想象力是创造力的重要营养供给，人在幼年时期虽谈不上具备了创造能力，但绝对是具备了创造能力培养的潜质。随着成长不断接受应试教育的培养，也伴随着想象力与创造潜质逐渐消退。然而，设计是创新的过程，推陈出新是设计师的责任，那么创造能力就应是设计师必备的能力，创造力的培养也正是设计师培养的重要目的。

那么如何唤醒我们随着成长而逐渐消退的创造能力呢？同样要依靠系统、科学的大学设计教育，通过专业的大学设计课程训练，使我们失去的想象力和创造力重新回到头脑中。这种能力也会随着由学校的设计课程训练到职业设计师的设计实践的转变而越来越旺盛、越来越充分，将永不枯竭地伴随设计师成长、成熟。

洞察能力、表达能力与创造能力是职业设计师所必备的能力，其实也可以简单概括为眼、手、脑的协调能力。但是要想成为优秀的职业设计师只具备这些还是远远不够的。即将毕业的设计专业的同学们还要坚定自己的设计师信念，不断学习，不断努力，就一定能够实现成为优秀职业设计师的梦想。

Some Thoughts on the Education Reform of the Environmental Design Major

The environmental design major is a comprehensive major focusing on practice and application. In the context that the society is changing rapidly and constantly, environmental design education at universities and colleges nationwide is confronted with multitude of problems, such as indistinct characteristics, ambiguous direction of talent development, and mismatch between education and market needs for talent. Therefore, it's worth thinking about how to redefine the characteristics of the major, and develop a talent development program according to market needs based on the disciplinary features of the major.

1. Reform of teaching & research methods

(1) Interdisciplinary teaching

The environmental design major involves various disciplines, such as architecture, landscape architecture, history, geography, ecology and landscape, which impose a higher requirement on students and corresponding teaching system. A challenge it is, interdisciplinary teaching is also the highlight of the major. It requires the teachers to focus on offering professional instructions from the perspectives of various fields.

(2) Teaching method innovation

To develop skilled talent, the curriculum should be designed by referring to social demands for talent, so that students really needed by the market can be developed. During the graduation project period, tasks required to be done in real career life shall be given to students, helping them develop abilities required by jobs. Besides, students shall carry out projects in groups by simulating the "project team" model in companies, thus having better practical experiences. This process helps students to put theories into practice, and develop the awareness of professionalism.

(3) Focusing on practice

The development of modern design disciplines now shows the trend toward teaching-research-practice integration. Design disciplines as applied disciplines must keep up with the pace of social and economic development. For example, organizing students to conduct site surveys and simulate construction at teaching bases is an effective way to improve student's ability to think and practice. Practical teaching should be involved in different courses, helping students gradually accumulate practical experience in a long term. In this process, students can complete their graduation projects, and the teaching objectives can be achieved.

(4) Innovation of characteristic courses

One current problem facing the environmental design major is the lack of innovation in some courses. From the perspective of disciplinary background, the major includes many architecture and landscape architecture courses. However, from the perspective of innovation, aside from these courses, practical courses should be designed in line with the characteristics of the major, to train students' practical ability, team work ability, organization ability, and the ability to gain an overall control of design works. Consequently, better teaching with the characteristics of the major can be offered.

(5) The importance of early-stage survey

According to the teaching program, early-stage surveys on design themes have always been a part of professional design courses; however, it turned out that the role played by surveys can hardly be seen in the final design schemes. The reason is that the students failed to well understand the necessity of the survey, and its relationship with the later design. The early-stage survey of a project, which is the foundation and premise of design, is a necessary part in the early stage of the project. Therefore, more survey tasks should be arranged based on professional design courses, with the view of developing students' ability to survey. Each professional design course should contain a social survey lasting for at least two weeks, creating an opportunity for students to effectively analyze and evaluate the natural and cultural factors regarding the theme and background of a project, and to observe the changes of such factors outside the campus. Before starting a design, students should have a comprehensive understanding of design content using such analysis methods as questionnaire surveys, visits, behavior analysis, photography, and measuring, for the purpose of laying a solid foundation for the design.

(6) Close course-practice integration

Design projects shall be conducted by cooperating with companies with topics closely related to real projects, creating a real work environment in which students will seriously carry out the design projects and learn in the process of practice. Advisors shall direct students to conduct site surveys, and directly communicate with persons in charge of the projects, with the view of developing students' ability to design and communicate, and helping students know about and learn to adapt to the society in advanced in the course of carrying out real projects.

(7) Research-teaching matching

Teacher development programs should encourage teachers to research what they teach, so that their research and teaching activities and results can effect and promote each other. In addition, teachers should also pay attention to helping students develop the ability to research in the course of teaching.

2. Reform of teaching methods

(1) University-business cooperation

The 2016 CIID "Interior Design 6+1" University-Business Graduation Projects Competition is a typical case of university-business cooperation. Its success indicates that university-business cooperation is an active teaching model or method for the development of skilled talent. In the case of the environmental design major, aside from graduation projects, main professional design courses should also adopt the method. In the course of cooperation, universities can incorporate new information like new materials, new technologies and skills, and new aesthetic trends into curriculum design and teaching; and at the same time, companies can know more about university teaching and excellent students, having an opportunity to

谈环境艺术专业教学改革

西安建筑科技大学

环境艺术专业是综合性、实践性很强的应用型学科专业，随着社会新形式的不断变化，各地方院校环艺专业环境艺术设计教育面临诸多困境，专业特色不明确、人才培养定位模糊、教育和市场的对接不顺畅等问题。如何基于本专业学科特性，重新定义专业特色、研发符合市场要求的人才培养计划是值得思考的问题。

一、教研手段的改革

（一）多学科知识的融入

环境艺术设计专业教学设计多个学科领域的知识，需要除建筑学、园林学外，历史、地理、生态、景观等跨学科的多面知识融入，这也对环境艺术专业的学生和教师教学体系提出了更高的要求。当然，这一学科的特色就在于它是多个学科的交叉与结合，在教学过程中应注重对学生多面知识的专业性指导。

（二）教学方式的创新

为培养应用型设计人才，在课程设置中，可依据实际社会岗位所需的能力作为教学内容的参考，有针对性、适用性的培养学生。在项目设计阶段，将实际岗位设计任务转化为课程要求的学习内容，使学生进行职业岗位的实际能力训练。以工作组为单位，模拟实际工作中“项目组”的模式，让学生体验实训。在这一过程中，使学生将理论知识与实践进行有效整合对接，培养学生在社会中的专业素质意识。

（三）注重实践性

现代设计学科目前在走教学、科研、实践一体化之路。设计学科作为应用学科，必须与社会发展、经济发展紧密对接。教学过程中带学生到教学基地实地考察、模拟施工，能够提高学生的思考、实践能力。在多个课程中加入实践性教学方式，长期渐进的带来学生不断实践考察，最终达成设计成果和教学目的。

（四）特色课程的创新

现有环境艺术设计专业课程中部分课程无新意，从学科背景角度来说，环艺专业基础课程中诸多建筑学、园林学课程，除此之外，应有针对性的开设符合本专业特色的实践性课程，实践性课程训练学生的实践能力、团队组织能力、对设计作品的整体把控能力，能有效提高专业特色的教学质量。

（五）注重设计前期调研

以往的专业设计课在教学计划中会安排学生针对设计主题进行前期调研活动，但往往反映在最终设计作品中的前期调研效果不佳，作用不大。归其原因是学生对设计前期调研的必要性和调研工作与后期设计之间的关系理解不足。设计前期调研是进行设计的前提与基础，是设计前期所做的必要工作，鉴于此，应使学生更加注重调研工作，结合专业设计课，开发多个调研小课，培养学生的调研能力。专业设计课中至少开设2周的社会调研，走出校园，对设计主题、设计背影的自然、人文因素进行有效分析和评估，了解各种因素的演变过程。通过问卷调查、走访、行为分析、拍照摄像、测量等手段，以各种分析法完成对所设计内容的全方位理解，为设计深入打下结实基础。

（六）课程设计紧密结合实际项目

在专业设计课程教学选题时，尽力做到“真题真做”，选题紧密结合真实项目，与企业合作，使学生“身临其境”，以专业严肃负责的学习态度对待设计题目，分组以老师指导带队的方式带学生实地了解项目情况，设计中与项目负责人直接接触、沟通，提前培养学生在实际社会环境中进行项目设计的能力、与人沟通设计方案的能力、提前了解、初步适应社会现状。

（七）研究与教学相匹配

在教师队伍人才培养的计划中，应提倡每位教师的研究与教学相匹配、相呼应，教师自身的专业研究方向应与教学方向保持吻合，研究与教学同步推进、互相影响、互相促进。同时在教学过程中引导学生的研究能力。

二、教学方法的改革

（一）校企联合

CIID“室内设计6+1”校企联合毕业设计活动是一个典型的成功案例，反映出校企联合是符合应用型人才培养的积极教学模式和教学方法。在环艺专业教学阶段中，除毕业设计之外，应在多个较大专业设计课程中采用此方法。与课程课题密切相关的企业，能够把行业中的新材料、新技术工艺、设计审美等带入课程教学中，同时让企业更加了解学校教育、了解优秀学生，为此后学生就业进行良好铺垫。

通过实际项目的依托，校企协作，将实际项目任务带入教学内容，实现产学结合，培养应用型人才。

（二）注重启发学生的感性与理性思维

select talent that they need.

Students' participating in real projects under the model of university-business cooperation is an effective way to develop skilled talent.

(2) Enlightening students on perceptual and rational thinking

In many universities and colleges, some students majoring in environmental design are recruited through the artistic entrance examination. Such students are always good at divergent and perceptual thinking, but weak at rational thinking – they tend to focusing too much on forms, colors and shapes during design, ignoring practical problems and the real value of design. They also lack the ability to have an overall control of design. For these students, developing their ability to analyze rationally and view things perceptually is very important, especially in the stage of basic knowledge teaching. Given the current situation that only three fundamental courses are offered, it's necessary to develop basic courses focusing on improving students' rational analysis ability while creating courses aiming at training the perceptual cognizance and creative thinking of students. The most important thing is directing students to combine conceptual thinking with rational analysis, so that conceptual thinking will not be unrealistic ideas blindly focusing on forms and colors, but an effective force with works together with rational thinking to create designs.

Small classes should focus on developing students' conceptual thinking and rational analysis abilities. Early-stage surveys at project sites are important for students to perceive and feel various environmental factors, and to record information in a perceptual way. In this stage, teachers should direct students to develop their ability to perceive using their sensory organs and intuition, and to record things that they perceive directly at project site.

The second stage is developing students' rational analysis ability. Rational analysis comprises two parts: rationally analyzing information perceived, and effectively analyzing objective information such as site conditions, potential possibilities, and users' requirements using such methods as static state recording, dynamic behavior recording, and questionnaire user surveys.

The final objective of site surveys is helping students directly combine survey and analysis results with design objectives in the process of repeatedly discovering and solving problems, and progressive training. In the stage of basic knowledge teaching, enlightening students on developing their perceptional and rational thinking is critical for their later study of professional courses.

Universities and colleges should design and offer environmental design courses according to current social development demands, and their advantages, so as to designing curriculums and teaching systems with their own characteristics, and developing professional environmental design talent with solid basic theoretical knowledge, diverse design skills, professional competence and ethics, and the spirit of team work.

诸多地方院校的环境艺术专业招收艺术类学生，艺术类学生的特点往往发散性思维和感性思维较强，而理性分析能力不足，进行设计时容易落入形式、色彩、造型的误区，忽略设计本应解决的实际问题，缺乏对整体设计的把控能力，对设计价值的思考不足。因此培养这类学生的理性分析能力和感性认知是关键的。尤其是在基础教学阶段，基础课的设置过于单薄，趋于常规，除基础三大构成课以外，应增设专门培养学生进行理性分析的基础课程，同时增设培养学生感性认知的创造性思维课程。更重要的是如何使学生的感性思维与理性分析结合起来，使感性思维不成为天马行空，不陷入形式、色彩的泥潭，而是有的放矢的与理性思维结合以服务于设计。

在调研小课中着重培养学生的感性认知和理性分析能力，在设计前期调研阶段，学生来到设计项目基地，用感性去认知、感受场地中的各种环境要素，用感知的方式进行记录。在这一阶段，要注重培养学生自我感觉的能力，培养学生能够利用五感，依靠直觉，记录场地带给自己的最直观感受。

第二个阶段，用理性思维进行分析。一部分由感性认知转化为理性分析。另一部分是对设计项目基地的场地限制条件、潜在可能性、和使用者需求进行有效分析，方法如场地静态现状记录、行为动态记录、使用者问卷调查等。

在不断地发现问题和解决问题之间来回往复，在进行多次循环往复的渐进式教学训练后，最终目的是使学生将调研分析内容与设计目标直接联系起来。在基础教育阶段启发培养学生的感性与理性思维能力对学生专业设计课程的学习是非常关键的。

环境艺术设计专业教学应紧密结合当前社会发展的需要，结合自身优势，形成自己的教学特色，为社会培养基础理论扎实，设计能力全面，具有专业素质、专业精神、团队协作精神的新型环境设计专业人才。

Bathroom Well-design Exploration

With the development of the housing industry, residential bathroom can not meet the needs of the people in the space and humane care. This paper is based on the well-design of bathroom space, to determine clearance axis by product and human behavior , and then according to structural space of equipment and partition wall to determine the building axis which control the applicability of the space, though modular coordination for social collaboration, factory production, assembly construction. So as to achieve save-material, adapted to the development of housing industrialization.

1. The common problems in the design of residential toilet

(1) The space design is not perfect

With the development of requiring for residential bathrooms , continue to highlight the problem of aging, old-age pension mode has become the ideal, put forward higher requirements on the space design of nursing care of the toilet, the toilet should be controlled in a reasonable size, to meet the different age generation center. From space in terms of size, insufficiency caused by unreasonable size of toilet building axis function; the designer of the pipeline, construction space equipment control is not accurate, the selection of wall tile is not conducive to the greatest degree of space saving; the reserved hole size is not enough; the toilet in front of the lack of space rotary people, open the door of the shower is not conducive to the way the old man entered the dielectric support, enter the bathroom door opening of the ground elevation effect wheelchair equipment; the installation position of the handrail space without consideration of the issue, consider the elderly Risk Report The installation position of police equipment; did not fully consider the disabled elderly using crutches, wheelchairs or via external force in the case of behavior problems in the bathroom. From the space perception factors, lack of light induction toilet elderly wheelchair use at night; the construction of sewer odors caused by details ignored and noise problems.

(2) The lack of space details of human care

In order to put the toilet facilities did not fully reflect the use of the old man, a reserved space size on the old behavior trajectory inconsiderate, did not consider the habits of the elderly and children toilet; no handrails according to body function characteristics and the direction of the force used elderly reserved position; not consider the elderly toilet auxiliary equipment use height; height and depth below the vanity is not considered into the wheelchair and the elderly physical space; bath without considering the safety of the elderly and children.

Because the clearance size is not fixed, the change of wall and floor tiles, lack of general modular series, leading to the scene must be cut according to the actual situation; the lack of toilet facilities "suit" combination, to influence the behavior of people lack of handrails, folding seats and other auxiliary equipment; the facilities between people are converted to the convenience of the account is not in place the bathroom; lack of construction design, integration of resources, products; to toilet function demand increasing, product diversification in the market lack of universal connection components; not fully these problems related to humanistic care for the elderly the bathroom. Residential toilet should meet the aging population, the maximum demand, promote the standard design of products, factory processing, on-site assembly, to achieve social cooperation.

2 . The optimization of residential bathrooms Countermeasures

(1) The use of facilities and human behavior to determine axis clearance

Headroom refers to the line axis of building space in a series of clearance size, can finely reflect the use of space, the axis control clearance can effectively ensure the facilities and space needs, as the dividing line clearance and partition structure space, and building axis parallel, as a common building important reference line.

Many generations in the bathroom is mainly the elderly, children in the family space, the traditional toilet in space planning, facility selection and human behavior space already can not meet the demand of all ages, needs to reflect the behavior characteristics of the elderly and children in space, to care the elderly, children can reasonable use of equipment enter the bathroom and convenience. From the human point of view, the behavior of the elderly activity has been determined in size and facilities track bathroom facilities, compact and orderly for memory and save energy. As the mainstream mode of old-age pension, must pay attention to the influence of the bathroom for the elderly, taking into account the elderly may need referral help or care who assisted wheelchair space, so the bathroom must be reserved the rotation radius of 750 mm, the width must be greater than 830 mm, to ensure that the old man Guan Shijie Helps to protect people or the use of auxiliary facilities can meet the requirements.

The gate in order to meet the different levels of the elderly care use wheelchairs, crutches or referral to help nursing people entered the bathroom, the hole must be increased according to the most wide size; sliding rails are arranged on the top of the door from the ground elevation by the toilet toilet in front; in order to facilitate the nursing safety of the elderly living space in front of the rotary the need to set aside the nursing auxiliary space or wheelchair; shower door to enter the same dielectric support the elderly, take sliding form, save space and increase the opening size; the old man alone in the bathroom to avoid dangerous situations, should be on the ground about 500 mm to install any state can touch the danger alarm equipment the old man in the wheelchair; night cannot safety contact switch should be installed in the induction lamp panel, side skirting; toilet smell problem is mainly a sewer pipe is not in place, Wash Taiwan water hose diameter is less than PVC of the ground pipe diameter, to prevent the smell not only to be sealed and water pipes need to return water bends into the toilet is not only noisy and deodorization function and siphon type is different.

By the standard size of the product and the behavior of people in the space scale is derived using the size of the space, to meet the different levels of the elderly and children of different ages and the use of children of different ages.

(2) The clearance axis and the construction space determine building axis

适老化卫生间设计探索

随着住宅产业的发展，住宅卫生间在空间与人文关怀方面不能完全满足多代居人群的需求。文章结合卫浴空间的精细化设计，以产品和人的行为空间确定出净空轴线，再依据设备、隔墙等构造空间确定出建筑轴线，以此控制了空间的适用性，通过模数协调进行社会化协作，工厂化生产，装配化施工，从而实现了节材减排，促进了住宅产业化的发展。

杨琳
北京建筑大学副教授

朱宁克
北京建筑大学讲师

一、住宅卫生间设计中存在的通病

（一）空间设计不完美

人们对住宅卫生间的需求不断提高，老龄化问题不断凸显，居家养老成为现阶段理想的养老模式之一，对介护卫生间的空间设计提出了更高的要求，介护卫生间应该控制在合理尺寸之内，以满足多代居中不同年龄层的人群使用。从空间尺寸来讲，对卫生间功能考虑不全造成建筑轴线尺寸不合理；设计师对管道、设备等构造空间控制不精确，隔墙砖的选型不利于最大程度的节省空间；预留洞口尺寸不够；坐便器前方缺乏人的回转空间，淋浴间门的开启方式不利于介护老人的进入，卫生间门洞口的地面高差影响轮椅等设备的进入；对扶手的安装位置考虑不足，空间未考虑老人危险报警设备的安装位置；没有全面考虑失能老人利用拐杖、轮椅或借助外力的情况下在卫生间的行为问题。从空间感知因素来讲，卫生间缺乏轮椅老人夜间使用的感应灯光；施工时忽略下水管道细部处理造成异味及噪音问题严重。

（二）空间细节缺乏人文关怀

卫生间设施的摆放次序没有充分体现老人的使用规律，空间预留尺寸对老人行为轨迹考虑不周，没有考虑老人及儿童对卫生间的使用习惯；扶手没有根据老人的身体机能特征及使用力度的方向预留位置；没有考虑老人坐便器辅助设备的使用高度；洗面台下方净高及进深未考虑到轮椅及老人的肢体空间；浴缸未充分考虑老人及儿童使用的安全性。

由于净空尺寸的不固定，现场变化性大，墙地砖缺乏通用的模数系列，导致现场必须根据实际情况进行切割；卫生间设施缺乏“套装”组合，导致影响人行为的扶手、折叠座椅等辅助设备不足；对设施之间人进行相互转换的便利性考虑不到位；缺乏对卫生间中设计、施工、产品的资源整合；对卫生间功能需求的不断增多，市场中出现的多样化产品缺乏连接构件的通用性；这些相关问题不能充分体现卫生间对老年人行为的人文关怀。住宅卫生间应该最大限度的满足老龄化人群的需求，促进产品的标准化设计、工厂化加工、现场化装配，实现社会化协作。

二、住宅卫生间的优化对策

（一）利用设施与人的行为确定出净空轴线

建筑净空轴线是指建筑空间中一系列净空尺寸的标注线，可以精细的体现使用空间，控制净空轴线可以有效的保证设施及人对空间的需求，作为净空与隔墙等构造空间的分界线，与建筑轴线相平行，共同作为建筑中重要的参考线。多代居卫生间主要是老人、儿童合居家庭使用的空间，传统卫生间在空间规划、设施选型以及人的行为空间上已经不能同时满足各年龄层的需求，需要体现老人及儿童在空间中的行为特征，促使介护老人、儿童都能合理的进入卫生间并便利的使用设备。从人性化的角度考虑，老人的行为活动已经决定了设施尺寸与人在卫生间的行为轨迹，设施的紧凑有序便于记忆和节省体力。居家养老作为养老的主流模式，必须关注卫生间对老人的重要影响，考虑到老人可能需要介助轮椅或介护人等辅助空间，所以卫生间内必须预留 750 毫米的旋转半径，通过宽度必须大于 830 毫米，以保证老人不管是介助介护人还是使用辅助设施都能满足要求。

门口为了满足不同介护程度的老人利用轮椅、双拐或介助介护人等情况进入卫生间，其洞口必须根据设备最宽尺寸进行加大；卫生间采用的推拉轨道安置在门上方免去地面高差；坐便器前方为了方便介护老人安全如厕，前方需留出介护人的辅助空间或轮椅的回转空间；淋浴间门同样为了介护老人的进入，采取推拉门的形式，节省空间又加大了开启尺寸；避免老人单独在卫生间出现危险情况，应该在地上 500 毫米左右安装任何状态都可触碰到的危险报警设备；由于轮椅老人在夜间无法安全接触开关面板，应该在踢脚侧方安装感应灯具；卫生间异味问题主要是下水管处理不到位，洗面台下水软管管径小于地面 PVC 管的管径，为防止异味不但要进行密封处理而且下水管需要返水弯，直冲式坐便器不仅噪音大而且防臭功能与虹吸式也有一定的差别。

通过产品的标准尺寸及人在空间的行为尺度推导出使用空间的尺寸，以满足不同介护程度的老人及不同年龄层的儿童共同使用。

（二）利用净空轴线与构造空间确定出建筑轴线

构造空间包括建筑隔墙、管道空间及墙体饰面三部分。墙体饰面根据不同

Construction space including building walls, pipe space and wall facing three parts. According to the construction wall facing the different ways of the occupied space of different sizes, the bathroom wall with dry hanging stone, installed steel in the reinforced concrete structure wall or block wall, after the water treatment, the dry hanging stone angle fixed in space. About 100 mm, the marble with stainless steel pendant although the construction method is simple, but compared to the wet paste occupy a large space, the wet paste by cement mortar plaster wall, 25 mm; wall tiles should pop up horizontal and vertical reference line and 90 degree or 45 degree angle with the door frame when paving, typesetting must be on, symmetrical. In tile before the construction, should determine the modulus of tiles through a row of brick laying line, according to the module.

The bathroom wells within the toilet usually clearance axis, take the use of space the size of the bathroom is well water pipes and ventilation, ventilation is usually prefabricated square does not need treatment, water pipes need plastering brick, and ventilation with outside tiles, takes about 500 x 250 mm space all the facilities, water pipes under the floor and the top of the main water pipe is connected on the water pipes installed hidden within the wall, the laying of water pipes is larger than the diameter of the groove spacing, hot and cold water must maintain a certain distance between the tube in tube without connecting pipe and tube fittings, avoid overlapping, piping should be horizontal and vertical to reduce the leakage point.

The bathroom wall if the concrete pouring, the minimum wall thickness of 240 mm, while the ratio of block wall strength is high, but the construction difficulty, high cost, serious pollution and loss of space is relatively large, cement brick masonry wall instead using environmentally friendly advantages, but also has the space consumption is small, light weight advantages.

From the behavior of people, products and construction space derived building axis, through the two steps of the backstepping, the bathroom either in the facilities size, the behavior of people in space or toilet water saving, space environmental problems have reached an ideal result.

(3) clearance modulus conform to product factory

From the perspective of considering the aging, the toilet can be installed by lifting facilities supporting the old man hard to stand up the old behavior, the use of the toilet is usually around 430 mm above, in order to reduce the burden of the elderly children toilet seat legs; can be installed in the adult toilet, chassis with affinity for PP material, body and seat are easy to install removed shower; configuration of lifting and folding stool, installed in the wall of U type or L type armrest; using a walk-in tub, during in the bottom of the bathtub; nursing toilet handrails should use some waterproof materials such as wood, plastic or resin mild, diameter 30 mm, maximum 35 mm.

Modular design is one of the methods of green design, the different specifications of the facilities according to the functional analysis of the function modules in the bathroom, the members can freely combined, shorten the production cycle of facilities, increase the facility type, according to different combinations of features to meet different needs. Modular applications to the bathroom has a certain degree of difficulty, but diversification of products through the commonality among the components, but also eliminates the difficulty of construction workers and the adverse impact on the environment, convenient after disassembly, maintenance, recycling and disposal; on this basis, put forward the method of "standard catalog", adding a modular series of facilities, facilities for the ergonomic requirements, and modulus and the axis of the relative clearance, all manufacturers to each product in accordance with the classification and size into the "directory", the designer selects products in the catalog, implementation General products, combined with building axis standards, improve the factory production to a certain extent, assembly of construction process; to reduce the amount of sales of blank Housing, the developers of large-scale decoration conditions, standard product component size, gradually realize the socialization cooperation.

In order to promote the development of residential industrialization, the standardization of design, based on general products, a complete set of system is to realize the renovation of residential washroom, two is the use of new, green decoration materials, to make full use of Recyclable, easy decomposition, regeneration and reuse and harmless to the environment and people's constantly using energy-saving materials. The toilet, the light off, intelligent health products, decoration garbage treatment and other measures to achieve emission reduction in industrial materials based on the toilet facilities, to the environment, the harmony of man.

3. The ideal bathroom design analysis

The toilet is not suitable for the aging of the size of the problem, using the above reasoning method is derived in the multi generation bathroom clearance size, the entire bathroom in accordance with a set of four dimensions of reasonable layout, in the limited space as much as possible to meet the different levels of the elderly care demand space.

Taking into account the care needs of the elderly wheelchair or care people need auxiliary space in the bathroom is left turning radius of 750 mm; the size of the door in order to meet the old man borrowed wheelchair and other auxiliary tools into the bathroom door need around the advantages more suitable for the elderly to use concession, door clear width is 850 mm, with transparent opaque glass as to the safety of the door, the wheelchair must be installed in the door below 350 mm kickplates, to avoid the influence of ground track placement on the height above the door.

The product in accordance with the selection and installation of multi generation in the condition of using the side door form a walk-in bathtub, width 700-900 mm, length according to the actual human scale but not longer than 1600 mm; 300 mm is arranged in the bathtub and bathtub handle arranged above the platform, can better meet the needs of people. The aging of the toilet the front rotary space need not less than 500 mm, 100 mm installed on both sides of the handle; vanity under the need to have a 650 mm height, and the old man into the wheelchair body, provisions in its front to keep 1100 x 800 mm rotary wheelchair space; design the mirror face Taiwan above can become free adjust the angle of the old man in the bathroom shower; because the ground is wet and slippery, decreased exercise capacity attenuation and perception, very prone to danger, the height of the folding seat configuration is less than or equal to 5 00 mm is appropriate, depending on the L type handrail wall at the side of the lowest ground distance of 650 mm, the highest is 1400 mm from the ground, the

的施工方式，占据空间大小不同，卫生间墙体采用干挂石材的做法，在钢筋混凝土结构墙体或砌块墙体上安装方钢，防水处理完成以后，把石材干挂在角钢上固定，占用空间约为100毫米，这种大理石配合不锈钢挂件虽然施工方法简便，但相比湿贴占用空间大，湿贴采用水泥砂浆外贴墙砖，需要25毫米；墙砖铺贴时应该弹出横纵向基准线并与门框呈90°或45°夹角，排版一定要对缝、对称，在贴砖施工之前，应该通过铺线确定墙地砖的模数，按照模数进行排砖。

卫生间管道井通常在卫生间净空轴线以内，占用使用空间的尺寸，卫生间管井是下水管道及通风道，通风道通常是预制好的正方形不需要处理，下水管道需要用砖砌进行抹灰，与通风道一并外贴瓷砖，大约占用了500毫米×250毫米的使用空间，所有设施的下水管道都在楼板下的顶部与主要下水管道相连接，上水管安装在墙体之内进行隐藏，铺设水管大于管径凹槽，冷热水管间距必须保持一定的距离，管于管之间不用连接配件，避免水管与电管重叠，管道走向应该横平竖直，减少渗漏隐患点。

卫生间隔墙如果采用现场浇筑混凝土，墙体厚度最低为240毫米，虽然比砌块墙体强度高，但是现场施工难度大、成本高、污染严重并且空间损耗比较大，相反砌块墙体所利用的水泥砖符合环保的优点，而且具有空间损耗小、质量轻的优点。

从人的行为、产品及构造空间推导出建筑轴线，通过两大步骤的反推，使得住宅卫生间无论是在设施尺寸、人的行为空间还是在卫生间节能节水、空间环境问题上都达到一个理想的结果。

（三）净空模数符合产品工厂化生产

从老龄化角度考虑，坐便器可通过安装升降设施支撑老人难以起立的行为，老人使用坐便器通常在430毫米以上，以减轻老人腿部负担；儿童马桶圈可以安装在成人马桶上使用，底盘采用亲和力的PP材料，本体及座圈都易于安装取下；淋浴间配置升降折叠坐凳，在墙体安装U型或L型扶手；采用步入式浴缸，在浴缸底部进行防滑处理；介护卫生间扶手应该选用一些防水木质、塑料或者树脂等温和的材料，直径30毫米左右，最大不能超过35毫米。

模块化设计作为绿色设计方法之一，在卫生间内把不同规格的设施根据功能分析出功能模块，促使构件之间能够随意组合，缩短设施的生产周期，增加设施的类型，根据不同的功能组合满足不同的需求。模块化应用到卫生间虽有一定的难度，但是通过零部件之间的通用化实现产品的多样化，也消除了工人施工难度以及对环境的不利影响，方便日后拆卸、维修、回收和处理；在此基础上，提出采用“产品标准目录”的方法，增加设施的模数系列，使设施符合人机要求，并且与净空轴线的模数进行相对应，所有厂家把每件产品按照分类与尺寸放进“目录”中，设计师在目录中选择产品，实现产品的通用化，与标准的建筑轴线相结合，在一定程度上提高工厂化生产、装配化施工的进程；减少毛坯房的大量销售，利用开发商大规模装修的条件，规范部品构件的尺寸，逐渐实现社会化协作。

为了推进住宅产业化的发展，在设计标准化、产品通用化的基础上，一是实现住宅卫生间成套化装修系统，二是采用新型、环保装修材料，尽量利用可回收、易分解、能再生使用的而且对人及环境无害的材料。不断采用节能坐便、轻质隔断、智能化产品、装修垃圾健康处理等措施，在产业化的基础上实现节材减排，使得住宅卫生间达到环境、设施、人的和谐统一。

三、理想卫生间设计解析

针对卫生间不适合老龄化尺寸的问题，利用以上推理方法推导出了多代居卫生间的净空尺寸，整个卫生间按照四件套的规范尺寸进行合理布局，在有限的空间内尽可能的满足不同介护程度的老人对空间的不同需求。

考虑到介护老人需要轮椅或介护人的辅助空间，需要在卫生间留有750毫米的回转半径；门口尺寸为了满足老人借用轮椅等辅助工具进入卫生间，推拉门不需要前后左右退让的优点更适宜老人使用，门净宽是850毫米，用透光不透明的玻璃作为门板，为了轮椅者的安全必须在门扇下方350毫米处安装护门板，把轨道安置在门上方避免地面高差对人的影响。

产品在符合多代居的条件下进行选择与安装，浴缸采用步入式的侧开门形式，宽度700~900毫米，长度根据实际人体尺度但是基本不得长于1600毫米；浴缸内设置300毫米的坐台以及浴缸上方安置把手，更能满足老龄化人群的需求。坐便器前方需要留出不小于500毫米的回转空间，在其两侧100毫米以内安装把手；洗面台下方需要留有650毫米的净高，便于轮椅及老人肢体的进入，规定在其前方必须留有1100毫米×800毫米的轮椅回转空间；洗面台上方的镜子设计成为可以自由调整的角度；老人在卫生间淋浴时由于地面湿滑、运动能力衰减以及感知能力下降，很容易发生危险，配置折叠座椅的高度以不超过500毫米为宜，靠墙一侧的L型扶手最低处距离地面650毫米，最高处离地面为1400毫米，淋浴间门口利用推拉门加大门口尺寸。在介护卫生间中危险报

shower door with sliding door to increase size. In the bathroom care risk alarm equipment installation height between 400-800 mm, can set the button or rope to form shapes and colors easy to identify; starting from the behavior of different facilities and the establishment of appropriate aging space are derived for 2700 axis clearance x 3000 mm.

According to the facility clearance size modulus form, and the use of environmentally friendly construction space partition occupies small space cement brick, pipeline according to the thickness of plastering block was hidden in the wall, plastering cement brick is 15 mm, 240 x 115 x 53 mm specification of cement standard brick. Mm mortar and wall finish is 25 mm, with lateral wall closed pipeline wells water pipe is 110 mm, 250 mm square with vents, water around the block and ventilation wall. The overall effect to achieve the visual appearance space; through behavior and products, building axis construction space deduced for 2900 x 3200 mm.

Figure 1 is derived using the bathroom layout (Figure 2), from the plane map can be seen in the facilities compact orderly arranged in the space around the door opposite the vanity and the mirror in the visual magnification of the use of space, to meet the different needs of the elderly care of the bathroom.

4. Conclusion

From the product and human behavior analysis of the residential toilet, have different needs of the elderly and children; through the support and care of the people by using dielectric behavior space, facilities should be consistent with the modulus is deduced according to the size of the axis clearance, derived method of building axis of the space structure, to achieve the fine the design of the toilet, can meet the future generations in the bathroom and the development trend of nursing toilet, including the needs of the family structure and the life style of space, with the age increasing in home care conditions; to vigorously promote the bathroom clearance line method, hope the designer through the standardized design of residential, humanized products the establishment of "product standards, manufacturers directory", further improve the level of production and product quality, promote industrial production, construction, promote socialization Collaboration, the ultimate realization of environmental protection, material saving policy, to contribute to the aging society.

警设备安装高度在 400~800 毫米之间，可以设置成按钮或拉绳形式，形状颜色要易于辨认；从设施和人的不同行为入手建立适老化空间，推导出的净空轴线为 2700 毫米 ×3000 毫米。利用图 1 推导出卫生间的平面布置图（图 2），从平面图中可以看出设施紧凑有序的安装在空间四周，门口正对着洗面台及镜子，在视觉上放大了使用空间，能够满足不同介护程度的老人对卫生间的需求。

四、结论

从产品及人的行为入手分析出的住宅卫生间，已经具备了老年人及儿童的不同需求；文章通过利用介护及被介护人的行为空间、设施应该符合的模数推导出净空轴线，再依据构造空间的尺寸推导出建筑轴线的方法，实现了卫生间的精细化设计，能够满足未来多代居卫生间及介护卫生间的发展趋势，包括家庭结构和生活方式对空间的需求以及随着年龄的不断增长在居家养老的条件；从而大力推行卫生间净空轴线法，希望设计师通过对住宅的标准化设计、产品的人性化选用，厂商对“产品标准目录”的设立，进一步提高生产水平与产品质量，推动工厂化生产、装配化施工，促进社会化协作，最终实现环保、节材减排的方针，为老龄化社会做出贡献。

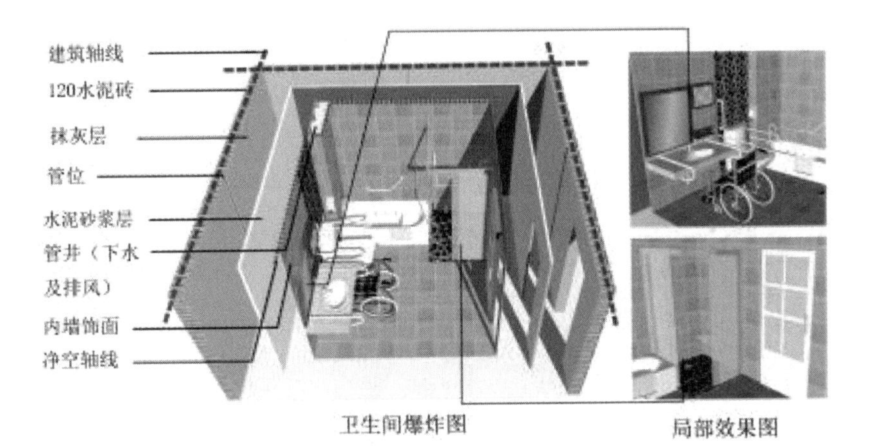

图 1

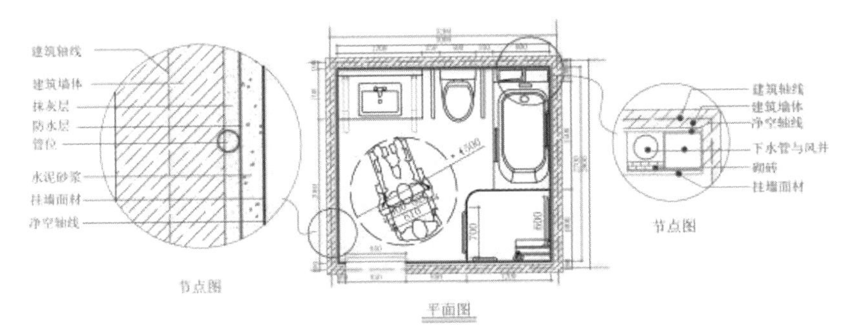

图 2

Professional Competencies Required to be Exhibition Designers in China: A Study

In China, exhibition design is an "invisible" profession with relation to exposition design, curating, and exposition art & technologies. It has no official designation, and there is no corresponding industry association; however, it's a fact that it has been widely recognized by relevant sectors. Then, what are the features of exhibition design? What's its content? How to learn exhibition design? What are the standards to judge an exhibition designer? What are the competencies required to be professional exhibition designers?

These questions can be answered through a method raised by Bobbitt, J. Franklin, that is, to analyze the activities and occupational stigma of people in a domain to find out the knowledge, abilities, habits and attitudes that they need, and then determine the targets and requirements of relevant talent development.

I. Definitions and requirements of relevant professions

Currently, there are five officially recognized professions closely connected with exhibition design: display & exhibition designer, exposition designer, and curator announced by the National Labor and Social Security; display designer defined by the China's Ministry of Commerce China General Chamber of Commerce; and exposition image designer defined by the SASAC Commercial Food Service Skills Identification and Development Center. The profession of exhibition designer has not been officially recognized yet so far.

The profession of "curator" is defined quite ambiguously, covering two sectors of economic management and exposition design. A "display & exhibition designer" is a person who designs products, showcases, shelves, mannequins, lights, music, and POP posters for the purpose of increasing product sales and promoting brand image. An "exposition designer" designs and builds creative, artistically impressive and visually attractive spatial environments for small, middle and large sized expositions. An "exposition image designer" is a professional in the planning, organization, execution, reception and design of expositions, primarily the planning and management of expositions. A "display designer" can be a clothing display designer or a decoration display designer. The former, obviously, designs the display of clothing, and the latter designs the display of home furnishings (mainly furniture and decorations) which is actual a part of interior design. The job of a "display designer" also includes the design of museums, large-sized commercial activities, and even personal image.

From the professional requirements, responsibilities, fields of service, technical qualifications, and qualification examinations of the five professions, we can see that curators and exposition image designers do similar work, mainly the planning, management and partial design of expositions; display & exhibition designers and display designers provide design service to commercial displays and small-sized exhibitions; and exposition designers provide design service to expositions and festival activities. Obviously, the definitions of the five professions are indistinct and overlapped – they are not really five different professions, but sub-professions under the profession of exhibition design.

The interesting thing is that to the exclusion of education backgrounds and working years, these professions have almost the same requirements on job duties, skills and knowledge. For example, aside from fundamentals of drawing and design (e.g. artistic calligraphy, plan design, 3D design, and color composition), they all require basic knowledge of design positioning, design creation, design expression, design execution management, as well as exhibition design, brief history of exhibition design, exhibition design methods and procedures, CAD, design practice and relevant laws and regulations. Abilities required by these professions include the ability to learn, to express, to calculate, to act, as well as good sense of space and color, and flexible fingers.

The professional requirements above actually present some problems in professional requirements. Firstly, some necessary requirements are not included. For example, the professional requirements of "display & exhibition designer" and "display designer" do not include CI and VI, and those of "exposition designer" fail to cover customs and performance. Secondly, the differences of requirements on different professions are not highlighted. Thirdly, there are no essential differences between them and the professional requirements of other design professions (e.g. interior design, industrial design and visual communication design) which make them independent principles and professions. Lastly, general basic abilities and qualities required by modern design are not included.

II. Characteristics and trends of exhibition design

Exhibition design is a process of information delivery through objects and human behavior in a certain spatial environment. It includes exposition design, museum exhibition design, commercial exhibition design, and festival exhibition design, featuring the integration and overlap of multiple disciplines.

Exhibition design delivers information and creates certain atmospheres. The Society for Experiential Graphic Design (SEGD), a US non-profit international environmental graphic design organization, defines exhibition design as the combination of information delivery and environment. Exhibition design is narrative, performative and experiential design in a given space. Narrative design refers to displaying exhibits in a way of telling a story, or, in a broader sense, to give a shared background to exhibits. When plots are given to an exhibition like telling a story, it will be easier-to-understand, striking a chord with the visitors, and presenting new meanings and value. Performative design emphasizes on the dialogues between space, people and time. When the behavior (bodies and body movements) of visitors constitutes a part of an exhibition, the boundaries between visitors and performance, or visitors and exhibits become fuzzy, which actually makes the exhibition more interesting and lively, and realizes more direct and effective information delivery. Experiential design is about using spaces, scenarios, and technologies like light and audio technologies to create an impressive atmosphere in which visitors feel amiable, horrible, sacred, depressive, happy, interesting, and so on. At present, brand experience has replaced product experience as a trend of commercial exhibition design, making the experiential memory of visitors a motive power that improves brand image and value.

我国展示设计人才职业要求研究

朱飞
南京艺术学院教授

展示设计在我国是个隐形的行业，栖身于会展设计师、陈列展览师等职业和会展艺术与技术专业之中，没有正式名称，更没有一个全国性的权威性管理机构，但展示设计这个名称得到了业内的广泛认可。那么，展示设计的行业特征是什么，涉及哪些内容，如何开展学习，其人才标准和职业要求又是怎样的呢？

我们可以按照博比特所提出的分析人们从事的活动、职业特征，找出所需知识、能力、习惯与态度，确定人才培养目标和具体要求 [Bobbitt, J. Franklin. (1924). How to Make a Curriculum. Boston:Houghton Mifflin.] 来开展本论题的研究。

一、相关职业的职业定义和要求

国内有相关部门公布的五个职业和展示设计密切有关：国家劳动和社会保障部颁布的“陈列展览设计师”、“会展设计师和会展策划师”，中国商务部中国商业联合会的“陈列设计师职业”，国务院国资委商业技能鉴定与饮食服务发展中心的“会展形象设计师职业”。但没有展示设计职业。

“会展策划师”的职业定义比较模糊，横跨经济管理和会展设计两大领域。陈列展览设计师其职责是在充分理解商品内涵的基础上，对产品、橱窗、货架、模特、灯光、音乐、POP 海报进行设计，以达到促进产品销售、提升品牌形象的目的。“会展设计师”需从事大、中、小型会展、各种节事活动的空间环境展示设计、施工，并提供具有创造性和艺术感染力的视觉化表现的人员。“会展形象设计师”是指从事展会策划、组织、实施、接待、设计等展会服务的专业人员，主要是从事会展策划和管理方面的工作。“陈列设计师”分为服饰类陈列设计师和装饰类陈列设计师，服饰类设计师主要从事服装商品的陈列，装饰类陈列设计师是指家居陈设设计，主要内容为家具布置和软装设计，实际上是室内装修的一个延伸。但其工作内容又注明可以从事博物馆设计、大型商业活动布展，甚至个人形象设计等。

综合分析以上所谓五个职业，从职业要求、工作职责、需求领域到专业技术资格条件、考试认证的情况来看，“会展策划师”和“会展形象设计师”比较接近，主要从事会展的策划、管理和部分设计工作，“陈列展览设计师”和“陈列设计师”是从事商业展示设计和一些小型展览展示设计，“会展设计师”的工作面在会展和节庆设计方面。这些职业的定义是模糊、交叉的，它们并不是五个不同的职业，而应该同属于展示设计，都是展示设计的一个组成部分。

很有意思的是，除了学历、从业年限等要求以外，它们在工作内容、技能要求和相关知识上几乎是一致的。在基本知识要求中，除常见的绘画基础、设计基础（如美术字，平面、立体、色彩构成等），均含有涉及设计定位、设计创意、设计表达、实施管理的内容，罗列了展示设计基础知识、展示设计史概述、展示设计方法和程序，以及计算机辅助设计和设计实务、相关法规等知识要求。能力要求一般为：学习能力、表达能力、计算能力、空间感觉、肢体能力、色觉和手指灵活性。

这样的行业知识技能要求反映出几个问题：首先，内容十分不完善，如从事商业展示设计的“陈列展览设计师”和“陈列设计师”居然没有 CI 或者 VI 设计方面的内容，也没有能够在会展的节庆礼仪中看到涉及到民俗、表演等方面的内容；其次，没有凸显展示设计范围内不同部分的内涵差别；另外，与其他诸如室内设计、工业设计、乃至视觉传达设计在本质要求上没有多少区别，缺乏成为独立学科和职业的核心内容。同时，也缺少现代通识性基础能力和素质要求。

二、展示设计的特征和发展趋势

展示设计就是在一定的空间环境中，通过物介和行为，进行信息传递。它包括会展设计、博物馆展示设计、商业展示设计、节庆礼仪设计。它具有多种学科融贯、交叉的特征。

展示设计融合信息的设计和氛围的营造，是有目的“场所与信息”的创造。因此，美国 SEGD（美国非盈利性国际环境平面组织）将展示设计的定义概括为信息传达与环境的融合 [（美）简·洛伦克，（美）斯科尼克，（澳）伯杰编著 . 邓涵予，张文颖，朱飚译 . 什么是展示设计 [M]. 北京：中国青年出版社，2008]。展示设计又具有空间的叙事性、述行性和体验性特征。叙事性就是以讲故事的方式有次序的陈列，从广义上看，叙述空间是指展示物融入背景。强调故事性、情节性和通俗性叙述方式可以激发观众的情感，使展品呈现出一种新的诠释与价值。述行性强调空间、人、时间之间的对话，观众的行为（身体和移动）构成了展示的要素，观众与表演、观众与展品之间的边界已经模糊，这增强展示设计的趣味性、生动性和信息传播的直接性与有效性。体验性是利用空间场景、声光电等技术渲染主题、表现意境，使人产生亲切、恐惧、神圣、

With expression methods and skills incorporating and immersing into each other, current exhibitions see fuzzier and fuzzier boundaries, becoming a complicated media system. When architectural spaces meet resultant images, simulation environments, and plots, exhibition design is linked increasingly closely with films & TV, lighting design, graphic design, installation art, performance art, and music & sound effects. Besides, with the participation of new media and fashion, exhibition design will be more interactive and attractive.

III. Analysis of professional exhibition designers' competencies

Professional exhibition designers' competencies include knowledge, abilities and qualities.

The knowledge of an exhibition designer should cover the six facets of text planning, space construction, visual communication, exhibition skills, aesthetics and engineering construction in the fields of culture, science and art at the four layers of planning, design, engineering and management. With various application areas such as architectural design, interior design, aesthetics, communication, information technology and marketing, exhibition design is a knowledge-intensive profession integrating various principles.

The abilities that an exhibition designer should have include a variety of work readiness such as active attitudes and actions, the sense of responsibility, oral and written communication skills, thinking & judgment, the ability to constantly learn, create and adapt, and teamwork skills. Such abilities as 1) innovation, 2) appreciation & aesthetic judgment, 3) expression, 4) multicultural thinking, 5) teamwork, 6) planning & organization, 7) communication, 8) computer-aided expression, 9) research & analysis, and 10) execution are also required.

Exhibition design is the most effective way to express creative ideas, and conversely, innovations, which cannot be copied or produced in mass, act as the main driving force of exhibition design. The opening ceremony of the Beijing Olympics must be different from that of the Athens Olympics, and the China Pavilion at the Shanghai Expo must be unique. Even a simple wedding ceremony should have its characteristics. Multicultural thinking advocates understanding and respecting different ideologies and ways of act existing in the human society, and learning to view things from the perspectives (e.g. psychological and cultural perspective) of others. In today's world, trans-regional and cross-border exhibition activities are happening every day. In such a case, we need to think how to introduce the history, cultures and customs of a country in a way that people from other countries can understand, thus realizing the interaction and communication between different countries.

Exhibition designers with an interdisciplinary education background, versatile abilities, professional ethics, strong sense of social responsibility, and physical and mental health are talent needed by the society.

VI. Learning modules and thoughts on the "6+1" competition

Exhibition design is a multidisciplinary profession covering the design of a wide variety of activities, primarily commercial activities, festival activities, conventions, expositions, performances and competitions. Since its knowledge and technical boundaries are relatively fuzzy, many important festival activities and celebrations are actually planned and designed by film and TV directors. For the study of exhibition design and corresponding performance assessment, students are required to, in a limited period of time, learn fundamentals and basic theories of modeling and design (e.g. history of Chinese and foreign architecture and interior design, brief history of Chinese and foreign arts, artistic design, introduction to exhibition design, ergonomics, and museology), and professional knowledge concerning text planning, space construction, visual communication, exhibition skills, aesthetics and engineering construction. Special attention should be paid to core knowledge such as spatial organization & expression, information delivery design, themed exhibition design (narrative, performative and experiential design), and exhibition technologies, as well as related knowledge like modern arts and graphic design. Besides, cutting-edge knowledge such as innovative interdisciplinary research fields like human behavior and brand experience should also be taught, for the purpose of developing exhibition designers that can adapt to and even challenge the times.

This year marks the 4th ceremony of the 2016 CIID "Interior Design 6+1" University-Business Graduation Projects Competition. With inspiring achievements, the event provides a great opportunity to learn and exchange ideas for both the teachers and students. Exhibition design award is offered each year at the event. Then, does the award just mean a design award or a quality element evaluation system established based on different design characteristics? What is an exhibition design evaluation system? In my view, it is a connotation and value system which advocates exhibition design works taking spatial design and information delivery as the foundation, features narrative, performative and experiential ideas, and stresses on innovation, cultural understanding and aesthetic.

压抑、欢快、趣味等不同的心理感受。品牌体验也已取代商品体验，成为商业展示设计的发展方向，观众的体验记忆正是品牌形象和价值发展的原动力。

现在，各类展示类型的界限已越来越模糊，很多表现方式和技巧都相互借鉴、渗透，成为一个复杂的传媒和相互关联体。合成影像、模拟环境、故事情节与建筑空间并置，展示设计与电影电视、灯光设计、图形设计、装置艺术、行为艺术、音乐音效的关系日益密切，并更加与新媒体、时尚为伍，互动性和参与性越来越得到重视。

三、展示设计的人才要素分析

人才要素包括知识、能力和素质三个方面。

在知识要求方面，展示设计结合了人文、科学、艺术三大领域，涉及策划、设计、工程、管理四个层面，需要具有策划文案、空间营造、视觉传达、展示技术、艺术审美、工程施工等六方面的知识，涉及建筑设计、室内设计、美学、传播学、信息技术、市场营销等应用领域，是一个知识高度密集型、融贯型的职业。

在能力要求方面，要具有积极的态度、行为和责任心，口头和书面交流能力，思维判断能力，不断自我学习、创造与适应能力，以及团队工作的技能等前能力要求［ 加拿大行业工会委员会颁发的所有从业人员必备的前能力要求 ］。以及：①创新能力；②审美、鉴赏能力；③表达能力；④多元文化思维能力；⑤团队协作能力；⑥策划、组织能力；⑦表达交流能力；⑧计算机表现能力；⑨调研分析能力；⑩动手能力。［ 这个展示设计人才能力排名是 2010 年南京艺术学院工业设计学院与江苏省会展办进行的一项问卷调查，对列举的 20 个展示设计专业能力重要性进行的评估 ］

展示设计是表现创意的最有效方式，创新能力是展示设计的主要原动力，它不可以复制，不可以成批量生产。我们绝不能容忍北京奥运会开幕式表演和雅典奥运会的一样，也不可能放弃对上海世博会中国国家馆的特色要求。就连一个简单的婚礼，也会尽可能做到有自己的特点。多元文化思维就是要了解并尊重人类社会存在的多样思想观念和行为方式，学习用他人的眼光、心态、文化视角来看待事物。现在，跨地区、跨国界的展示活动已成为常态，如何体现本国的历史文化传统，并以别人的习惯解读方式去展现，实现互动和交流，这个问题非常值得研究。

展示设计专业人才培养应该是知识融贯型、能力复合型，同时，还要具有良好的职业道德、强烈的社会责任感和健康的身心素质。

四、学习模块和“6+1”活动的思考

展示设计是多重学科的组合，包括了商、节、会、展、演、赛等活动，其知识和技术边界比较模糊，很多重要节庆礼仪活动的策划和设计都是由电影、电视导演来负责。但作为一项职业学习和考核来说，在有限的时间内，既要打好造型基础、设计基础和相关的理论基础（中外建筑及室内设计史、中外美术简史、艺术设计学、展示设计概论、人体工程学、博物馆学等），也要搭建策划文案、空间营造、视觉传达、展示技术、艺术审美、工程施工六方面的专业学习知识模块，特别需要强调其中的空间组织和表现、信息传达设计、展示专题设计（叙事性、述行性、体验性）、展示技术等核心内容，以及现代艺术或者图形设计等关联内容，并要设置诸如人的行为方式和品牌体验等交叉学科的创意研究等前沿课题学习，使得展示设计师的培养更具有时代性和挑战性。

CIID 举办的“室内设计 6+1”校企联合毕业设计竞赛活动已经举办了四年了，取得了令人鼓舞的成绩，学生与老师都获得了交流和提高。在每一期评比中都设有展示设计的奖项，那么，我们的活动是多设立了一个设计奖项呢？还是多了一个基于不同设计特征之下的素质要素评价体系呢？展示设计类的评价体系又是什么呢？我认为，应该是基于展示设计空间 + 信息本质特点的，具有空间叙事性、述行性和体验性特征的，强调创新、文化理解力和审美的内涵考量和价值推崇。

Research-Oriented Graduation Projects Under the "Meridian Theory – Survey throughout the Entire Project" Innovative Teaching Model

——Some Thoughts on the CIID "Interior Design 6+1" University-Business Graduation Projects Competition

The graduation project, which is the core of the undergraduate development program, acts as an important standard to assess the teaching quality and curriculum continuity rationality of the four-year undergraduate education, and also a key indicator to assess the quality of innovative undergraduates. The teaching model adopted to carry out the graduation project is one of the main factors affecting teaching quality and results during the graduation project period.

Current graduation project teaching model and research-oriented graduation project topics

In the original undergraduate education model, graduation project topics generally fall into three categories: topics studied by teachers, topics determined by teachers according practical market needs, and topics determined by students. The current teaching model focuses on the process of students' practice and the final project results, but fails to direct students to think and develop multiple solutions to one topic.

The determination of topics should not only consider increasingly diverse design types and requirements, but also current social concerns. Bachelor's degree graduation projects play an important role in developing innovative talent needed by the society. Therefore, it's necessary to introduce research-oriented topics, such as problems faced by target groups and the environment in different times, and hot-button social issue like harmonious man-environment coexistence, into graduation projects, realizing the combination of practical topics and research-oriented topics. To develop art students with innovation and originality, it's necessary to combine study-oriented topics with practical and conceptual topics. Design originates from life. Aside from developing students' ability to think and judge independently, graduation projects should also focus on solving hot-button social issues with design, for the purpose of providing effective, forward-looking and feasible solutions, and constructive suggestions and advices to such issues and related target groups.

Practical significance of the "Meridian Theory – Survey throughout the Entire Project" teaching model in graduation project teaching practice

Meridians and parallels help us determine locations and directions. Through meridians and parallels, we can accurately locate a place on a globe or a map. For the "Meridian Theory – Survey throughout the Entire Project" teaching model, multidimensional survey methods applied in the three stages (thesis determination stage, middle stage and integrated design stage) form meridians offering main design clues; and design ideas, as methods and approaches to solve problems, form parallels. According to the results of surveys conducted from multiple perspectives based on multiple data and samples, design methods are adjusted constantly. Based on the design clues provided by meridians and adjusted design methods, design strategies and thoughts can be accurately determined, and consequently, rational and appropriate design schemes can be developed.

"Meridian Theory – Survey throughout the Entire Project" teaching model features a harmonious combination of conceptual thinking and rational thinking. Under this model, teachers will, based on the objective conclusions made in the three stages mentioned above, try their best to direct and stimulate students to use their conceptual and rational thinking in the entire project process, to learn and adjust thoughts actively, and to apply design approaches according to the survey results at various stages. In this way, students' ability to discover, analyze and solve problems can be really improved.

1. Design starts with surveys – thesis determination stage

The determination of graduation project topics forms an important part in the comprehensive practical teaching of undergraduate education. How to give full play to it in graduation project teaching is an urgent problem needing to be discussed. This paper explores how to carry out research-oriented graduation projects under an innovative teaching model, and discusses how to make innovations in terms of teaching programs, teaching goals, instructional mechanisms, and process supervision. The effectiveness of the teaching model has been proved by the practice results achieved during the competition.

Conduct surveys, analyze findings, and discover problems. In the thesis determination stage of a project, an early-stage survey should be conducted after the design specification is formulated, to study the design task, content, subject and target on a large scale, for the purpose of having a deeper understanding of the project, including its site characteristics, surrounding environment, spatial characteristics, target group, and redesign area. In this stage, students must look up relevant materials to find out proper survey methods, develop survey models, analyze survey data, and select and analyze samples. Then, based on the above achievements, the survey methods and content of existing research-oriented topics should be determined. In the case of Beijing Yaoyang International Retirement Community, similar retirement communities should be surveyed and analyzed, relevant domestic and foreign redesign projects should be studied and analyzed, and the current situations and site of the community should be surveyed and analyzed. In the process of site surveying and data analysis, the entry point of design and current problems faced by the project will become clear and clear. Only when problems are discovered, analyzed and solved, can the intersections of meridians and parallels be accurately found. Think outside the box of reference images. After the above surveys and analysis are completed, many students would start to search and study excellent cases online to collect reference images. However, in this process, their design thoughts and ideas tend to be restricted by what

“经线论——调研贯穿设计始终”创新教学模式下的研究型毕业设计课题实践

——由CIID“室内设计6+1”校企联合毕业设计教学活动引发的思考

任彝
浙江工业大学副教授

毕业设计是高等院校本科人才培养方案的核心组成部分，也是检验本科教学四年过程中课程教学质量、课程衔接关系合理性的重要评价标准。同时也是评价创新型本科人才培养质量的重要指标，而影响毕业设计教学质量和教学效果的主要因素之一就是本科毕业设计的教学模式。

本科毕业设计教学模式现状及研究型毕业设计课题

原有的本科教学模式中毕业设计课题大致分为三类：结合教师科研课题、教师收集的结合生产实际的课题、学生自立课题。在现有的教学模式中，通常仅注重学生对实际课题的实践过程把握和设计表现，并未对同一课题可能产生的多种设计思考和设计结果进行专门的引导和研究。

面对日益多元的设计类型和设计要求，结合社会当下关注的热点问题。本科毕业设计教学作为为社会输送创新设计人才的同时，更需要增加一些研究型的毕业设计课题，将受众与生存环境在不同时代所面临的问题、人与环境如何和谐共生等热点关注的选题，在实际课题中结合一些具有研究型方向的毕业设计课题。在当前注重培养艺术类学生创新能力和原创设计能力的前提下，无论在实际课题或概念设计课题中都应结合研究型课题的方向和切入点。再者，设计来源于生活，除了扎实培养学生在毕业设计环节中有独立思辨的能力，还能形成与当下社会生活发生紧密关系的设计结合点，旨在为这些当下的社会问题或不同的受众群体，提供有效的具前瞻性的解决方法、可实施的解决手段、建设性的意见和建议。

“经线论——调研贯穿设计始终”教学模式在毕业设计教学实践中的现实意义

经线和纬线是人们为了在地球上确定位置和方向的。通过经线和纬线的准确定位可以在地球仪和地图上找到准确的地理位置。“经线论——调研贯穿设计始终”教学模式是指在毕业设计教学过程中，经线是由设计开题、中期到最终的设计整合阶段三个阶段运用的多维的调研方法组成主要线索，而设计作为手段和解决问题的途径可以视为纬线，通过调研方法的多角度、多数据、多样本的综合分析，不断调整设计手段和方法，围绕经线所提供的线索，在与经线交叉点精准的定位设计策略和设计思想，最终实现设计的合理性和适宜性。

感性与理性的有机结合是“经线论——调研贯穿设计始终”教学模式的总特征。“经线论”式的教学是教师在教学中，依据研究过程的三个阶段调研形成的客观结论，最大限度地调动学生思维的教学方式，引导学生在做研究型毕业设计课题时如何既有感性的发散思维，也有理性的逻辑思维，主动学习和调整研究思路，融会贯通得将设计手段与表现结合各阶段调研成果，真正有效地提高学生发现问题、分析问题和解决问题的能力。

一、设计始于调研——开题阶段

毕业设计课题是本科生培养过程中重要的综合性实践教学环节。如何使其在本科毕业设计课题教学中充分发挥积极作用是迫切需要探讨的问题。本文旨在讨论面对研究型毕业设计课题，探索如何在教学模式中创新，制定相应的指导教学计划、教学目标、指导机制、过程监督等诸方面进行了探讨和创新，在本次竞赛的教学实践中证明，该模式取得了优良的教学效果。

学习设计调研方法——分析调研结果，发现问题。在毕业设计的开题阶段，设计前期的调研即在确认设计任务书后，对设计任务、设计内容、设计主体及设计目标进行较广范围的研究。主要目的是加深对项目的理解及场地特征、周边环境、空间特点、受众群体及设计改造范围进行广度的调研。在这个阶段的学习过程中，学生必须自觉查阅与项目相关的文献资料，对如何设计针对课题的调研方法，做调研模型及分析调研数据、如何选择有用的样本进行分析等一系列的内容理解和掌握。在此基础上对现有研究型的课题进行针对性强的调研方法及调研内容，例如对相关老年公寓建成案例的实地调研分析、相关国内外典型案例的资料分析、北京曜阳国际老年公寓的现状分析及场地分析等。设计的切入点和项目现状中存在的问题，都将在实地调研结果的数据分析、样本模型中逐渐清晰，分析问题，发现问题，才能准确找到设计线索中经线与纬线的交叉点。走出意向图误区。大部分的学生进行到这个阶段时，就开始进行大量的网上调研及典型优秀案例的收集，于是约定俗成的设计思想和设计概念就逐渐在意向图的收集过程中，逐步形成。甚至不为少数的本科毕业设计完全依靠设计意向图，东拼西凑的初步设计概念完全背离了具体设计的基础条件和客观性。这类设计与原创设计最大的区别就是完全不能追根溯源到设计的最初，呈现出的设计概念都是短期的拼凑和逻辑不通的表象，经不起推敲。意向图和参

they've found. Some students even directly use collected images to develop their preliminary design schemes, regardless of the basic and objective conditions of the project. The biggest difference between such schemes and original schemes is that the originals of the former cannot be traced, and the design concepts contained therein are just temporarily combined together in an illogical way. Reference images are only inspiring materials, rather than the results of conceptual and rational thinking. Brainstorm original design concept and the logical structure of design philosophy (research thoughts). Aside from multidimensional surveys conducted in the early stage, brainstorms are an effective way to expand the way of thinking which can hardly be realized by individual divergent thinking. Brainstorms are the best way of communication among the designer, target group, and other relevant parties for the purpose of finding new ideas. Students should be divided into several groups for brainstorms. The associative thinking of a group of people is much powerful than individual thinking. Brainstorms produce great innovations and ideas, remove bottlenecks, and solve difficulties. In the process of brainstorms, students can learn from each other, and the process of discussion transforms into a process of study featuring interactive learning. In this process, the project progresses from the stage of individual surveying, reporting and thinking to the next stage of observation, recording, discussion, criticism, summarizing and evaluation. With the logic of thinking becomes more and more clear, the original conceptual scheme naturally comes into being, the logical mind map of design ideas shapes up, and design content fills the logical structure of design philosophy. All the design methods and expressions are included in the conceptual framework and logistic structure of the design.

2. Conduct multidimensional surveys based on the behavior habits, traffic flows and ergonomics of the target group in the middle stage

In the middle stage, students have already had clear design ideas and concepts. According to the "Meridian Theory" teaching model, in this stage, corresponding survey methods should be applied to further study and analyze the behavior, characteristic classification, traffic flows, habits, the priority of the five senses, and ergonomics of the target group which is disabled and partially disabled seniors in the case of the competition. Based on the analyze results, design objectives, task, direction and content can be determined substantially. In the process of mid-term survey, problems like uncertainties and improper design strategies occurring in the thesis determination stage should be solved immediately. Approaches such as diagrams, freehand drawings and spatial renderings should be adopted to demonstrate the design content. Special attention should be paid to the design of key spatial nodes. Design is not the art of average, or a game of filling. Many students tend to focus on all the facets in spatial redesign, which leads to the lack of highlights featuring unique design ideas and impressive structural design. A fine painting always features densely colored areas and blank areas. Likewise, a good design work should also have highlights and also "blank areas" for the target group to develop.

3. Target group evaluation and information feedback surveying in the integrated design stage

Following the middle stage is the integrated design stage in which previous achievements are adjusted and integrated. In this stage, many students feel confused about how to judge correctness, appropriateness and feasibility of such achievements. Since there is no unified standard and value orientation, it's hard for advisors to find a completely objective evaluation system, and an interdisciplinary evaluation standard. Under the "Meridian Theory" teaching model, the most effective evaluation method is that people living in the design space, or the target group comprehensively evaluate the design results and give their feedback, and the students then discuss the feedback, screening the most direct, objective and constructive suggestions and advices. For example, in case of exhibition design, the feedback of visitors can be classified according to the types of visitors, and then analyzed.

Conclusion:

From the thesis determination stage to the middle stage to the integrated design stage, multidimensional surveys, like meridians on the globe, help students integrate the knowledge they've learned in the four-year undergraduate education system, and have a clear understanding of the entire design process and the importance of logical thinking structure. All design methods and expressions are like parallels crossing meridians, forming intersections. When such intersections are linked, a complete and logically structured design process comes into being. As an innovative teaching model featuring the organic combination, proper separation and harmonious co-existence of conceptual thinking and rational thinking, the "Meridian Theory – Survey throughout the Entire Project" teaching model is an effective way to control the entire graduation design process.

考图只是设计过程中启发思维的资料，完全不能代表感性与理性有机结合的设计过程。头脑风暴结合集体汇谈，生成原创设计概念原型，设计理念（研究思想）逻辑结构形成。除了设计初期的多维调研方法外，组织学生进行若干次头脑风暴和集体汇谈，有效的解决了发散思维中靠个人力量最难解决的思维的广度和跨度。头脑风暴是一种设计师与受众，各环节不同设计方共同完成一项设计最好的拓展思路的沟通方式。学生分成几组，进行集体汇谈的过程中产生的联想式思维要远远优于个体的思考能力，通常好的创意和设计构想在小组汇谈的过程中灵感闪现，苦苦思索或遇到瓶颈的设计难点也在讨论声中迎刃而解。旨在思维的交流碰撞中促进合作型学习，将交谈的过程演变上升为一种具有互动特征的学习型空间。将以往仅停留在每个个体完成考察调研报告、个体思辨的阶段，进一步向观察、记录、讨论、批判、总结、评价的综合阶段延伸。随着思维逻辑的逐渐清晰，方案的概念原型应运而生，设计理念的逻辑思维导图整理成形，整体的设计内容和设计环节都紧扣逻辑结构进行展开，所有的设计手段和表现都统一于设计整体概念和逻辑结构。

二、受众行为习性、行动轨迹、人体工程学等多维调研方法的实施——中期阶段

进入毕业设计的中期阶段，学生已有了明确的设计思路和理念。“经线论”教学模式的第二阶段的调研方法开始对空间主体对象的行为方式、受众特性分类、跟踪行动轨迹、自理和半自理老人的生活习性、五感的优先性、针对老年群体的人体工程学进行更进一步的分析和调研。根据更进一步深入的调研结果的分析和总结，介入对项目设计目标、设计任务、设计方向及设计内容的实质性设计阶段。在进行中期调研的过程中，及时对开题阶段的一些不确定或设计策略上的偏颇及时调整。运用大量的分析图解、手绘及空间渲染表现将设计内容层层推进。更重要的是在设计过程中要注重各空间的重要节点设计。设计不是平均的艺术，更不是一种填充的游戏，为数不少的学生在毕业设计阶段会对所有面对的空间问题事无巨细，逐一进行改造设计，导致整体设计缺少深入，缺少既有设计想法又有设计深度和结构做法实施的重点设计。如同艺术作品中的重彩和留白，所有的好设计都是有的放矢，既有重点也留有余地的待受众主动参与去创造和开发更多元的空间。

三、受众评价及信息反馈调研——整合设计阶段

历经设计的中期阶段，最后进入研究性毕业设计课题的整合设计阶段。此阶段要将设计过程中阶段成果和内容进行全局的设计调整和整合。但大部分让学生困惑的是如何去判断设计中的对错、适度性和可行性，设计没有绝对的判断的标准和价值取向，也同样很难让指导教师给予完全客观的评价体系和多元化的评判标准。“经线论”的教学模式最有效的参考范式即把设计结果交给空间主体——受众进行综合评价及信息反馈，学生们在进行受众评价及体验信息反馈调研后，会得到对自己的设计最直接最为客观的建设性建议和意见。例如做展示设计的设计评价就可以根据受众群体的分类，进行展示设计受众评价调研，互动设施评价及意见反馈。

结语

从开题阶段——中期阶段——整合设计阶段贯穿始终的多维调研方法如同地球仪中的经线，能真正帮助学生在本科阶段的最重要设计课题中，融汇贯通所有专业课程的知识点，清晰整个设计过程及逻辑思维结构的重要性，所有的设计手段和设计表现如同纬线一样，紧密围绕横向的经线，精确捕捉到与经线在不同阶段的交叉点，将各个交叉点连接即成为整个严密逻辑构成完整的设计过程。同时，“经线论——调研贯穿设计始终”成为把控整个研究型毕业设计课题过程中客观有效的方法；也成为设计过程中感性思维与理性思维有机交织、适当分离、共生融合的创新教学模式。

零玖

09

活動印記

Activity Mark

CIID“室内设计 6+1”2016（第四届）

校企联合毕业设计

CIID"Interior Design 6+1"2016(Fourth Session)University and Enterprise Joint in Graduation Design

鹤发医养卷

——北京曜阳国际老年公寓环境改造设计

White hair volume medical support

—Beijing Yao Yang International Apartments for the elderly environmental reconstruction design

卓培
北京城建设计发展集团股份有限公司建筑院高级工程师

作为“北京密云曜阳国际老年公寓”原项目的设计主持人，我在2015年10月刚刚接触到CIID“室内设计6+1” 2016（第四届）校企联合毕业设计“北京密云曜阳国际老年公寓——环境改造设计” 活动的命题时，感觉这个题目对于大学四年级的毕业生来讲是有很大设计难度的。

时间的推移、社会的发展、产业的变迁、功能的转变，都给同学们怎么开题，怎么解题，怎么结题，提出了严峻的挑战。

北京密云曜阳国际老年公寓的原C区建筑是按照年轻人群的普通宿舍来设计的，并不符合老年公寓的设计标准。同时，老年建筑的设计并不仅仅是满足国家规范设计标准的要求，而是要考虑到“不需要介护老人”“半介护老人”“全介护老人”的不同居住人群是有各自独特的心理特征和行为特点的。

其中，自理型老人是不需要介护或半介护的人群，养生的理念非常重要；半介护老人是不需要全介护的人群，专业护理的理疗概念非常重要；全介护老人是最特殊的人群，专业医疗的设施非常重要。

同学们在指导老师用心地全程辅导下，开题时认真地准备前期资料和社会调研；中期时积极地开拓和整理设计构思，奇思妙想层出不穷；答辩时精心地绘制完成了自己的作品和汇报PPT文件，最终每组同学都交出了一份令评委们满意的答卷。不过，同学们要注意一点：在设计答辩或者汇报方案时，要事先充分地准备、熟记于心，做到脱稿汇报或者半脱稿汇报，一定不能低头念稿，要抬起头来用眼神和动作与评委和听众交流互动。

另外，在这次活动中我也深刻地体会到了中国建筑学会室内设计分会（CIID）和CIID“室内设计6+1”活动的各个大学的优秀组织能力和良好合作氛围，我也看到“团结互助、学术交流、共同提高”的宗旨贯穿着整个毕业设计活动的全程。

“最重要的是过程，而不是奖项的高低”，本届活动对于参与其中的每个人都是一次有意的“再思考”过程。

最后，我也在此感谢为本题目提供大力支持的北京城建设计发展集团股份有限公司的我的同事们，他们是公司副总工金路（建筑院总建筑师）和藏艳（建筑院建筑师）。

This year, the CIID “Interior Design 6+1” University-Business Graduation Projects Competition set the topic of “Beijing Yaoyang International Retirement Community Environmental Redesign”. As the project leader of the original Beijing Yaoyang International Retirement Community Project, I think the topic is quiet difficult for those senior students.

For this topic, many factors like the passage of time, social development, industry change, and change of functions all pose challenges to the students in determining theses, exploring solutions and giving solutions.

The buildings in Area C of Beijing Yaoyang International Retirement Community were originally designed as the common dormitories for young people, not specially designed for seniors. The design of buildings for seniors, to the exclusion of meeting relevant national design standards, should also take into account the psychological and behavior characteristics of not disabled, partially disabled and fully disabled seniors.

For not disabled seniors who need no nursing service, healthcare is very important. For partially disabled seniors, professional physical therapies are important. And for fully disabled seniors who are the most special group, professional medical equipment matter a lot.

Thanks to the guidance from the tutors, the students carefully prepared early-stage materials and conducted social surveys after determining their theses, came up with and discussed many creative ideas in the process of design, and finished high-quality drawings and PPT reports at last. The jury is very satisfied with their results. But there is still one thing that I'd like to point out: when defending theses or making reports, it's important to communicate with the audiences with eyes, body postures and gestures during the entire process based on a good command of the text, rather than reading the text.

What else impressed me is the excellent organizing ability of CIID, and its successful cooperation with the universities participating in the CIID “Interior Design 6+1” University-Business Graduation Projects Competition. The principle of “solidarity, academic exchange and shared growth” was successfully demonstrated in the entire process of the event.

Process is more valuable than prizes. For everyone taking part in the event, he experienced the meaningful process to “rethink”.

At last, I'd like to express my thanks to my colleagues at Beijing Urban Construction Design & Development Group Co., Ltd. They are Deputy General Manager and Chief Architect Jin Ru, and Chief Architect Zang Yan.

这是一次指向精准的命题与涉猎宽泛的解题之间的有益回应，是一次“教、学、用”三方资源较充分互动、张力凸显的教育实验，是一次精彩纷呈、活力满载的设计实践。

对于教师而言：

不同专业背景的（六校）同学，围绕同一命题而展开的设计，自然呈现出合而不同的生动且鲜活面貌；而对它们的集约化展示，则显现了关注面的差异、切入路径的不同、处理问题的多样性、以及价值评判体系的区别。

对于学生而言：

——由空间改造而切入设计的同学，通过尺度朝向界面区域等的建构，显现出自身的设计逻辑；

——由功能安排而切入的同学，则通过行为学及设施流程等路径，表明自身的设计侧重；

——由运营管理而切入的同学，则通过形象塑造话题演绎品牌营销等线索，展现自身的设计策略……

今后应注重的：

——教与学双方似应强调“问题意识”，力戒“美化式”设计；

——教与学双方似应设定“限利条件”，力戒无限定的“想像力”设计；

——教与学双方似应注重“设计策略”，避免“兵来将挡 、水来土掩”式的无择重设计。

This is a vigorous and meaningful response where a specific topic is answered by various solutions. This is an instructional experiment jointly performed by teachers, students and the project owner based on active communication. This is also a design practice giving rise to diverse wonderful ideas.

For the teachers:

This time, students from different professional backgrounds of six universities gave their answers to a same question, and naturally, their answers take on different looks, yet containing the same vitality. During the intensive presentation, from different perspectives and through different entry paths, these answers solve problems in different ways based on different value evaluation systems.

For the students:

—Those viewing the topic from the perspective of space reconstruction focus on the reconstruction of space elements such as dimensions, orientations, borders and areas which reflect their logic of design.

—Those viewing the topic from the perspective of functions focus on functional aspects like behavioral science, equipment, and procedures.

—Those viewing the topic from the perspective of operations management focus on project operating strategies like image building, theme presentation, and brand marketing.

Teachers and students should:

—Develop the “problem awareness” to avoid “self-deception”.

—Set “restrictions” to avoid unrestricted “imagination”;

—Pay attention to “design strategies” to avoid “aimless design”.

赵健

CIID 资深顾问

广州美术学院教授

CIID“室内设计 6+1”校企联合毕业设计活动创办四年来，历经四年的实践，越发趋向完善成熟。它不仅仅是一次毕业设计答辩活动，也是一次高校与高校间的学术研讨交流的盛会。每次的设计选题始终是建筑行业及社会密切关注的热点问题为导向，依托校企合作、开展联合毕业设计的教学模式，体现教育服务于人才，人才服务于社会培养需求的原则。

作为一名即将走入实践的设计师，需要一种兼具“普世性”的职业精神。正确建立起自然、空间、人三者之间的关系，使之能够长久地共融发展，有效地解决人类的生存和居住问题。让弱势群体能够获得设计所带来的经济机会降低能源消耗，并提升人们的生活空间质量，为社会良性发展做出应有的贡献，这是设计师应当具备的责任。

祝愿在学会及各院校的共同努力下， CIID“室内设计 6+1”校企联合毕业设计活动越办越成功。

This year marks the fourth anniversary of the CIID “Interior Design 6+1” University-Business Graduation Projects Competition. The past four years witnessed its growth to an influential event for graduation project defense, and for the academic exchange among universities. Each year, the topic was carefully chosen so that a social concern can be solved by architecture under the mode of university-business cooperation. This event well demonstrates the principle that

王兆明

CIID 副理事长

黑龙江建筑职业技术学院教授

"education should be offered in such a way that talent required by the society is developed".

A real designer should involve the spirit of universality in his design, that is, to meet man's needs for living without breaking the harmonious co-existence of nature, space and man, to bring the economic opportunities brought by design to vulnerable groups, and to create better living spaces at a low energy cost, thus contributing to the healthy development of the society. This is the responsibility of a designer.

I wish the CIID "Interior Design 6+1" University-Business Graduation Projects Competition greater success in the future with the joint efforts of CIID and the universities involved.

马本和
CIID 理事
齐齐哈尔大学美术与艺术设计学院环艺系主任教授

很荣幸能够被邀请参加在哈尔滨工业大学建筑学院举办的，由中国建筑学会室内设计分会（CIID）组织的，2016 年第四届“室内设计 6+1”校企联合毕业设计的答辩会，同时我也见证了时下中国最具权威性的建筑室内外环境设计方面的优秀毕业生设计作品。

本次答辩环节，7 所高校的 13 组设计作品充分展示了每个高校的教学优势与教学特色，同时针对医养建筑室内外空间设计，结合各自院校不同专业的特点，展开了深入并有设计指导意义的理论研究。

CIID“6+1”校企联合毕业设计的模式非常适合当下中国高校环境设计学科在毕业设计教学环节方面的教学模式，我想，若能将这一模式与经验，通过 CIID 学会平台，向全国各高校推广，那么全国环境设计专业的毕业设计水平与能力都会得到非常大的提高。

建议将整个“6+1”联合毕业设计的开题、中检、答辩各环节制作成视频文件，通过互联网进行更广泛的传播，这样更有利于边远地区的高校师生分享，也更充分实现CIID学会引领室内设计行业与室内设计教育进步与发展的初衷，进而也会提升 CIID“6+1”校企联合设计的品牌价值与影响力。

I'm honored to be invited to the 2016 CIID "Interior Design 6+1" University-Business Graduation Projects Competition organized by CIID at the School of Architecture, Harbin Institute of Technology. This event provides a stage to showcase the graduation projects developed by excellent students from the most authoritative universities in interior and exterior environmental design.

13 design schemes presenting the instructional advantages and characteristics of 7 universities showed up at the project defense. Taking the interior and exterior space design of a retirement community as the topic, these schemes express ideas and discuss theories profoundly and instructively.

The university-business cooperation mode adopted by the CIID "6+1" competition sets a good example for the environmental design departments of various universities in graduation project implementation. If we introduce the mode to universities nationwide through the platform of CIID, the professional level and capacity of environmental design students in implementing graduation projects will be greatly improved.

I suggest that we can make videos of the thesis proposal presentation, mid-term reporting and defenses included in the CIID "6+1" competition, and share them online with other schools, including schools in remote areas. This is a good way to realize CIIS's vision of promoting the development of China's interior design industry and interior design education, and to promote the brand value and influence of the CIID "6+1" competition.

2016 第四届 CIID“室内设计 6+1”校企联合毕业设计答辩搭建了校企之间、兄弟院校之间、相关专业之间的联合教育平台，也让我们看到了现如今高校毕业生设计的能力，通过对企业给定的课题从建筑结构、空间设计、景观设计和心理学等多个角度，分析总结了《鹤发医养——北京曜阳国际老年公寓环境改造设计》并根据得出结论进行改造和设计。

参赛的各高校从命题、开题、调研到答辩，共同完成了这次多方、三地的联合教学系列活动。

整个答辩过程中十三组同学的精彩解说展示了各自的设计方案，各位专家评委的透彻点评总结升华了方案的设计理念，推动了专业发展，提高了学术水平，拉近了校企合作关系，达到了各院校间相互交流、共同学习的最终目的。

建议更多院校的师生参与到“CIID 室内设计 6+1 校企联合毕业设计”系列活动中，从而丰富活动内容，扩大其影响力，增强其形象力，同时使高校实践课题得到推广，让学生得到更多锻炼的机会。

张红松
哈尔滨师范大学美术学院常务副院长兼环艺系主任、教授

The 2016 CIID "Interior Design 6+1" University-Business Graduation Projects Competition builds a platform of joint education for businesses and universities, and also a stage for university students to show their design capacity. From the perspectives of architectural structure, space design, landscape design and psychology, the submitted design schemes roll out different ideas on the project of Healthy Aging – Beijing Yaoyang International Retirement Community Environmental Redesign.

From topic determination, thesis proposal presentation, conducting surveys, to defending theses, the universities and business involved made concerted efforts to carry out the event which covered three cities.

During the defenses, 13 design teams presented their schemes using wonderful arguments, and these schemes, thanks to the comments, summaries, and suggestions of the jury, were then pulled to a higher level. The event not only improved the professional and academic level of the universities, but also promoted business-university cooperation and the communication among universities. Put simply, all the parties involved experienced improvements through the event.

I suggest more universities to participate in the CIID "Interior Design 6+1" University-Business Graduation Projects Competition. In this way, the event will be more substantial, influential and high-profile, and at the same time, university practices can be promoted and more practice opportunities can be created to the students.

“室内设计 6+1”校企联合毕业设计项目在 CIID 的领导下，深化设计教育改革，丰富高校探索设计人才培养思路，引领与推动中国环境艺术设计教育方式的多元化，促进探索自己的教育特色，为中国高校室内设计人才培养做出了积极的贡献。祝“室内设计 6+1”校企联合毕业设计项目越办越好！

Under the leadership of CIID, the CIID "Interior Design 6+1" University-Business Graduation Projects Competition has been playing an active role in deepening the design education reform, injecting new ideas into university talent development, and promoting universities to develop diverse and characteristic modes in environmental art design education. It has made a positive contribution to the development of interior design talent in China's universities. I wish the CIID "6+1" competition a better future.

董赤
CIID 理事
吉林艺术学院设计学院副院长

陈天力
日本大阪市立大学院生活科学研究科博士唯美同想设计咨询机构设计师

首先感谢主办方的邀请，有幸参加CIID“室内设计6+1”校企联合毕业设计评审活动。在国外工作学习了多年，对于国内的教学、学术研讨交流没有过多的参与，但因为在留学期间主要的研究方向是高龄者的生活空间方面的研究，所以对于老年人建筑的设计等各方面一直抱有浓厚的兴趣。

本次CIID“室内设计6+1”校企联合毕业设计的选题非常恰当精准，将目前社会密切关注的热点问题作为设计题目，对于提升学生的主观积极性及实践性起到了大力的推动作用。在学校与社会对接的教学机制下，学生可以更好地了解具体项目运作时的重点及相关的注意事项。通过较长周期的毕业设计，从题目的分析、现场勘测、对业主方的要求的理解、设计方案的构思、图纸的绘制、汇报文件的制作，到最后答辩汇报。在这个过程中需要学生综合运用各项专业知识去发现、分析、论证，系统地解决实际问题，是从学校到社会实践的一次全面提升。

历经四年的实践，不仅仅是一次高水准的毕业设计答辩活动，也是一次高校与企业间的学术研讨交流的盛会，探索设计的真正社会价值的盛会。

First of all, I'd like to express my thanks to the organizer for inviting me to be a jury member of the 2016 CIID "Interior Design 6+1" University-Business Graduation Projects Competition. I spent many years abroad, so my experiences in participating in domestic instructional or academic activities are kind of limited, but since I focused on studying the living spaces of seniors during my school days abroad, I have been very interested in architectural designs related to seniors.

This year's CIID "Interior Design 6+1" University-Business Graduation Projects Competition adopts a social concern as the topic which greatly aroused the enthusiasm of the students and stimulated them to put what they know into practice. Under the mode that universities and businesses work together in education, the students can have a better understanding of the emphases and matters needing special attention in the process of carrying out a project. It takes a long time to complete a graduation project. From topic analysis, site survey, understanding of the owner's requirements, to schematic design, preparation of specifications and drawings, preparation of the report, and the final oral defense, the students are required to integrate their specialized knowledge to discover, analyze, demonstrate and systematically solve various practical problems. This process is a process of turning knowledge into practice.

After four years' development, the Competition has grown into a high-level event for not only graduation project defense, but also the academic exchange between universities and businesses, and the exploration of the real social value of design.

关注CIID“6+1”活动是从第一届杨琳老师出题“鸟巢”再利用改造开始，鸟巢的再利用改造涉及很复杂的背景和跨学科设计研究，从选题上来看，“6+1”活动是有自己的定位和活动价值思考的。对大型体育赛事基础设施指点“江山”，试图产生扭转乾坤的影响力，能感觉CIID对这个赛事的“企图和野心”，即6+1活动定位不是简单地凑几个不错的设计院校在一起做个关于设计与教育的交流和分享会，而是就当前的重要建造现象和热点拿出自己的独特思考和观点。

今年这次养老地产项目的设计研究汇报，我有幸参与了同济大学的中期汇报和哈尔滨工业大学的终期汇报，受益匪浅——尤其是北京建筑大学在同济大学中期汇报和终期哈尔滨工业大学的方案进步之大，同济大学和哈工大等团队都体现出了很好的设计思维和很高的设计素养。使我认识到，6+1活动无疑可以承担更大的社会责任，她已经积累了重要的经验，并得到大家的认可，也已经在产生重要的社会影响力。

“6+1”活动的未来如何走？可能是一个更大的挑战，随着全球化的发展，一个设计项目的设计坐标都已经不是一个简单的地域性的某一个学科的设计思考，涉及更复杂的全球化设计思维和跨学科基础研究，设计要解决的问题的维度和层次也在发生巨大的变化。

祝“6+1”活动在未来的发展中，勇于迎接各种挑战，整合行业力量，富于探索和实验，取得更大的成绩。亚洲城市与建筑联盟和亚洲设计学年奖组委会将始终陪伴“6+1”的发展，为行业和产业的发展做出自己的贡献！

姚领

6+1特邀观察员
亚洲设计学年奖
亚洲城市与建筑联盟AAUA秘书长

The topic of the first CIID “6+1” competition is the transformation of the Bird’s Nest which involves complicated backgrounds and interdisciplinary design research. Obviously, such a topic reflects the positioning of the event and the value it pursues. This is an ambitious topic – to “do something” on a large stadium, exerting dramatic influence on the use of it. In a broad sense, it aims at not just gathering universities to exchange and share ideas about design and education, but encouraging the students to roll out their unique ideas and opinions on current key phenomena and hotspots in the field of architecture.

This time, I had the privilege of participating in the mid-term reporting of Tongji University and the final reporting of Harbin Institute of Technology of the retirement real estate design project, and I did marvel at what I saw, especially the big improvements made by the two universities in their design schemes. I saw impressive design ideas and high professional ability of their design teams. Undoubtedly, the CIID “6+1” competition is able to assume more social responsibilities, for it has accumulated important experience as an industry-recognized event, and most importantly, it has started to exert significant social influence.

Then, what does the future hold for the CIID “6+1” competition? This may be a bigger challenge faced by the event. In the context of globalization, what design needs is a global perspective and interdisciplinary research, rather than a regional perspective and single principle research in the past. What’s more, problems occurring in the process of design need to be solved from completely different dimensions and in quiet different degrees.

I wish the CIID “6+1” competition a great success in the future. With the joint efforts of all parties involved, the challenges ahead will be addressed in the course of exploration and experiment.

The Asia Architecture and Urbanism Alliance and Annual Asia Design Award Committee will keep supporting the growth of the CIID “6+1” competition, making contributions to industry development.

命题研讨过程

开题现场考察

方案交流现场

师生讨论交流

表彰颁奖仪式

专家导师合影

10

單位介紹

Unit Introduction

CIID“室内设计 6+1”2016（第四届）校企联合毕业设计

CIID"Interior Design 6+1"2016(Fourth Session)University and Enterprise Joint in Graduation Design

鹤发医养卷

——北京曜阳国际老年公寓环境改造设计

White hair volume medical support

—Beijing Yao Yang International Apartments for the elderly environmental reconstruction design

China Institute of Interior Design

Six colleges display differently and this reflects a certain difference between them.The predecessor of China Institute of Interior Design (CIID) is China Institute of Interior Architects. Since itsestablishment in 1989, CIID has been the only authorized academic institution in the field of interior design in China.CIID aims to unite interior architects of the whole country, raise the theoretical and practical level of China's interior design industry, pioneer the Chinese characteristics of interior design, help interior architects play their social role, preserve the rights and interests of interior architects, foster professional exchanges and cooperation with international peers, so as to serve and facilitate the construction of China's modernization.

Since its foundation 20 years ago, CIID hold abundant and colorful academic exchanges every year, building aplatform for designers to communicate and to study meanwhile update designer information of design industry, related competitions and business promotion, to enhance the better and rapid development of interior design industry of China.

Members of CIID are composed of individual members (including student members, associate members, fullmembers, advanced members, foreign members) and group members. By now, CIID has a large membership of more than ten thousands prominent designers who are from all over the country and passed the strict assessment by CIID.

Every year CIID will organize various types of competitions which include Institute Award of China Interior DesignAward, Influential People of China Interior Design, "Renewal Design" Original Competition, National-level Interior Design Competition for Young Students, China Hand-drawn Art Design Competition and so on.

CIID Secretarial is located in Beijing, taking charge of institute work. CIID secretariat publishes membership periodical china interior, Collection of Entries of China Interior Design, periodical Ornament and Decoration World,Home Adornments, ID+C. CIID website: www.ciid.com.cn.

Tongji University

Tongji University, established in 1907, is a top university of China Ministry of Education. During the time of restructuring of the university and college systems in 1952,the Department of Architecture was formed at Tongji University , and in 1986 was renamed as the College of Architecture Urban Planning (CAUP). Currently CAUP has three departments: the Department of Architecture, the Departmentof Urban Planning, the Department of Landscape Design. The undergraduate program covers:Architecture, Urban Planning, Landscape Design, Historic Building Protection and Interior Design. CAUP is one of China's most influential educational institutions with the most extensive programs among its peers, and the largest body of postgraduate students in the world. Today, CAUP hasbeen recognized as an international academic center with a global influence in the academic fields.

Tongji University's interior design education originated from the Department of Architecture which started to conduct interior space research in the 1950's.In 1959, it applied for the establishment of the "Interior Decoration and Furniture Specialty" within Architecture Discipline. In 1986,approved by the Ministry of Education and the Ministry of Construction, the "Interior Design Discipline" was formally founded. Starting to admit undergraduate studentsin 1987,Tongji University was one of two earliesthigh education institutions in mainland China to train interior design professionals in a University of science and technology. In 2011.

"Interior Design" officially became the secondary discipline of the Architecture Discipline. In the same year, the "Interior Design Research Team" was established,providing even broader room for subject development. Tongji University's interior design education crystallizes its own characteristics,emphasizing rational thinking and proposing the interior design concept of"human centric, ecological consciousness, overall environmental perspective,equal time and regional characteristic significance,technology and art integration".

CIID 中国建筑学会室内设计分会

中国建筑学会室内设计分会（简称 CIID），前身是中国室内建筑师学会，成立于 1989 年，是在住房和城乡建设部中国建筑学会直接领导下、民政部注册登记的社团组织。CIID 是获得国际室内设计组织认可的中国室内设计师的学术团体，是中国最具权威的室内设计学术组织。

学会的宗旨是团结全国室内设计师，提高中国室内设计的理论与实践水平，探索具有中国特色的室内设计道路，发挥室内设计师的社会作用，维护室内设计师的权益，发展与世界各国同行间的合作，为我国现代化建设服务。

CIID 成立 20 多年来，每年举办丰富多彩的学术交流活动，为设计师提供交流和学习的场所，同时也为设计师提供丰富的设计信息，提供各类大型赛事信息，提供各项商务帮助，促进中国室内设计行业更好更快地发展。
CIID 设有个人会员（包括学生会员、准会员、正式会员、资深会员、外籍会员）和团体会员。目前，已有会员 1 万余名，均是经过严格资格评审的精英设计师，遍布全国各地。CIID 每年举办各类赛事，包括中国室内设计大奖赛“学会奖”，“中国室内设计影响力人物”评选，“设计再造”创意大赛，“新人杯”全国青年学生室内设计竞赛，中国手绘艺术设计大赛等一系列奖项。

CIID 秘书处设在北京，负责学会相关工作。秘书处定期出版会员会刊《中国室内》以及《中国室内设计年刊》，同时学会拥有会刊《装饰装修天地》《家饰》《室内设计与装修 ID+C》。CIID 的官方网站为中国室内设计网。

同济大学

同济大学创建于 1907 年，教育部直属重点大学。同济大学 1952 年在国家院系调整过程中成立建筑系，1986 年发展为建筑与城市规划学院，下设建筑系、城市规划系和景观学系，专业设置涵盖城市规划、建筑设计、景观设计、历史建筑保护、室内设计等广泛领域。同济大学建筑与城市规划学院是中国大陆同类院校中专业设置齐全、本科生招生规模最大，世界上同类院校中研究生培养规模第一，具有全球性影响力的建筑规划设计教学和科研机构，是重要的国际学术中心之一。

同济大学室内设计教育起源于建筑系，同济大学建筑系于 20 世纪 50 年代就开始注重建筑内部空间的研究，1959 年曾尝试在建筑学专业中申请设立“室内装饰与家具专门化”。1986 年经国家建设部和教育部批准，同济大学建筑系成立了室内设计专业，1987 年正式招生，成为中国大陆最早在工科类（综合类）高等院校中设立的室内设计专业。1996 年原上海建材学院室内设计与装饰专业并入同济大学建筑系；2000 年原上海铁道大学装饰艺术专业并入同济大学建筑系。2009 年同济大学开始恢复建筑学专业（室内设计方向）的招生工作。2011 年建筑学一级学科目录下，设立“室内设计”二级学科。

同济大学建筑城规学院的教学理念为以现代建筑的理性精神为灵魂，以自主创造、博采众长的学术品格为本色，以当代技术与地域文化的并重交融为导向，以国际学科前沿的跟踪交流为背景。室内设计教学突出建筑类院校室内设计教学特色，强调理性精神，提出“以人为本、关注生态、注重环境整体观、时代性和地域性并重、融科学性和艺术性于一体”的室内设计观。

South China University of Technology

South China University of Technology(SCUT), located in Guangzhou city, Guangdong province, was founded in 1934. It is a well-known Chinese university which has a long history and enjoys a high reputation. It is a national key university directly under the Ministry of Education of the People's Republic of China, one of the first national "211 project" and "985 project" key construction of colleges and universities.

The design institute of SCUT was founded in June 2010, with majors including industrial design, environmental design, information and interaction design, clothing and apparel design. The design institute closely relies on the strong advantage of technology and deep cultural heritage of SCUT and actively explores the way of highly industry integration and international cooperation, aiming to create famous heights for domestic and foreign design innovative talent training as well as design practice and service.

At present, the design institute grasps the development opportunity in design creativity industry, constantly renews education idea and makes bold exploration in design innovation personnel training mode. With "leading technology innovation, leading culture innovation, leading industry transformation and sustainable development" as the construction goal, a series of production and research platform including "creative and sustainable design and research institute" "space of contemporary art" "public platform of design experiment and practice" "interdisciplinary top creative talents cultivation test area" and "Tenglong research&development center" "cultural art and creative industry research center" "Chinese folk art research center" "ceramic culture research institute" has been built, striving to become the domestic leading design instutite with international influence, so as to support and lead the design industry development in guangdong and across the country.

Harbin Institute of Technology

Harbin Institute of Technology affiliates to the Ministry of Industry and Information Technology, and is among the first group of the national key universities to enter the national"211Project""985 Project" and to start the collaborative innovation "2011 plan". In order to train engineers, the Mid east railway authority founded the Harbin Sino Russian school in 1920, the predecessor of Harbin Institute of technology, which becomes the cradle of China's modern industry and technical personnel.The School has evolved into a distinctive, powerful, first class national key university, which is multidisciplinary, open, researchful and with international influence.

The discipline of Architecture in Harbin Institute of technology is one of the earliest architectural subjects in China, with more than 90 years'ups and downs. The school of Architecture has 4 undergraduate disciplines, including Architecture, Urban Planning, Landscape Architecture,Environmental Design, and 3 first-level disciplines, including Architecture, Urban and Rural Planning,Landscape Architecture, and secondary master's disciplines in Design. We have the first-level doctorate and master's authorization in Architecture, Urban and Rural Planning and Landscape Architecture, and secondary-discipline master's authorization in Design, and Post-doctoral Research Institute on architectural first-level discipline.With the cultural spirits of rigor and diligence,The school of Architecture has created a devoted, distinctive, qualified and dedicated teachers' team.We have gained distinctive and outstanding achievements in undergraduate teaching, postgraduate education and scientific research, and have formed our own academic characteristics in the Design of Public Buildings in Cold Region, Regional Architecture, Building Technology in Cold Region, Architectural History and Theory, Urban Planning and Designing in Cold Region and Environmental Design in Cold Region.

Xi'an University of Architecture and Technology

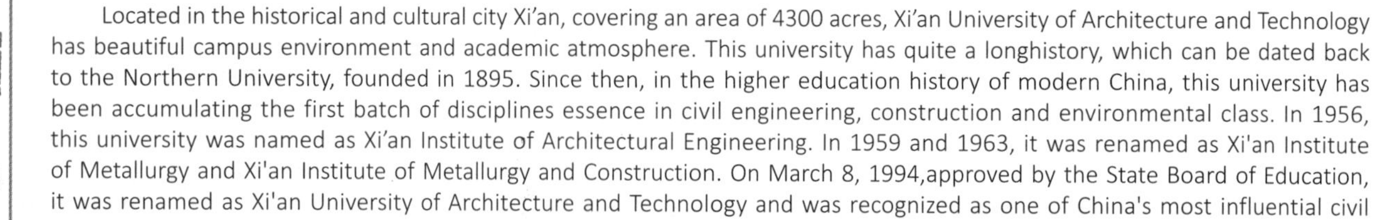

Located in the historical and cultural city Xi'an, covering an area of 4300 acres, Xi'an University of Architecture and Technology has beautiful campus environment and academic atmosphere. This university has quite a longhistory, which can be dated back to the Northern University, founded in 1895. Since then, in the higher education history of modern China, this university has been accumulating the first batch of disciplines essence in civil engineering, construction and environmental class. In 1956, this university was named as Xi'an Institute of Architectural Engineering. In 1959 and 1963, it was renamed as Xi'an Institute of Metallurgy and Xi'an Institute of Metallurgy and Construction. On March 8, 1994,approved by the State Board of Education, it was renamed as Xi'an University of Architecture and Technology and was recognized as one of China's most influential civil

华南理工大学

华南理工大学位于广东省广州市，创建于1934年，是历史悠久、享有盛誉的中国著名高等学府。是中华人民共和国教育部直属的全国重点大学、首批国家“211工程”“985工程”重点建设院校之一。

华南理工大学设计学院组建成立于2010年6月，现有工业设计、环境设计、信息与交互设计、服装与服饰设计等4个系。设计学院紧密依托华南理工大学雄厚的理工优势和深厚的人文底蕴，积极探寻与产业高度结合和国际化合作的道路，旨在打造享誉国内外设计创新人才培养和设计实践与服务的研究高地。

当前，设计学院紧紧把握设计创意产业的发展契机，不断创新 教育理念，大胆探索设计创新人才培养模式，以“技术创新引领、文化创意引领、产业转型引领、可持续发展引领”为建设目标，拥有“创意与可持续设计研究院”以及“当代艺术空间”“设计实验与实践公共平台”“跨学科拔尖创新人才培养试验区”和“腾龙研发中心”“文化艺术与创意产业研究中心”“中国民间艺术研究中心”“陶瓷文化研究所”等一系列产学研平台，力争建设成为国内领先、有国际影响的设计学院，从而支撑、引领国家和广东设计产业发展。

哈尔滨工业大学

哈尔滨工业大学隶属于国家工业和信息化部，是首批进入国家“211工程”“985工程”和首批启动协同创新“2011 计划”建设的国家重点大学。1920年，中东铁路管理局为培养工程技术人员创办了哈尔滨中俄工业学校——即哈尔滨工业大学的前身，学校成为中国近代培养工业技术人才的摇篮。学校已经发展成为一所特色鲜明、实力雄厚，居于国内一流水平，在国际上有较大影响的多学科、开放式、研究型的国家重点大学。

哈尔滨工业大学建筑学学科是我国最早建立的建筑学科之一，历经90余载风雨砥砺。建筑学院建筑学科现有建筑学、城市规划、景观学、环境设计4个本科专业和建筑学、城乡规划学、风景园林学3个一级学科和设计学二级学科硕士点。已获得建筑学、城乡规划学和风景园林学一级学科博士、硕士授予权，以及设计学二级学科硕士授予权，还设有建筑学一级学科博士后科研流动站。建筑学院始终秉持严谨治学、精于耕耘的文化精神，打造了一支朴实敬业、有特色、有能力、肯奉献的优秀教师团队。在本科教学、研究生培养及科学研究方面，特色鲜明，成绩显著。在寒地公共建筑设计、地域建筑设计、寒地建筑技术、建筑历史与理论、寒地城市规划与城市设计、寒地环境艺术设计等诸多方向上，均形成自己的学术特色。

西安建筑科技大学

西安建筑科技大学坐落在历史文化名城西安，学校总占地4300余亩，校园环境优美，办学氛围浓郁。学校办学历史源远流长，其办学历史最早可追溯到始建于1895年的北洋大学，积淀了我国近代高等教育史上最早的一批土木、建筑、环境类学科精华。1956年，时名西安建筑工程学院。1959年和1963年，曾先后易名为西安冶金学院、西安冶金建筑学院。1994年3月8日，经国家教委批准，更名为西安建筑科技大学，是公认的中国最具影响力的土木建筑类院校之一及原冶金部重点大学。

engineering colleges and the key university of the former Ministry of colleges and the key university of the former Ministry of Metallurgical.

Featured by civil engineering, construction,environment and materials science, engineering disciplines as the main body, Xi'an University of Architecture and Technology is a multidisciplinary university also with liberal arts, science, economics, management, arts, law and other disciplines. The university has 16 departments, 60 undergraduate programs so it can launch the first batch of undergraduate enrollment. It also has the right to recruit students by recommendation and the right of implementation of Accelerated Degree. Undergraduate art and design program is the featured major in Shaanxi Province.

Founded in April, 2004, Xi'an University of Architecture and Technology was established by the undergraduates from the major of art design and photography and from mechanical and electrical engineering industrial design and the relevant teachers from newly established sculpture and other specialties. The current undergraduate majors in this college include art and design, industrial design, photography, sculpture, exhibition art and technology, with more than 1,200 undergraduate students. Art Design was named "national characteristic specialty","provincial famous professional". This university has gathered many multidisciplinary researchers,including architecture, planning, landscape, etc. All these research teams have a long history of working towards the research of western region cultures, through undertaking many national and provincial funds subjects. The Arts College has actively organized (or as the contractor) the national academic, discipline-building meetings; inviting international and domestic famous professors to come for academic exchanges. It also has developed management approach, and set up a special fund to encourage young teachers and outstanding doctoral students to carry out academic exchanges and international (inside) collaborative researches. In the meantime it has established friendly and cooperative relations with the universities in Europe, Asia and other countries.

The university has taken the overall development of students as its training objectives, the improvement of theoverall quality of them as the aim to focus on. Relying on various student organizations carrier and platforms,the university has carried out various forms of extracurricular activities. And also it has focused on strengthening academic exchanges and interaction, inviting scholars, experts and celebrities to come to listen to the lectures and presentations,which can broaden the students' horizons, improve their knowledge structure and culture their spirits of science, technology and humanities. In other ways, the university organized the students to actively participate in academiccompetitions, and guided or encouraged students to engage in research activities, and many students have published various papers in the national magazines. The college has transferred departments, libraries, laboratories and paid multi-interactive efforts or work together to build a teaching-research-student trinity open experiment (work) platform. The graduates trained by the college have been welcomed by employers and the graduates are in short supply.

Beijing University of Civil Engineering and Architecture

Beijing University of Civil Engineering and Architecture is a university in Beijing is the only architectural colleges. Civil Engineering Division, began in the early 1936's liberation Beiping Li Industrial Vocational High School for Beijing Architectural College (when he was vice mayor of Beijing, a famous historian Wu Han as president), approved by the State Council in 1977 as a university, renamed Beijing construction Engineering College, 1982 was identified as the first country to grant bachelor units, in 1986 allowed for a master's degree granting unit. 2011 was identified as the Ministry of Education "Excellent Engineers Training Education Program" Universities. April 2013 was renamed the Ministry of Education approved the Beijing University of Civil Engineering and Architecture, was named "Green Capital University campus demonstration school." Beijing 2014 was rated "party building and ideological and political work of art colleges and universities."

Since the founding of New China, especially since the reform and opening up three decades, Beijing University of Civil Engineering and Architecture positive development of the capital and the capital to meet the needs of urban and rural construction of higher education, "based in Beijing, for the country, relying on the construction, service urbanization" and the future, times, continue to broaden educational horizons and continue to strengthen the professional disciplines, and continuously improve teaching quality and ability to serve the community, which has become an engineering, engineering, management, science, law, arts and other subjects of mutual support and coordinated development the distinctive characteristics of the multidisciplinary university, is an important force in Beijing and the country's urban and rural construction.

The school has 11 two colleges and three basic teaching units, College of Architecture and Urban Planning, Civil and Transportation Engineering, Environmental and Energy Engineering, School of Electrical and Information Engineering, Economics and Management, Engineering, Surveying and Urban Space information, electromechanical and Vehicle Engineering, and Law, College of Continuing Education, international Education, Department of computer Education and network information, ideological and political theory teaching and research department and Sports Ministry. Schools existing more than 12,500 people,

西安建筑科技大学是以土木、建筑、环境、材料学科为特色，工程学科为主体，兼有文、理、经、管、艺、法等学科的多科性大学。学校现有 16 个院（系），其 60 个本科专业面向全国第一批招生，有权招收保送生，实行本硕连读。艺术设计本科专业为陕西省特色专业。

西安建筑科技大学艺术学院成立于 2002 年 4 月，是由建筑学院的艺术设计专业和摄影专业本科生、机电工程学院工业设计专业本科生和新成立的雕塑专业及各专业关教师组建而成。学院现有艺术设计、工业设计、摄影、雕塑、会展艺术与技术 5 个本科专业，在校本科生 1200 余人。艺术设计专业被评为“国家级特色专业”“省级名牌专业”。学院集聚了包括建筑、规划、景观等在内的多学科的研究人才，学科团队长期致力于西部地区地域文化研究，承担了多项国家、省部级基金课题。艺术学院积极主办（承办）国家级学术、学科建设会议；邀请国际、国内知名教授来我校进行学术交流；制定管理办法，并设立专项基金，鼓励青年教师和优秀博士生开展学术交流、国际（内）合作研究，与欧洲、亚洲地区的多所大学建立了友好合作关系。

学院以学生全面发展为培养目标，注重学生综合素质提高，依托各类学生组织载体和平台，开展形式多样的课外活动。注重加强学术交流与互动，邀请学者、专家和社知名人士来我院举办讲座和专题报告，开阔学生视野，改善学生知识结构，培养学生的科技、人文精神。组织学生积极参与学科竞赛，指导、鼓励学生从事科研活动，在国内刊物上发表各类论文。学院调动教研室、资料室、实验室，多方互动，通力合作，构建了教学、科研、学生三位一体的开放性实验（工作）平台。学院培养的学生深受用人单位欢迎，毕业生供不应求。

北京建筑大学

北京建筑大学是北京市与住房和城乡建设部共建高校、教育部“卓越工程师教育培养计划”试点高校和北京市党的建设和思想政治工作先进高校，是一所具有鲜明建筑特色、以工为主的多科性大学，是“北京城市规划、建设、管理的人才培养基地和科技服务基地”“北京应对气候变化研究和人才培养基地”和“国家建筑遗产保护研究和人才培养基地”，是北京地区唯一一所建筑类高等学校。

学校肇始于 1907 年的京师初等工业学堂。1982 年被确定为国家首批学士学位授予高校。1986 年获准为硕士学位授予单位。2011 年被确定为教育部“卓越工程师教育培养计划”试点高校。2012 年“建筑遗产保护理论与技术”获批服务国家特殊需求博士人才培养项目，成为博士人才培养单位。2013 年 4 月经教育部批准更名为北京建筑大学。2014 年获批设立“建筑学”博士后科研流动站。2015 年 10 月北京市人民政府与住房和城乡建设部签署共建协议，学校正式进入省部共建高校行列。2016 年 5 月，学校“未来城市设计高精尖创新中心”获批“北京高等学校高精尖创新中心”。

学校有西城和大兴两个校区。目前，学校正按照“大兴校区建成高质量本科人才培养基地，西城校区建成高水平人才培养和科技创新成果转化协同创新基地”的“两高”发展布局目标加快推进两校区建设。与住建部共建中国建筑图书馆，是全国建筑类图书种类最为齐全的高校。

学校学科专业特色鲜明，人才培养体系完备。学校现有 10 个学院和 2 个基础教学单位，另设有继续教育学院、国际教育学院和创新创业教育学院。现

including more than 7,500 full-time undergraduates, doctoral and master more than 1,900 people, adult education college students, more than 3,000 people, more than 120 students. Existing 34 undergraduate majors, including the Ministry of Education specialty 3 - architecture, civil engineering, building environment and equipment engineering; Beijing specialty 7 - architecture, civil engineering, building environment and equipment engineering, water supply and drainage engineering, project management, surveying engineering and automation. The school has "architecture" doctoral-level disciplines, a post-doctoral research stations, 12 master's a discipline, a master's interdisciplinary point five professional degree authorization points and 8 categories Master's degree in the field of engineering authorization points. Beijing has a key disciplines three disciplines - architecture, civil engineering, surveying and mapping science and technology, a discipline in Beijing two key disciplines - management science and engineering, urban planning study.

School has a staff of 1,000 people, of which 671 full-time teachers. 381 teachers with senior professional and technical positions, including 105 professors, part-time tutor 20 people. 1 person has Changjiang Scholars, a national outstanding teachers who enjoy special government allowance of eight people, innovation and entrepreneurship students Special Contribution Award winner a Ministry of Education in the new century a man of talent, a leading talent and Technology Beijing one hundred people, heavenly Wan talent Project municipal candidate four people, the Great Wall scholar 3, teachers teaching seven people, three high-level personnel, academic creative talents of five people, technology nova eight people, young talents of 20 people, 20 young talents, Beijing Excellent team teaching, academic innovation team, management innovation team 25. Schools are divided into West and Daxing two campuses. West Campus area of 123,000 square meters, the building area of 202,000 square meters; Daxing campus covers 501,000 square meters, the building area of 265,000 square meters. School library books paper 1,529,000 (of which the Department of Housing and China to build China Architectural Culture Center Library building, sharing 360 000), 1.22 million e-books, a large electronic literature database 46, and China and the Ministry of Housing and Urban architecture Library, a national architectural books broadest range of colleges and universities.

Schools adhere to the center of the teaching work, efforts to improve education quality, won the national teaching achievement award 1, 21 municipal teaching awards, including first prize in nine. 2013 approved a national virtual simulation teaching center and National Experimental Teaching Center. In order to meet the capital needs of urban modernization of senior personnel, and actively carry out educational reform, focusing on improving students' ability to learn, practice and innovation ability, 119 schools have been built outside of practice teaching base. For five years, I have students in the country and the Capital University "Challenge Cup" and other scientific and cultural activities, access to provincial and ministerial awards 374.

The past five years, more than 1,900 school research various research projects, of which 863 national, provincial and ministerial level National Science and Technology Support more than 390 research projects in; provincial and ministerial level science and technology achievement award 58, which won the National Science and Technology Progress Awards, awards a total of 10 scientific and technological inventions, 2010,2011,2012 three consecutive years in the first unit of the State presided Technology Progress Award. 2014 "efficient low-temperature flue gas condensing heat corrosion depth of the use of technology," the State Technology Invention. 2014 Technology Services funds 280 million, eight consecutive years, Beijing is in the forefront among the universities.

Nanjing University of the Arts

Nanjing University of the Arts is one of the earliest arts institutions in China. It consists of 14 schools, 27 undergraduate majors and 50 major directions. It has Master's and Doctoral degrees and Post-doctoral stations in 5 subdisciplines under the first-class discipline of thearts: Arts Theory, Music and Dance, Drama and Film, Fine Arts Theory and Design Theory.

The professional background: in 2005, Display Design was set up as a major direction in undergraduate level and a research direction in master program in Nanjing University of the Arts; in 2008, it was incorporated into industrial design major as its one direction in School of Industrial Design; in 2011, it was approved by the Ministry of Education as an independent sub-discipline in then national disciplinary classification in undergraduate education; in 2012, it was classified into the first-class discipline of design with a new major name of "Art and Technology ".Through nearly 10 years of efforts by adhering to the principle that is students centered, academy-oriented, practice focused and development-guided, Art and Technology (Display Design) major has formed a coherent and open modular curriculum system of coherent knowledge and rational structure supported by modernization and globalization oriented course contents. The major is to cultivate professional design talents for the cultural sector,the museum sector, medium and large exhibition halls, the design community, the tourism sector, exhibition and other institutions. The graduates of this major are to have the ability to do design and research in the manner of integrating question, market and culture. And they are also to be cultivated as talents with the capacity of high-level artistic formation and excellent expression, as well as rational expertise structure and outstanding professional features.

有 34 个本科专业，其中国家级特色专业 3 个、北京市特色专业 7 个。学校设有研究生院，有 1 个服务国家特殊需求博士人才培养项目，1 个博士后科研流动站，12 个一级学科硕士学位授权点，有 1 个硕士学位授权交叉学科点，5 个专业学位授权类别点和 8 个工程专业学位授权领域点。拥有一级学科北京市重点（建设）学科 5 个。

学校名师荟萃、师资队伍实力雄厚。现有教职工 1000 余人，其中专任教师近 700 名。学校坚持质量立校，教育教学成果丰硕。学校 2014 年获得国家教学成果一等奖，并在近三届北京市教学成果奖评选中获得教学成果奖 21 项，其中一等奖 8 项。近五年来，学生在全国和首都高校“挑战杯”等科技文化活动中，获得省部级以上奖励 336 项。

学校坚持立德树人，培育精英良才。现有各类在校生近 12500 人，已形成从本科生、硕士生到博士生和博士后，从全日制到成人教育、留学生教育全方位、多层次的办学格局和教育体系。多年来，学校为国家培养了 6 万多名优秀毕业生，他们参与了北京 60 年来重大城市建设工程，成为国家和首都城市建设系统的骨干力量。学校毕业生全员就业率多年来一直保持在 95% 以上，2014 年进入“全国高校就业 50 强”行列。

学校坚持科技兴校，科学研究硕果累累。近五年以来，在研各类科研项目 1900 余项，其中国家 863、国家科技支撑等省部级以上科研项目 390 余项；获省部级以上科技成果奖励 58 项，其中荣获国家科技进步奖、技术发明奖共 11 项，2010、2011、2012 连续三年以第一主持单位获得国家科技进步二奖，2014 年以第一主持单位获得国家技术发明奖。科技服务经费连续 8 年过亿，2014 年达到 2.8 亿元。

学校面向国际，办学形式多样。学校始终坚持开放办学战略，广泛开展国际教育交流与合作。目前已与美国、法国、英国、德国等 24 个国家和地区的 41 所大学建立了校际交流与合作关系。

站在新的历史起点上，学校正按照“提质、转型、升级”的基本发展策略，围绕立德树人的根本任务，全面推进内涵建设，全面深化综合改革，全面实施依法治校，全面加强党的建设，持续增强学校的办学实力、核心竞争力和社会影响力，以首善标准推动学校各项事业上层次、上水平，向着“到 2036 年建校 100 周年之际把学校建设国内一流、国际知名的有特色、高水平、创新型大学”的宏伟目标奋进。

南京艺术学院

南京艺术学院是我国独立建制创办最早并延续至今的高等艺术学府。下设 14 个二级学院，27 个本科专业及 50 个专业方向。拥有艺术学学科门类下设的艺术学理论、音乐与舞蹈学、戏剧与影视学、美术学以及设计学全部 5 个一级学科的博士、硕士学位授予权及博士后科研流动站。

南京艺术学院从 2005 年开设了展示设计本科专业和硕士专业研究方向；2008 年该专业并入工业设计学院，2011 年会展艺术与技术专业作为独立的二级学科获得国家教育部的正式批准，2012 年该专业又被归为设计学类，成为“艺术与科技”专业。南京艺术学院工业设计学院的艺术与科技（展示设计）专业以学生为中心，以学术为导向，以实践为手段，以发展为目标，通过近 10 年的发展，已经逐步形成知识融贯、结构合理、连贯而开放的模块化专业课程体系和走向现代化、全球化的课程内容。旨在为文化部门、博物馆部门、大中型展馆、设计团体、旅游部门、会展机构等单位培养具有一定的理论素养，专业知识合理，专业特点突出，具备问题导入、市场导入和文化导入的整合设计和研究能力，以及高度艺术造型及表达能力的专业设计人才。

Zhejiang University of Technology

Zhejiang University of technology is a key comprehensive college of the Zhejiang Province; its predecessor can be traced back to the founding in 1910 as Zhejiang secondary industrial school.After several generations' hard working and unremitting efforts, the school now has grown to be a comprehensive University in teaching and researching which is very influential. The comprehensive strength ranks the top colleges and universities. In 2009, Zhejiang province people's government and the Ministry of education signed a joint agreement; Zhejiang University of Technology became the province ministry co construction universities.

In 2013 Zhejiang University of Technology led the construction of Yangtze River Delta green pharmaceutical Collaborative Innovation Center which was selected for the national 2011 program,to become one of the first 14 of 2011 collaborative innovation center. There are 68 undergraduate schools; 101 grade-2 subjects of master's degree authorization; 25 grade-2 subjects of doctor's degree authorization; 4 postdoctoral research stations. Subjects include philosophy, economics, law,education, literature, science, engineering, agriculture, medicine, management, arts and other 11 categories. School teacher is strong. There are 2 Chinese academicians of Academy of Engineering,sharing 3 academicians of Chinese Academy of Sciences and Academy of Engineering; 6 national young experts with outstanding contributions; 3 National Teaching Masters, 3 winners of national outstanding youth fund, 2 people were selected to central thousand person plan, the Ministry of education,1 professor of the Yangtze River scholars, 1 innovative team of Ministry of Education, 2 national teaching teams, and 26 person were selected to all kinds of national personnel training plans.Zhejiang University of Technology adheres to its motto "Profound accomplishment and invigorating practice. Accumulate virtues and good practice." To improve the quality of education in a prominent position, and strive to cultivate to lead, promote Zhejiang and even the country's economic and social development of elite talent.

Beijing Urban Construction Design and Development Group Co., Limited

Beijing Urban Construction Design and Development Group Co., Limited (01599.HK) is a scientific engineering company jointly established by Beijing City Construction Group and a number of large state-owned enterprises,to provide professional services for city construction,its operation cover fields of city mass transit,comprehensive traffic junction,development of underground space,industrial and civil buildings municipal engineering,bridge and road,providing customers with full process professional high quality service of project initial period consultancy,planning,investment and financing,survey and mapping,design,project management,project EPC,system integration,project evaluation and economic analysis.The Group has the qualification of comprehensive Grade A for engineering design,Level I for overall contracting for building project construction and for municipal utility project construction,and specialized contracting qualification for city mass transit projects,and many other qualification for engineering consultancy,urban and rural planning preparation,project cost,comprehensive project survey,project measurement,geological disaster evaluation and construction drawing review.

The predecessor of Beijing City Construction Design and Development Group Co.,Ltd. Is Beijing City Construction Design and Research General Institute Co.,Ltd., founded in 1958 specifically for the construction of Beijing Metro Line 1,the first metro line of the New china.It has been development for over half a century,and today,about 4000 employees in more than 40 branches are working hard for the dreams of beautiful city with the mission of city design to build the future,to promote the harmony and sustainable development of people,cities and environment.We push forward the construction and development of cities with our wisdom and mission,and have become outstanding in the field of city mass transit construction.

We do everything with complete sincerity and fineness.We have completed many world-attractive projects over half an century and more,our services have been well recognized by the government and customers,with many honors such as the National "May 1" Labor Certificate of Merit and Capital Labor Certificate of Merit awarded to us.The Group was listed with IPO in Hong Kong Stock Exchange on July 8,2014,as a comprehensive service provider in city construction with design as the leading sector.

浙江工业大学

浙江工业大学是一所教育部和浙江省共建的省属重点大学，其前身可以追溯到1910年创立的浙江中等工业学堂。经过几代工大人的艰苦创业和不懈奋斗，学校目前已发展成为国内有一定影响力的综合性的教学研究型大学，综合实力稳居全国高校百强行列。

2013年浙江工业大学牵头建设的长三角绿色制药协同创新中心入选国家2011计划，成为全国首批14家拥有“2011协同创新中心”之一的高校。目前学校有本科专业68个；硕士学位授权二级学科101个；博士学位授权二级学科25个；博士学位授权一级学科5个；博士后流动站4个。学科涵盖哲学、经济学、法学、教育学、文学、理学、工学、农学、医学、管理学、艺术学等11大门类。学校师资力量雄厚，拥有中国工程院院士2人、共享中国科学院和中国工程院院士3人、国家级有突出贡献中青年专家6人、国家级教学名师3人、国家杰出青年基金获得者3人、中央千人计划入选者2人、教育部长江学者特聘教授1人、教育部创新团队1个、国家级教学团队2个、各类国家级人才培养计划入选者26人次。浙江工业大学坚持厚德健行的校训，把提高教育质量放在突出位置，努力培养能够引领、推动浙江乃至全国经济和社会发展的精英人才。

北京城建设计发展集团股份有限公司

北京城建设计发展集团股份有限公司（01599.HK）是北京城建集团等多家大型国企共同发起成立的，为城市建设提供专业服务的科技型工程公司，业务范围涵盖城市轨道交通、综合交通枢纽、地下空间开发、工业与民用建筑、市政、桥梁、道路等领域，为客户提供工程前期咨询、规划、投融资、勘察测绘、设计、项目管理、工程总承包、系统集成、项目评价、经济分析等专业化高质量的全程服务。拥有工程设计综合甲级、房屋建筑工程施工总承包一级、市政公用工程施工总承包一级、城市轨道交通工程专业承包资质，以及工程询价、城乡规划编制、工程造价、工程勘察综合类、工程测量、地质灾害评估、施工图审查等多项资质。

北京城建设计发展集团股份有限公司前身是北京城建设计研究总院有限责任公司，成立于1958年，是专门为新中国第一条地铁北京地铁1号线建设而成立的。半个多世纪以来，分布于40多家分支机构的近4000名员工秉承设计城市、构筑未来的使命，致力于构建美丽城市的梦想，促进人与城市、环境的和谐、可持续发展。我们用智慧和使命推动着城市的建设和发展，业已成为城市轨道交通建设行业的翘楚。

为人至诚，为业至精。历经半个多世纪，我们完成了众多令世人瞩目的工程，我们的服务获得了政府、客户的认可，荣获全国“五一”劳动奖状、首都劳动奖状等多项荣誉。我们于2014年7月8日震撼登陆香港股市，致力于成立以设计为引领的城市建设综合服务商。

致　谢
Acknowledgement

北京城建设计发展集团股份有限公司

同济大学建筑与城市规划学院

华南理工大学设计学院

哈尔滨工业大学建筑学院

西安建筑科技大学艺术学院

北京建筑大学建筑与城市规划学院

南京艺术学院工业设计学院

浙江工业大学

CIID第二（上海）专业委员会

CIID第十一（哈尔滨）专业委员会

北京市民政局副局长 李红兵先生

CIID学术委员会主任，清华大学美术学院教授 周浩明先生

中国养老健康国际联盟副主席（首席医养顾问），中国老龄事业发展基金会老年健康基金管理委员会副秘书长 胡晶女士

英智康复健康管理集团总裁，中国医促会健康养老专委会副主任 关晓立女士

北京城建设计发展集团股份有限公司建筑院总建筑师 金路先生

北京城建设计发展集团股份有限公司建筑院项目负责人高级工程师 卓培先生

CIID副理事长华东建筑集团股份有限公司副总裁 沈立东先生

CIID资深顾问 来增祥先生

中国建筑学会室内设计分会副理事长，上海现代建筑装饰环境设计研究院有限公司总工程师　王传顺先生

中国照明学会副理事长，国际交流工作委员会主任 郝洛西女士

同济大学建筑城规学院副教授 戴颂华女士

中国建筑学会室内设计分会（CIID）副理事长，黑龙江建筑职业技术学院教授 王兆明先生

中国建筑学会室内设计分会（CIID）资深顾问，广州美术学院教授 赵健先生

中国建筑学会室内设计分会（CIID）理事，吉林艺术学院设计学院副院长教授 董赤先生

哈尔滨师范大学美术学院常务副院长兼环艺系主任教授 张红松先生

中国建筑学会室内设计分会（CIID）理事，齐齐哈尔大学艺术与设计学院环艺系主任教授 马本和先生

日本大阪市立大学院生活科学研究科，博士，唯美同想设计咨询机构设计师 陈天力先生

学会特邀观察员，亚洲设计学年奖、亚洲城市与建筑联盟（AAUA）秘书长 姚领先生